# ENGINEERING MODELLING AND ANALYSIS

This introduction to numerical analysis and computing covers a range of topics suitable for undergraduate engineering students with a particular focus on civil and environmental engineering.

Taking a balanced approach to teaching computing and computer methods at the same time, *Engineering Modelling and Analysis* satisfies the need to be able to use computers using formal languages such as Fortran and other applications such as Matlab and Microsoft Excel. Realistic and relevant problems are provided throughout the book in short accessible chapters which follow the structure of a degree course.

**David Walker**, **Michael Leonard** and **Martin Lambert** are in the School of Civil, Environmental and Mining Engineering, and **Andrew Metcalfe** is in the School of Mathematical Sciences, all at the University of Adelaide, Australia. They are all active in teaching and research and the content of the book reflects a strong belief that the one should complement the other.

# ENGINEERING MODELLING AND ANALYSIS

*David Walker, Michael Leonard, Andrew Metcalfe & Martin Lambert*

LONDON AND NEW YORK

First published 2009
by Taylor & Francis
2 Park Square, Milton Park, Abingdon, Oxon OX14 4RN

Simultaneously published in the USA and Canada
by Taylor & Francis
270 Madison Avenue, New York, NY 10016, USA

*Taylor & Francis is an imprint of the Taylor & Francis Group, an informa business*



This work has been produced from typeset copy supplied by the Authors
Printed and bound in Great Britain by
CPI Antony Rowe, Chippenham, Wiltshire



*British Library Cataloguing in Publication Data*
A catalogue record for this book is available from the British Library

*Library of Congress Cataloging in Publication Data*
Engineering modelling and analysis / Martin Lambert … [et al.].
p. cm.
Includes bibliographical references and index.
1. Engineering models. 2. Engineering – Mathematical models. I. Lambert, Martin. II. Title: Engineering modeling and analysis.
TA177.E54 2009
620.001'5118 – dc22

2008022437

ISBN 10: 0-415-46961-9 (hbk)
ISBN 10: 0-415-46962-7 (pbk)
ISBN 10: 0-203-89454-5 (ebk)

ISBN 13: 978-0-415-46961-6 (hbk)
ISBN 13: 978-0-415-46962-3 (pbk)
ISBN 13: 978-0-203-89454-5 (ebk)

# Contents

# Preface

The application of science and mathematics in engineering has developed to the point where a significant portion of design, analysis and assessment is now computer-based. While this has enhanced the ability of the engineer in his or her work there are some trends developing that should ring alarm bells. One such trend is that towards more and more user-friendly computer software. In many cases now it appears that it is simply not necessary to understand the theory of a particular solution procedure; the requested data are simply fed in to an appropriate program and an answer appears. It almost sounds too good to be true, and in many respects it is. While the experienced engineer, the one who understands the problem and its solution, is able to use such tools in an appropriate fashion, there is the potential for the inexperienced to use these tools incorrectly and without realising this. Such practice should be a serious concern.

This book, then, is aimed at educating engineers-in-training in the basics of computer modelling and analysis, with a clear emphasis on students developing their own code and procedures in a range of modes from a formal computing language (such as Fortran or C) to sophisticated spreadsheets (Microsoft Excel) and mathematical packages (Matlab). It is through such development that a deep understanding of the topics develops leading to an increased ability to apply the methods appropriately and to extend well beyond the confines of the enclosed chapters.

In designing the book a deliberate decision was taken to keep the chapters short and self-contained. In many cases a series of chapters together define an overall topic but there is nothing to suggest that all need to be followed, or that they should be followed in any particular way. Computer programming takes time, and there is likely to be a significant difference in the time required to master a topic from student to student and, indeed, for a particular student from topic to topic. As authors we would like to see a situation where the student spends twice as much time programming as reading and we see a short chapter as our part of the equation.

In many cases the methods covered will be those that were developed first. This is done quite deliberately for two reasons: firstly, the early methods are often the ones that can be used to explain the fundamental principles most simply; secondly, these methods often give a good feel for the processes at work in the solution and it is the *feel* for the methods that is so important in understanding the topics to the depth required.

The coverage in the book is clearly focussed on engineering applications, and in particular civil and environmental applications. We have designed the book to be well aligned with the strengths and interests of engineering undergraduates through the nature of the coverage of the theory of the methods and also through a wide range of examples of the application of such methods in current engineering practice.

A common topic when discussing numerical methods is the competition between engineers and mathematicians. One way of expressing the difference comes from paraphrasing Patrick Rivett and Samuel Eilon (Operations Research practitioners) (Rivett, 1981; Eilon, 1982):

> *the mathematician is concerned with exact solutions to approximate problems while the engineer is more likely interested in the provision of approximate solutions to exact problems.*

This idea of approximate solutions is central to much of this book and the methods covered. Approximate does not mean incorrect, nor does it mean a rough guess. Approximate means accurate to a known tolerance or level of error.

Another feature of the book is the interconnection of topics. Although it may be possible to learn all about finite difference modelling and the normal distribution independently (for example), that learning can be enhanced considerably by appreciating the links between the two topics and understanding how they are inter-related.

We have taken much care in trying to promote a deep learning environment by highlighting where possible *some* of the links between topics: students will no doubt find many more through their own efforts and reflections.

David Walker
Michael Leonard
Andrew Metcalfe
Martin Lambert

Adelaide, 2008.

Acknowledgment: Microsoft product screen shots reprinted with permission from Microsoft Corporation.

CHAPTER ONE

# Introduction (Engineering Modelling and Analysis)

## 1.1 INTRODUCTION

The word "model" is one that is used commonly in everyday language, and has a range of meanings. Dictionary definitions include "a three-dimensional reproduction of something, usually on a smaller scale", "a design or style of structure, e.g. this year's model", and "to design or plan (a thing) in accordance with a model, e.g. the new method is modelled on the old one". However, when the term is used in an engineering context it is often quite specific and conceptually different. To an engineer a model can be defined as an "abstraction of reality" (Izquierdo et al., 2004). This concept is illustrated in Figure 1.1 (Dandy et al., 2008) where a real world system (e.g. a bridge, highway, open channel, or dam wall) that needs to be understood sufficiently to make predictions about how it will behave under a range of situations is conceptualised in an idealised form, which is then analysed. This conceptualisation leads to the development of a model and forms an important part of engineering design. Based on the analysis that is then carried out, the properties of the real world system are inferred.

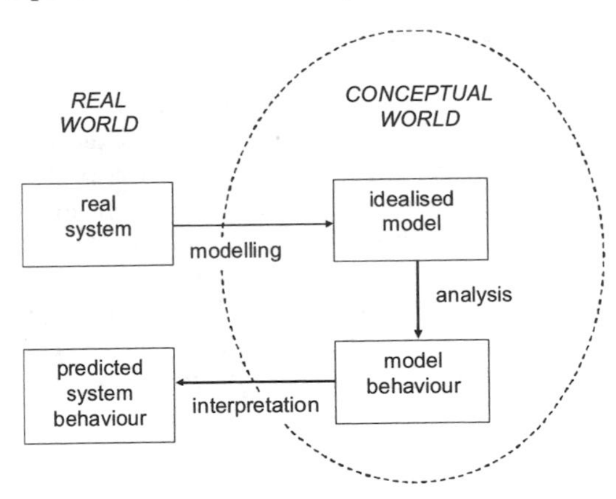


**Figure 1.1** Modelling and analysis in a systems approach. From Dandy et al. (2008).

It might be thought that once the model is developed the hard work has been done; however, this is only partly true. Certainly, a good model is a requirement for valid predictions of real-world behaviour but the analysis stage is important and will provide much of the focus of this book.

The generation of a model and subsequent analysis is best illustrated with an example. Consider a structural beam that has a set of loads applied to it, such as the one shown in Figure 1.2. In many situations the engineer may be required to predict the deflection of such a beam and to do this it may be necessary to develop a model of the beam and its loads. One approach is based on assuming that the curvature of the beam can be expressed as a second-order differential equation:

$$\frac{d^2y}{dx^2} = \frac{M}{EI} \tag{1.1}$$

where $y$ is the vertical deflection, $x$ is the distance along the beam, $M$ is the bending moment at any point along the beam, $E$ is Young's modulus which relates to the material properties of the beam (based on whether it is constructed of steel or wood or concrete, for example) and $I$ is a property of the beam based on its cross section. This equation, then, is a model of the actual beam and is intended to represent the beam's deflection under a known distribution of loads, given the material and section properties. While Equation (1.1) is a model, it is not the solution. For a solution it is necessary to solve the equation using some sort of analysis procedure.

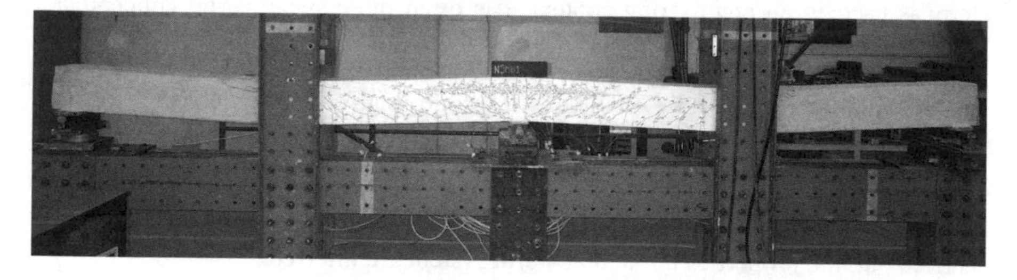

**Figure 1.2** A loaded beam in a structural test to failure.

In some situations an exact analytical solution may be known. For example, for situations where the load is a single unit load and the material and section properties are constant along the length of the beam it is possible to integrate the model equation twice and obtain an exact expression for the deflection, $y$. However, under many situations this will not be possible and the analysis becomes more difficult and requires some sort of numerical integration to approximate the solution. Assuming that the moment and material and section properties are known at all points along the beam it will be possible to plot the term $M/EI$. Such a plot is shown in Figure 1.3. With this it is possible to integrate Equation (1.1):

$$\frac{dy}{dx} = \int \frac{M}{EI} dx \tag{1.2}$$

where the terms are as defined previously. Now the problem is how to carry out the integration of the shape shown in Figure 1.3. An intuitive approach is to simply divide the shape into a large number of strips and assume each is a rectangle, the area of which can be calculated and summed to give the total. While this is intuitively appealing, is it the best approach? The short answer is no, and numerical integration will be covered in some detail in early chapters of this book.

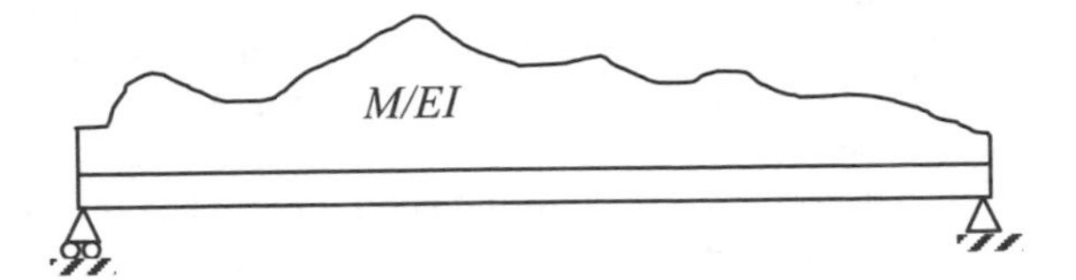


**Figure 1.3** A beam where a particular loading pattern has led to a distribution of $M/EI$.

The integration shown in Equation (1.2) leads to an expression for the slope of the beam everywhere. A second integration, again using numerical techniques, leads to an expression for the beam deflection. From here the engineer can transfer back to the real world and infer actual beam deflections based on a convenient model and a series of analytical and numerical procedures.

As a second example, consider a wide open channel with a pipe that discharges some sort of foreign material into the flow. Such a situation is illustrated in Figure 1.4. As the material is carried downstream it tends to spread out laterally and the question is: how would an engineer predict the concentration of this material at any point in the flow downstream of the discharge point?

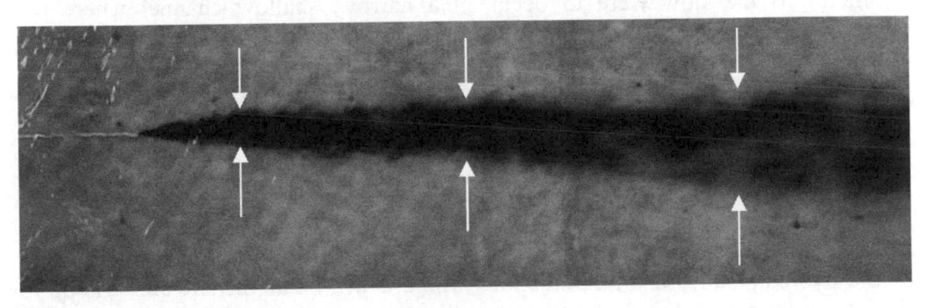

**Figure 1.4** Dye discharging into an open channel flow. Note the increasing width of dye which comes from the diffusion processes.

A knowledge of flows and the behaviour of a pollutant suggests that this is in fact a very complex system. The movement of the material is mainly due to the surrounding water and this is complex, with effects from the bottom and sides of the channel affecting the flow. For most flows of interest to an engineer there will also be turbulence, variations in the flow with a largely random nature. The spreading is influenced by all of these factors, plus the physical and chemical properties of the material itself. There is, however, a way to handle this situation. Experience has shown that the movement of the material, perhaps visualised as a dye, can be modelled in terms of the average open channel velocity. The spreading of the material can be modelled as a diffusion process and the combination of the two processes can be described (modelled) by the transport equation:

$$\frac{\partial C}{\partial t}+u\frac{\partial C}{\partial x}+v\frac{\partial C}{\partial y}+w\frac{\partial C}{\partial z}=K_x\frac{\partial^2 C}{\partial x^2}+K_y\frac{\partial^2 C}{\partial y^2}+K_z\frac{\partial^2 C}{\partial z^2} \tag{1.3}$$

where $C$ is the concentration of the material (in this case the dye), $x$, $y$ and $z$ are the three orthogonal axes that define the geometry of the situation, $u$, $v$ and $w$ are the three velocities in the three axis directions, and $K_x$, $K_y$ and $K_z$ are diffusion coefficients to explain the spreading of the dye due to turbulence.

Equation (1.3), therefore, is a three-dimensional model of the behaviour of the dye in the flow. It does not describe the processes exactly, but gives a way of handling them in a conceptual way. The "model" can be simplified further if appropriate. For example, suppose the flow is very shallow and relatively wide. In this case the variation that could be expected over the depth would be quite small compared to the variation that could be expected over the width and length of the channel, so the model could be simplified by ignoring the terms with a $z$ derivative in Equation (1.3). This leaves:

$$\frac{\partial C}{\partial t}+u\frac{\partial C}{\partial x}+v\frac{\partial C}{\partial y}=K_x\frac{\partial^2 C}{\partial x^2}+K_y\frac{\partial^2 C}{\partial y^2} \tag{1.4}$$

with the quantities defined as before. In this case it is assumed that the concentration will be uniform through the flow depth. This is also a model of the situation under investigation but one that relies on certain conditions regarding the geometry. If the flow were to occur in a narrow, shallow channel where the variation over the width and depth were small compared to the variation along the length of the channel, it would be possible to simplify the model further by removing from Equation (1.4) any derivatives in the $y$ dimension:

$$\frac{\partial C}{\partial t}+u\frac{\partial C}{\partial x}=K_x\frac{\partial^2 C}{\partial x^2} \tag{1.5}$$

In summary, each of the Equations (1.3) to (1.5) is a model of the dispersion of the dye pollutant. Each has situations where it will be applicable and it is up to the engineer or model designer to ensure that these conditions are adhered to. The model, though, is not the solution. For that it is necessary to go to the analysis phase to actually develop some solutions that allow predictions about real-world behaviour to be made. There are a number of ways of solving Equation (1.5): one uses an analytical approach, another uses the finite difference method. The actual solution will be dealt with later in the book. At this stage it is sufficient to understand the processes of modelling and analysis and to appreciate the need for a deep understanding of the real-world processes and the range of mathematical tools that will lead to solutions being developed.

In many cases there are several quite different models that can be developed to approximate a real-world situation. For the movement of material in a river it is also possible to model it as a large number of discrete particles that are carried with the mean flow but that also have a random component to their movement. This is called particle tracking and is a Monte Carlo approach. This powerful modelling approach will also be discussed in later chapters.

There is nothing to say that that a finite difference model based on the transport equation is any better than the Monte Carlo approach. Both are valid models and both allow the real world system to be understood and predictions made. Each has its own advantages and disadvantages and method of solution, but that is only natural. Different models require different analysis methods to obtain a solution.

## 1.2 ENGINEERING MODELS

Models, by necessity, have to make a range of simplifications and assumptions in their generation. Consider, for example, a desire to model flow in a large river system (e.g. Figure 1.5). The real world system comprises a main channel and its tributaries together with the thousands of smaller channels that feed these. The rivers and creeks all have banks that have been shaped by the flows, leading to a complex feedback mechanism. There will also be trees and other plants that stand in the flow and generate a range of eddies and other disturbances. During high flows the edges of the river will change, perhaps moving out over a vast floodplain.

In attempting to understand the total system the mechanisms that feed water into these waterways, the rainfall, the natural spring and groundwater systems, should also be considered. The system is also made more complex because the sediment that lies on the river beds can become part of the flow and, in doing so, will gradually change the shape of the river channel. In the flows there will also be a range of chemicals which may be important. Suddenly the idea of dealing with this vast and complex system is looking beyond the range of what is possible.

**Figure 1.5** An Australian river system, located in the Northern Territory.

At this stage the thing to do is to stop and consider what aspects of the river system are of crucial interest. If a flood map is to be drawn then it may be possible to dispense with much of the detail on the bottom sediments other than to take account of them through an average roughness coefficient. The inflows can also be approximated by a single flow or range of design flows. It may even be possible to ignore flooding and the effect of floodplains and consider the system as a series of connected one-dimensional channels. If, at the other end of the detail scale, the real interest is on the behaviour of a set of piers on a particular bridge somewhere in the system, then a model that looked at only the immediate surrounds in very fine detail and ignored most of the other aspects would be more appropriate.

In choosing any of these courses of action, what is actually being done is modelling: moving to a conceptual world where particular features are idealised and the total problem is simplified. There are, of course, two aspects to this. Firstly, it is necessary to understand the system sufficiently to know what can be ignored and what cannot. That is where the engineering knowledge and judgement comes in and the importance of this ingredient cannot be overemphasised. Secondly, it is then necessary to be able to develop a solution to the model. That is where the application of mathematical methods and approximations come in. The primary aim of this book is to deal with the latter issue: the application of mathematical tools and methods to the solution of problems relevant to engineering.

In the forthcoming chapters a range of modelling tools will be described. Some may be directly applicable to problems such as river flows or the analysis of structural systems. Others will provide the necessary background information required to develop and run the models. One of the interesting features of the topics will be the level of interconnection that exists between them and this in turn should make becoming familiar and confident with them just that little bit easier.

## 1.3 ENGINEERING MODELLING

Engineering modelling is concerned with understanding an entire system, be it a river or bridge or building, and identifying the key components that are the focus of the current investigation. It is important to realise that these will vary as the focus varies. As Dandy et al. (2008) point out, the colour that a bridge is painted will have little relevance in a study of its structural behaviour but will be crucial in a study of temperature effects. Similar issues arise across the spectrum of engineering modelling.

As an example from the field of hydraulic engineering, some of the computer programs available from the Danish Hydraulics Institute for modelling water flow in the environment are listed in Table 1.1. The suite of programs allows one-dimensional, two-dimensional or three-dimensional problems to be modelled. The program MIKE FLOOD perhaps illustrates the modelling philosophy best by being able to be set up to cater for a range of dimensions within the one model.

With appropriate levels of simplification and mathematical techniques that are able to solve the resulting equations in an accurate and reliable fashion, numerical models offer a range of benefits to the engineer. Computer simulation allows expensive and time-consuming laboratory modelling to be dispensed with, in favour of digital computer solution of the equations that describe the system under study. The time-consuming component now becomes developing the numerical model, generating the input for the model and interpreting the output. There is some economy with this, though. A computer simulation model can be easily modified and applied to a new or alternative designs. Computer simulation is, however, not a panacea for every situation as numerical models can suffer from a range of problems too. As an example, the model may be:

- too complicated, and obscuring a process of interest;
- too simplistic with unrealistic assumptions;
- not validated against real data;
- incorrectly or inappropriately applied;
- run with the wrong input data giving incorrect output;
- incorrect, with numerical or programming errors; or
- overly time-consuming, taking so long to compute that it will not finish the computation before the structure is to be built.

**Table 1.1** Description of the DHI computer models. From http://www.dhigroup.com/Software/Download/DHISoftware2008.aspx (downloaded 20/2/08).

| Program | Description |
|---|---|
| MIKE 11 | One-dimensional simulation of hydrology, hydraulics, water quality and sediment transport in estuaries, rivers, irrigation systems and other inland waters. Ignores variations in depth and width along the channel. |
| MIKE 21 | Solves the vertically integrated equations of continuity and conservation of momentum (the Saint Venant equations) in two directions and includes water surface slopes, bed shear stress, momentum dispersion, Coriolis forces and wind forces. |
| MIKE 3 | Handles three-dimensional free-surface flows and is applicable to simulations of flows, cohesive sediments, water quality and ecology in rivers, lakes, estuaries, bays, coastal areas and seas. |
| MIKE FLOOD | Integrates three of the most widely used hydrodynamic models, namely MIKE 21, MIKE 11 and MIKE URBAN, into one package where the appropriate spatial resolution is applied where needed, e.g. pipes and narrow rivers are modelled using one-dimensional solvers whereas the overland flow is modelled using two spatial dimensions. |

## 1.4 PHYSICAL MODELS

Prior to the widespread introduction and use of computer models, engineers relied to a large extent on physical models. In physical models a structure may be built to scale and tested with suitably scaled loads, or a hydraulic feature may be constructed to scale and flow patterns and other behaviours observed. In these cases the scaling must be carried out taking account of the most important features that are to be modelled, and taking account also of the physics of the system. In testing Formula One cars, for example, a 1/10$^{th}$-scale model must be tested at 10 times the expected wind speeds if the drag forces are to be reproduced correctly. This explains the desire for large-scale test facilities that can handle half-scale, or larger, cars. Models of dams and spillways generally require similarity in the Froude number, which leads to time and velocity being scaled to the square root of the physical scales. The details of scaling laws in engineering are beyond the scope of this book, but the key point to grasp is that physical modelling is much more than simply building something to a chosen physical scale and seeing what happens under some arbitrary conditions.

A key problem with physical models, other than the difficulty in scaling the relevant aspects correctly, is the cost of constructing and running the model. It is true that in the past vast models of river basins have been constructed; however, the size and cost of running these is now leading to increased interest in computer modelling, where the change in the size of the main channel or the main beam is achieved at the touch of a button.

Having outlined some issues with physical models it is important to understand that they still have a role. There are situations where numerical solutions are still not possible and for these physical models are a valid way forward. Also, a physical model can be useful in assisting the engineer to understand the system and the problem.

## 1.5 MODERN ENGINEERING COMPUTING

As a means of illustrating the all-pervading nature of numerical modelling and analysis, the technical papers in a recent edition of the ASCE *Journal of Environmental Engineering*, (December 2007, Volume 133, Issue 12) were searched for the use of modelling and analysis methods. The results demonstrate the importance of numerical modelling and analysis in current engineering research and practice. The papers are listed in Table 1.2 and a summary of the applications is given in Table 1.3.

The results highlight the importance of numerical modelling and data analysis, with a high reliance on statistical analysis evident in the papers. While it might be argued that these papers are, by nature, research-focussed and therefore possibly unrepresentative of the wider engineering scope, it is equally true that higher-level engineering students should be able to cope with the methods being applied in research and that, in addition, this year's research is likely to be common practice in the next five years. The need for high-level skills across a range of numerical and analytical methods has never been greater.

Even in situations where the methods are not applied explicitly there is often an underlying assumption that the terms and concepts are understood and need not be stated formally. The concept of standard deviations or skewness of data are so well entrenched that it would not be considered necessary to define the terms or say how they are used, and yet many of the conclusions drawn in papers and articles might rely to a large extent on the reader understanding at quite a deep level what these terms are and what they actually mean when applied to engineering.

**Table 1.2** The papers in the *Journal of Environmental Engineering*, Volume 133(12).

| Paper No. | Paper title |
|---|---|
| 1 | Retardation of Nonlinearly Sorbed Solutes in Porous Media. |
| 2 | Dispersion Modeling of Leachates from Thermal Power Plants. |
| 3 | Total Coliform Survival Characteristics in Frozen Soils. |
| 4 | Spatial Variation of Sediment Sulfate Reduction Rates in a Saline Lake. |
| 5 | Strategy for Complete Nitrogen Removal in Bioreactor Landfills. |
| 6 | Evolutionary Multivariate Dynamic Process Model Induction for a Biological Nutrient Removal Process. |
| 7 | Evaluation of Membrane Processes for Reducing Total Dissolved Solids Discharged to the Truckee River. |
| 8 | Biohydrogen Production by Mesophilic Anaerobic Fermentation of Glucose in the Presence of Linoleic Acid. |
| 9 | Prevention Efficiencies of Woven Straw to Reduce PM10 Emissions from Exposed Area. |

**Table 1.3** The modelling and analysis content of the papers in the *Journal of Environmental Engineering*, Volume 133(12).

| Applications / Paper No. | 1 | 2 | 3 | 4 | 5 | 6 | 7 | 8 | 9 |
|---|---|---|---|---|---|---|---|---|---|
| Finite difference modelling of PDEs | ✓ | | | | | | | | |
| Solution of ODEs | | ✓ | ✓ | ✓ | | | | | |
| Optimisation using genetic algorithms | | | | | | ✓ | | | |
| Time series modelling | | | | | | ✓ | | | |
| Linear regression and correlation | ✓ | ✓ | ✓ | | | | ✓ | | ✓ |
| Confidence intervals | | | ✓ | ✓ | | | | ✓ | ✓ |
| General statistical analysis | | | | | | | | ✓ | |

## 1.6 A WORD OF WARNING

Before launching into this book, a word of warning. In many situations the algorithms set out will not be the ones that are now used in practice. For many of the topics people have developed and are now using much more sophisticated algorithms that leave the relatively simple ones in this book very dated. However, the methods and algorithms in this book are still valid for a number of reasons.

1. In many cases the algorithms represent the earliest stages of the development of the topic and it is useful to have a realistic view of the history of the methods to appreciate the level of sophistication that now exists.
2. More importantly, the algorithms are simple enough to allow the novice programmer to write valid code that works, and works well. There is little point trying to understand the more complex codes before the simple ones, and in many cases the complex ones are simply extensions of the simple ones.

3. The development of sophisticated algorithms and computer programs presupposes the ability to develop and write code. The methods in this book fulfil this need to be able to develop the necessary skills.
4. A significant number of programs, for example Microsoft Excel, assume that users are familiar with a wide range of numerical techniques and the help given within the program is of limited use to those unfamiliar with the material. Excel, for example, has a Fast Fourier Transform capability in its Data Analysis Toolbox, but there is little assistance given to, say, what the transform means or indeed what to make of the results.

## PROBLEMS

**1.1** Consider the design of a simple truss and how the reality is transformed into a model that an engineer can then solve. What are the key modelling assumptions used in the solution of a statically determinate system of elements?

**1.2** In the design of an urban drainage system it is usual to assume that one of the critical states is a steady design flow. Why is this done and what are the alternatives? What are the modelling and analysis implications of the alternatives?

**1.3** Find any engineering journal volume and investigate the application of modelling and analysis in the contributions. Rather than trying to identify the various tools, try and find any work that does not use computer modelling and analysis. What fraction of the volume do such contributions represent?

**1.4** A new freeway is being built to improve traffic flow through a congested city. List as many aspects as possible that might be relevant to the freeway design. Which aspects will require detailed modelling?

**1.5** A colleague suggests that computer models cannot be trusted and physical models are much better. Also, computer models need maintenance too. Are they correct?

**1.6** A soil sample is taken to a depth of 3 m. A computer model assumes that the soil profile is the same across all locations where a new three-storey shopping centre is to be built. Discuss whether this assumption is reasonable.

**1.7** Some models focus on peak or extreme events while others look at long-term behaviour (and some do both). Consider examples of sediment erosion, reservoir levels, traffic flow, the design of wind farms, a ski lift, a bridge, or an electricity network. List whether extreme events or long-term behaviour is applicable to engineering modelling and analysis of the system. What advantages and disadvantages are there using two separate models for peak loading and long-term loading instead of one?

# CHAPTER TWO

# Introduction (Accuracy, Speed and Algorithms)

## 2.1 INTRODUCTION

From the relative safety of the post-2000 era it is hard to describe the level of anxiety that existed around the world as the year 2000 approached. More important than the debate over when the old millennium finished and the new one started (either the end of 1999 or the end of 2000) was the concern that the ticking over of the computer clocks from 1999 to 2000 on a range of electronic devices, from computers to elevator microprocessors, would cause pandemonium. The issue came about because in the days when early computers were first being developed computer programmers had decided on an abbreviated storage mechanism for the date that effectively meant that the computer only tracked the last two digits (assuming all dates would start 19) so that when it ticked over from 1999 it would effectively go back to an assumed date of 1900. This was what was referred to as the Y2K (Year 2000) bug. It may seem an odd thing to do now when memory is discussed in megabytes and gigabytes, but at the time the decision was made, computer memory was scarce and expensive and savings were considered favourably.

There was real concern that this might cause some significant problems. Some said the banking system would collapse and there was panic buying of food and provisions. Duggan (1999), in a paper directed at the emergency medicine fraternity, suggested that "hospitals should prepare to feed a large number of people (patients, families, and staff) and provide alternate means of transportation. Disaster plans should be reviewed and revised to accommodate weakened water, electricity, and supply infrastructure." Others predicted that planes would fall out of the sky, and many scheduled flights were actually cancelled (just in case). A lot of countries spent significant sums of money on software patches to get around the problem. Where the computers were too old to accept the patch the machines were simply scrapped. World-wide costs were estimated to be between US$580 billion and US$3 trillion (Pickering, 1999; Phillimore and Davison, 2002).

In the end, the transfer to 2000 and beyond went smoothly, and even countries that had not spent up big preparing for the change seemed to cope (Ravetz, 2000). Perhaps the danger was over-stated. Whatever the case, it's hard to imagine a time when there has been more focus on computer programmers and their craft.

Without being overdramatic, the code that computer programmers, software engineers, and graduate engineers are writing at the moment has some chance of causing similar flutterings at some time in the future. Perhaps not in the same way, but the reality that was Y2K is an example of the fact that the power and utility of computers can easily be turned around into a real problem. Numerical modelling, computer analysis and computer programming in general are vitally important and this book is aimed at setting out some of the basics of these three inter-connected areas. That task, however, can wait, and before introducing the main topics covered in the book, a brief history of computers and computing will be given as a way of setting the scene.

## 2.2 A BRIEF HISTORY OF COMPUTERS AND COMPUTING

In the days before the World War II computers were not coloured lumps of metal and plastic (with a few trace elements) that sat *on* a desk, they were humans who sat *at* a desk. Using mechanical calculators and perhaps look-up tables these humans computed; they added, multiplied, divided and subtracted and, by following strict lists of instructions (algorithms), they calculated complex sums and results. There were, however, a couple of problems: the computers were slow, and prone to making mistakes. Although the speed was an issue it was the latter problem of accuracy that was of more concern. In fact, much of the early interest in mechanical or electronic computing devices was as a means of overcoming the mistakes made by the human computers. Charles Babbage (1791–1871), widely credited with designing (and attempting to build) the first mechanical computer, set out, initially at least, to automate the process of calculating and printing the tables that were used for navigation (Swade, 2000). At the time of his first interest, in the 1820s, these tables would come with errors and extensive errata sheets and then further errata sheets for the original errata sheets! The whole process was bordering on the unworkable.

It is generally reckoned that Babbage was ahead of his time and certainly ahead of what was possible technically at the time, in terms of instrument precision and reliability. His computer, or difference engine as it was called, was never finished, although the London Science Museum built a working version of Babbage's design, completing it in 1991. The first computer programmer is often reckoned to be Ada Lovelace, who worked with Babbage on aspects of his difference engine. She wrote an algorithm to compute Bernoulli numbers and published an article on the machine. The article was partly a translation from an earlier paper written in French, but in preparing it she added extensive notes and explanations (Swade, 2000).

As time went by, the accuracy of calculations from human computers was still an issue but increasingly speed was also a concern. Goldstine (1990) gives details of a mechanical device used in the 1930s to calculate the flight paths of missiles. It was reckoned accurate to 5 parts in 10,000 and was able to determine a typical trajectory in 10 to 20 minutes. This would have taken a human computer 7 hours, so the time saving was significant. This particular problem then led to the development of the first electronic digital computer, ENIAC (Electronic Numerical Integrator And Computer). According to Goldstine (1990) the developers "were keen on the problem of automating dull tasks that could be done better by machine than by human". Speed was also an issue: the ENIAC, with its electronic components, was 100 times faster than human computers (Gear and Skeel, 1990). It is of interest to note that, due to the timing of these developments and the war that was being waged at the time, many of the early programmers were women.

At the time of the ENIAC (mid 1940s) a further challenge for computing was being raised: the need to be able to extend solutions from linear systems and approximations to the full non-linear versions of the equations. This was one particular field of endeavour that John von Neumann was keen to pursue and one in which serious advances were made. The breakthroughs came in the form of machines that were able to work on a range of quite different problems by being externally programmed (through the use of punched cards or punched tape). The development of the solid-state transistor in the 1960s by William Shockley, Walter Brattain and John Bardeen provided a significant boost for the electronic computer, mainly in terms of size, speed and reliability.

**The First Computer Bug**

Rear Admiral Grace Hopper was one of the early computer programmers, working at Harvard on an early machine. At the time computers took up entire rooms and one day an erroneous result was traced to a moth that was trapped in one of the electrical relays. Taylor (1984) wrote that Hopper taped the insect into her logbook which has been preserved. The entry states "first actual case of bug being found". Interestingly, this entry indicates that the term "bugs" had been associated with errors prior to this point, but this was the first case of an "actual bug". The expression, and the moth, have both been retained to this day.

In terms of what is required from modern computing the issues of the past point the way. The most basic need is one of accuracy: results from computers must be correct to a known tolerance. (This concept of accuracy which relies on an answer being sufficiently close to the exact value is crucial to the whole subject of numerical modelling and analysis and will come up frequently throughout the book.) At the next level there is speed: the sorts of tasks carried out on computers are becoming more and more complex and require more and more computing power. In many situations the job that can be run is limited purely by how long people are prepared to wait for an answer. Early meteorological models conceived by John von Neumann, for example, were deliberately designed based on their run times. He argued that "whether one does a simple 24-hour forecast in half an hour or in two minutes is not decisive. But in a 30-day hemispheric calculation it is very important whether one needs 24 hours or a month. If it takes a month one will probably not do it. If it takes 24 hours, one may be willing to spend several months doing it 20 times, which is just what is needed." (von Neumann, 1963, cited in Goldstine, 1990).

Finally, the highest level of computer program should also be sophisticated enough to deal with the mathematical and numerical complexity presented, and this will often mean sophisticated algorithms that have the ability to compute at the highest levels of accuracy and stability. These requirements of accuracy, speed and sophistication mean that computer programs must produce accurate results, run as fast as possible given the constraints of providing accurate results, and be based on sophisticated algorithms that are mathematically correct and have characteristics that are known to produce predictable behaviour.

The people who employ or engage professionals to produce computer code or automated routines are not interested in making the programmer's life easy, and expect the best that can be achieved.

**Computing in 1967**

An article in *The Australian* newspaper on 12 December, 1967 outlined the research being undertaken at the time to allow the home television set to be used as a display for a home or small business computer. As an aid to appreciating the benefit of owning a computer the paper outlined the general procedure for using one: *The operator simply types out on a keyboard the problem to be solved and the solution given by the computer is displayed instantaneously on the screen.* (If only it were so simple!)

## 2.3 TOPICS AND THEIR FOCUS

As the history of computers has developed, so too has the level of sophistication of algorithms run on these computers. These algorithms have not developed in isolation and the task of selecting a suitable range of topics for an engineering modelling and analysis text is an interesting one. The aim of this section is to set out the topics that have been selected and to do something that is seldom done: to show the rich web of connections that exist between what are, in many cases, apparently quite different topics. Table 2.1 lists the main topics in order as they appear in the table of contents. The order is based on a logical approach to the topics such that they generally increase in difficulty and complexity. The treatment of these topics is intended for an undergraduate audience, but it is important to note that any of these topics could be pursued in the literature to more advanced levels.

**Table 2.1** Topics as listed in the table of contents and an indication of their characteristics.

| Topic | Type | Level |
|---|---|---|
| Roots of Equations | Numerical | Introductory |
| Numerical Integration | Numerical | Introductory |
| Numerical Interpolation | Numerical | Introductory |
| Systems of Linear Equations | Numerical | Introductory |
| Numerical Solution of ODEs | Numerical | Intermediate |
| Finite Difference Modelling | Numerical | Intermediate |
| Probability and Statistics | Statistical | Intermediate |
| Probability Distributions | Statistical | Intermediate |
| Monte-Carlo Method | Statistical | Intermediate |
| Stochastic Modelling | Statistical | Advanced |
| Optimisation | Statistical | Advanced |
| Spectral Analysis | Numerical | Advanced |

The topics chosen for the book have been selected in such a way as to provide a reasonable introduction to statistical and numerical methods without attempting to provide an exhaustive list of options. Some of the topics have been presented in the context of modelling while others have been presented in the context of analysis. However, this does not imply that those topics are limited to that context. For example, in the design of a marina, ocean wave data may be analysed to determine an appropriate spectral model, which then requires the output of the model to be analysed and verified against the observed data.

In addition to the main topics there are a number of short appendices each covering a quite specific mathematical tool. The Taylor series, for example, comes up in a number of topics throughout the book and it was felt important that it was covered in a concise form somewhere.

## 2.4 TOPICS AND THEIR CONNECTIONS

Even before any of the topics are investigated it is worth realising that there are few that sit entirely on their own without connections to others covered in the book. Figure 2.1 sets out the topics and the links between them. Some key points to explain the links between topics are listed in Table 2.2. The list is not exhaustive.

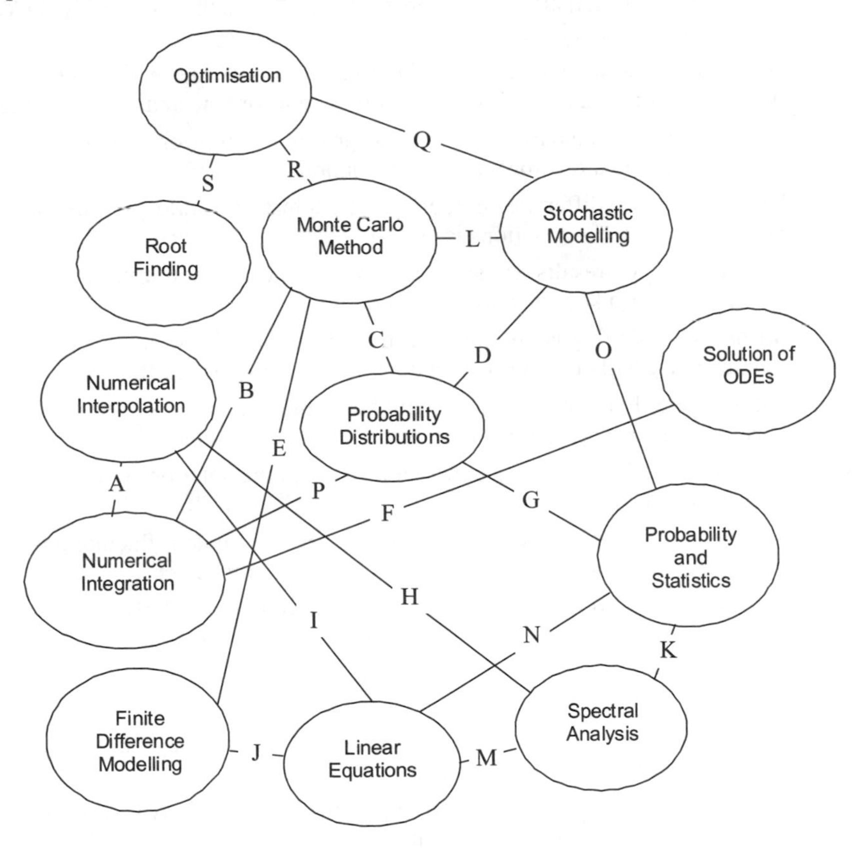


**Figure 2.1** Topics and their connections.

**Table 2.2** A brief description of the main links between topics.

| Link | Connection |
|---|---|
| A | Numerical integration uses the polynomial fitting employed in numerical interpolation as the basis for some of the higher-order methods. |
| B | A common demonstration of the Monte Carlo method is a numerical integration of a function. |
| C | The Monte Carlo method workings are largely based on the generation and application of random numbers with a particular probability distribution. |
| D | Stochastic modelling is largely based on the use and properties of a range of probability distributions. |
| E | Finite difference modelling and the Monte Carlo method are two quite different modelling approaches that can be used to tackle similar problems. |
| F | The solution of ordinary differential equations leads to a numerical method that can be applied to the integration of appropriate functions. |
| G | Probability and statistics and probability distributions are clearly linked through the common interest in probability. |
| H | Numerical interpolation and the Fourier transform are both centred on fitting data using standard functions: polynomials or sine and cosine waves. |
| I | Cubic splines, a standard curve fitting procedure, rely on the solution of simultaneous equations for their determination. |
| J | All implicit finite difference models require a matrix solution and the link to the systems of linear equations is strong. |
| K | Spectral analysis results are subject to statistical variation and the concept of confidence is a key issue to be considered. |
| L | Stochastic modelling is an application of the Monte Carlo method, usually in time, and usually to determine a complex property or quantity. |
| M | The discrete Fourier transform can be expressed as a series of linear equations and can be implemented using matrix operations. |
| N | The solution of multivariate regression requires the solution of a system of linear equations. |
| O | Time series models in stochastic modelling involve random fluctuations that are correlated in time and these are tied to probability issues. |
| P | Numerical integration is often required to determine cumulative probability distributions – for example, in the case of the normal distribution. |
| Q | The calibration of stochastic models is essentially an optimisation problem and this provides a strong link between these topics. |
| R | Some global optimisers use random jumps in a similar way that the Monte Carlo method randomly samples a probability distribution. |
| S | In "hill climbing" optimisation techniques the root of a derivative is used to identify a local maximum or minimum. |

Within particular topics there are also connections. One example, the normal distribution and its relationships to other probability distributions, is shown in Figure 2.2. Once again, there is a rich web of inter-connections and this should encourage ongoing learning as the reader investigates further topics.

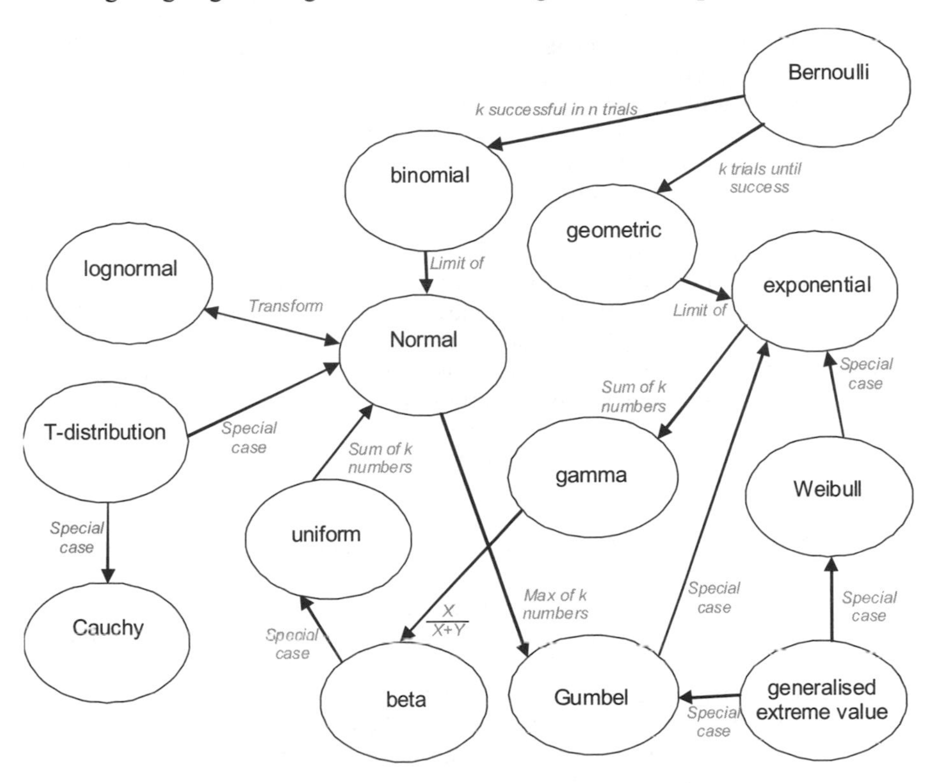


**Figure 2.2** The normal distribution and its connections.

## 2.5 DEVELOPING COMPUTING SKILLS

A fundamental theme that runs through all of this book is that the best way to learn computing is to do computing. It is all very well to know and understand the theory behind the topics, but that is not enough. It is useful to be able to recognise good code but, once again, that is not enough either. Computer programming comes from practice; lots of practice. In programming there will be few "eureka" moments: the sudden rush that comes when something suddenly becomes clear. It is more likely that progress will come in small steps in understanding, reinforced by successfully coding an algorithm or solving a particular problem.

The individual steps may be small but the satisfaction that comes from successfully coding an algorithm to solve a problem not otherwise achieved by hand is worth the effort. Modern desktop computers are extremely powerful and with "a few keystrokes" this power is at the fingertips of the diligent programmer.

## 2.6 NUMERICAL RECIPES

It might be thought that in this day and age there would be little call for programmers who can code solutions themselves in what might be regarded as old fashioned 'scientific' languages. Publications such as Numerical Recipes by Press et al. (1992) that provide callable routines in Fortran, C and Pascal might be thought to make obsolete much of the need to learn the basics. This may be partly true, but only partly. It may well be that in an engineering career, if a problem arises that can be solved with existing code, such as in Numerical Recipes, the best course of action is simply to use those routines. That is true, but it is unrealistic to expect an engineer who has little training in computer programming to be able to select the appropriate routine, or to know how to implement it, or to even be able to recognise when it has produced a realistic answer.

It may be that having completed this book, the engineer is then ready to use the high-powered and sophisticated routines available in Numerical Recipes. If that is so, then the authors will rest easy, sure in the knowledge that the training received working through this book has prepared the engineer to use and understand any pre-existing computing routine and to use them intelligently.

## PROBLEMS

**2.1** When designing a program, which is more important: accuracy or speed? Why?

**2.2** What are some advantages and disadvantages of electronic computers compared to human computers?

**2.3** The time someone is prepared to wait for a program to run depends on the importance of the results. Describe programs where the user will (i) want instant answers, (ii) wait for a few hours, (iii) wait for several days, (iv) wait for up to a month, and (v) wait up to a year.

**2.4** An algorithm is a list of instructions. Write a list of instructions on how to get from home to the city centre. Would the list be different if it were for a local person or a foreign traveller? How many additional instructions would be required to make the home-to-city algorithm 100% reliable?

**2.5** Investigate Moore's Law. What is it and is it still accurate in its predictions?

**2.6** The Chi-square distribution is most commonly used in verifying the fit of experimental data to a particular distribution. Investigate the distribution and see which of those shown in Figure 2.2 it can be linked to. Determine the nature of the links.

**2.7** Select a topic from those shown in Figure 2.1 and verify and expand, where possible, the links. It is likely that there are more than those shown although some will be stronger than others. Some may be quite tenuous.

CHAPTER THREE

# Roots of Equations (Introduction)

## 3.1 INTRODUCTION

In open channel flow an important concept is that of normal depth. This is the depth a given flow would reach under steady-state equilibrium conditions when the channel cross-section and slope are constant. The uniform flow which results is a balance between the driving gravity force and the resisting friction force around the perimeter of the channel. As an example, the expression that relates the key variables in a rectangular channel can be written:

$$Q = \frac{d_n B}{n}\left(\frac{d_n B}{B + 2d_n}\right)^{2/3} S^{1/2} \tag{3.1}$$

where $Q$ is the flow rate, $B$ the width of the channel, $S$ the slope of the channel and $d_n$ the normal depth. The friction is described using Manning's roughness coefficient, $n$, which may vary between 0.01 for very smooth surfaces and 0.1 for rough, heavily vegetated floodplains. A typical value for a concrete channel, such as that shown in Figure 3.1, is 0.014.

**Figure 3.1** Concrete drainage channel.

Normal depth is of great interest when solving open channel flow problems. As an added issue, in numerical models designed to solve for flow conditions Equation (3.1) will often have to be determined thousands of times during the course of a channel design leading to a need for speed of solution. There is, however, a problem: it is impossible to rearrange Equation (3.1) in such a way that a single unknown, $d_n$, is on the left hand side and the knowns ($Q$, $B$, $n$ and $S$) are on the right hand side; the form of the equation does not allow this. As way of a solution, it is possible to guess a value for $d_n$ and see if it satisfies the equation; then, if it doesn't, simply guess again. However, this is hardly a rapid or reliable way to a solution!

In engineering there are a number of situations where this sort of problem arises. This chapter introduces some of these problems and the techniques that may work. However, before that it is worth giving some background to the topic.

## 3.2 WHAT ARE ROOTS?

One of the early tools that students learn about in algebra is how to solve a quadratic equation of the form: $ax^2 + bx + c = 0$. The x values that arise from the solution of this equation are the roots. These are the location or locations where the value of the function is equal to zero. For a quadratic equation the solution can be written in terms of the coefficients $a$, $b$ and $c$ by completing the square:

$$x = \frac{-b \pm \sqrt{b^2 - 4ac}}{2a} \tag{3.2}$$

For example, the function where $a = 1$, $b = 5$, $c = -10$ is shown in Figure 3.2 and the solution of Equation (3.2) gives $x = 1.53, -6.53$. The graph is important because a deep understanding of the numerical methods comes from appreciating the form of the function and in particular what happens to the function in the vicinity of the roots, i.e. where $f(x) = 0$.

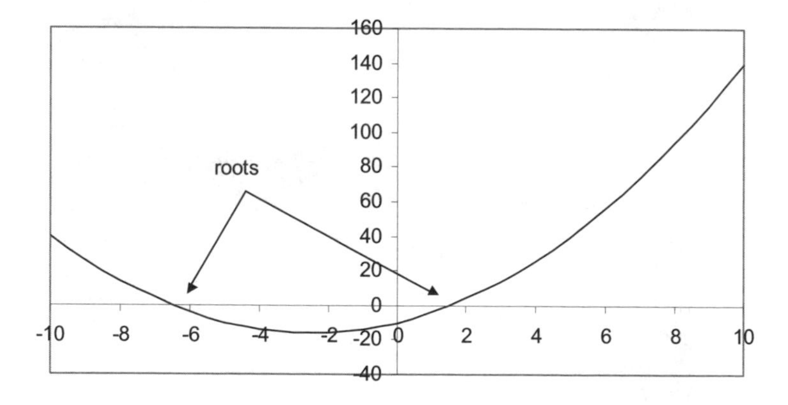


**Figure 3.2** The roots of a parabola.

Matlab allows for an easy solution of polynomials. The solution of an equation such as $f(x) = x^2 + 5x - 10$ can be determined by defining the polynomial and using the root command. An example is shown in Figure 3.3.

```
>> p=[1 5 -10];     % defines polynomial
>> r = roots(p)     % determines roots
r =
   -6.5311
    1.5311
```

**Figure 3.3** Matlab code to solve a polynomial root-finding problem.

This type of solution, using Equation (3.2), is called explicit since the solution can be determined directly with the unknown on the left hand side of the equation and all known quantities on the right hand side. Although there are applications for the solution of quadratic equations in mathematics – for example, in the solution of second order differential equations with constant coefficients – there are many more situations in engineering where it is required to find roots of a general equation $f(x) = 0$ where the function is not a simple linear or quadratic relationship with an explicit solution. In these situations it is necessary to be able to apply other methods to achieve a solution.

Returning to the normal depth problem, it is possible to re-arrange the equation for normal depth as:

$$Q - \frac{d_n B}{n}\left(\frac{d_n B}{B + 2d_n}\right)^{2/3} S^{1/2} = 0 \tag{3.3}$$

Since $d_n$ is the only unknown, this is of the form $f(d_n) = 0$ and, in general, one seeks methods to solve the equation $f(x) = 0$ where $x$ is the unknown. It should now be clear that if it is possible to find the root of this equation, this will give the solution for normal depth, $d_n$. Before looking at some simple solution methods it is worth investigating other equations where the same solution problem exists.

## 3.3 ENGINEERING PROBLEMS REQUIRING ROOT FINDING

Another example of a situation where root-finding methods are required is in the calculation of wavelength (see Figure 3.4) in water waves. Here the formula is:

$$L = \frac{gT^2}{2\pi}\tanh\left(\frac{2\pi d}{L}\right) \tag{3.4}$$

where $g$ is the gravitational constant, $T$ is the wave period, $d$ is the depth and $L$ is the wavelength. Notice that $L$ appears on both sides of the equation and, once again, it is not possible to isolate it. To solve using root-finding techniques the expression is re-written:

$$f(L) = L - \frac{gT^2}{2\pi}\tanh\left(\frac{2\pi d}{L}\right) = 0 \qquad (3.5)$$

or $f(L) = 0$. It is evident that the general method for setting up a root-finding solution is to re-arrange the equation such that all the terms are on one side of the equation and this is equated to zero.

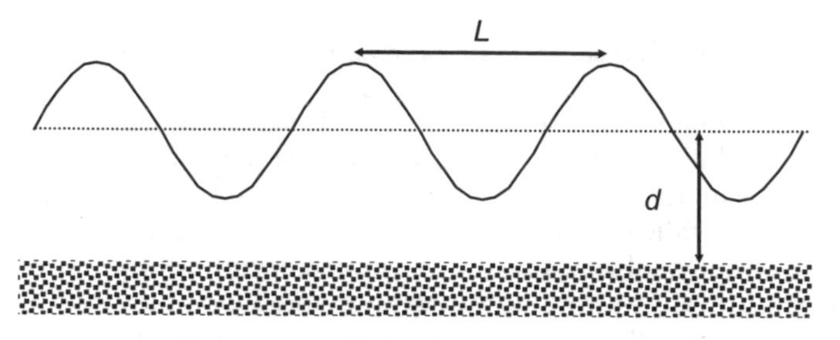


**Figure 3.4** Wavelength, *L*, on a sinusoidal ocean wave.

Another example of an equation that can be expressed as a root-finding problem comes from financial analysis where the quantity, Internal Rate of Return (IRR), can be calculated by determining a discount rate where the present value of all benefits exactly equals the present value of all costs (that is, the Net Present Value is zero). The need to calculate this quantity can arise in engineering project assessment analysis where a variety of possible schemes are being compared on economic grounds. The solution for the Internal Rate of Return can be written:

$$\sum_{t=0}^{n} \frac{B_t}{(1+r)^t} = \sum_{t=0}^{n} \frac{C_t}{(1+r)^t} \qquad (3.6)$$

where $B_t$ is the benefit in year $t$, $r$ is the unknown Internal Rate of Return and $C_t$ is the cost in year $t$. In a particular example, where there was an initial cost of \$400 m, a final cost after 5 years of \$620 m and annual revenue over the five years of \$200 m, the net present value (NPV) (which must equal \$0 for the IRR) can be written as:

$$NPV = 200\frac{1-(1+r)^{-5}}{r} - \frac{620}{(1+r)^6} - 400 = 0 \qquad (3.7)$$

Plotting *NPV* against $r$ in Figure 3.5 shows that there are in fact two solutions between 0 and 0.2, one showing an implied interest rate of approximately 4% and the other around 16%. Although the graphical approach is useful and provides the engineer with a good feel for the function and its behaviour, a numerical method to determine these values will often be required.

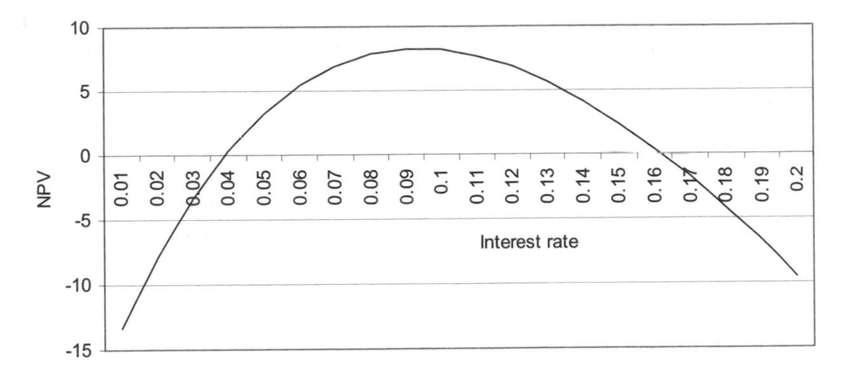


**Figure 3.5** NPV v discount rate, $r$.

## 3.4 WORKED EXAMPLE

It would be possible to look at a real engineering example of a root-finding problem like normal depth or wavelength, but for ease of understanding it is more convenient to use a simple function so that the problem of root finding is not lost beneath the problem of understanding the equation that is being solved.

*Problem*: Find $x$ such that $x = e^{-x}$ .

*Preparation:* Since $x$ appears on both sides of the equation and cannot be isolated, a root-finding technique is required. To formulate the equation in the required format it is re-arranged as:

$$f(x) = x - e^{-x} = 0$$

*Solution 1*: The first solution method is simply guessing and gradually refining the guess until the answer is sufficiently close. Guessing might proceed as shown in Table 3.1 and conclude that the required answer is $x \approx 0.56$. As a strategy for a solution it is worth noting that if a guess gives a positive value of $f(x)$ and another guess gives a negative value of $f(x)$, then the root will be between these two guesses. In this example, guessing $x = 0$ and $x = 1$ shows that the root is between these two values.

**Table 3.1** Root finding by guessing.

| $x$ | $e^{-x}$ | $f(x)$ |
|---|---|---|
| 0.00 | 1.00 | –1.00 |
| 1.00 | 0.37 | 0.63 |
| 0.50 | 0.61 | –0.11 |
| 0.60 | 0.55 | 0.05 |
| 0.56 | 0.57 | –0.01 |
| etc. | | |

*Solution 2*: It is also possible to plot the equation and determine the root from that. This is a good procedure for gaining an understanding of the function and what it looks like but it is difficult to read the root accurately in most cases. An example plot is shown in Figure 3.6.

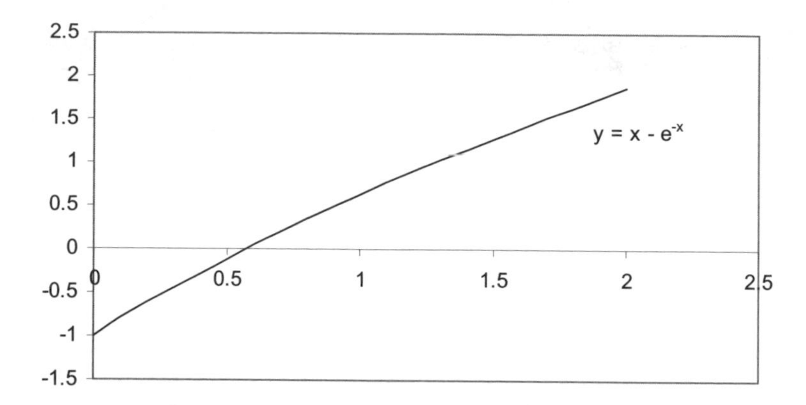


**Figure 3.6** A plot to determine where $x = e^{-x}$.

These methods are both acceptable but a bit slow, tedious and definitely not adaptable to fast efficient computer solutions. As mentioned earlier, they are important in understanding the way that the numerical methods work. In the next chapters fast and reliable computer methods will be outlined. However, before moving on to a range of methods it is worth looking briefly at the in-built capability of Excel for solving some of these problems.

Modern scientific calculators often have an excellent non-linear root-finding capability and it is worth the effort mastering this tool. It is also worth remembering that in a situation with limited technology available, the trial and error method can work quite effectively.

## 3.5 GOAL SEEK IN EXCEL

Excel has a function called "Goal Seek" which is under "Tools" on the drop-down menu. It allows the solution of these sorts of problems using an automatic iterative procedure. Figure 3.7 shows the two stages in a solution of wavelength. In Figure 3.7(a) $T$ and $d$ have been defined and a guess for wavelength has been given a nominal value of 1.0 m. A new cell has been set up that contains the function in a root-finding format, i.e. $L - gT^2/2\pi \tanh(2\pi d/L)$. In Figure 3.7(b) the solution has been obtained and is shown in the cell containing L. The error in convergence is also calculated.

Since only one value is required to start the iteration, the method cannot be a bracket method (discussed in Chapter 4). This has one major implication: a solution is not guaranteed. The Goal Seek process is usually applied manually on a worksheet but can be automated using a macro.

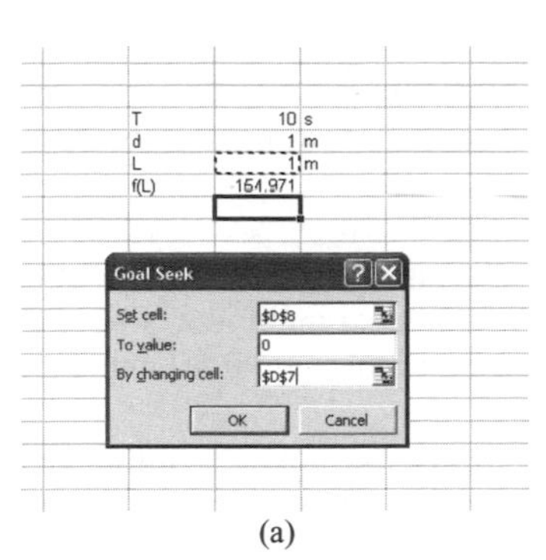


(a)

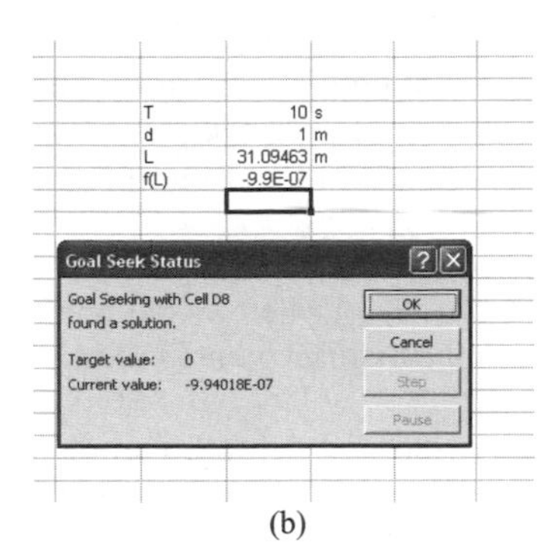


(b)

**Figure 3.7** A screen dump showing the operation of Goal Seek in Excel.

## PROBLEMS

**3.1** Find the roots of $x^2 - 6x + 5 = 0$ by:

a) Plotting the function and determining the approximate $x$ values where the function value is zero.
b) Using Equation (3.2).
c) Using a trial and error technique with initial guesses at $x = 0$ and $x = 3$.
d) Using a trial and error technique with initial guesses at $x = 0$ and $x = 10$. It is not possible to be certain if there are one or more roots between these to points. Why?

**3.2** The simple equation $x^2 + 1 = 0$ would confound a numerical search for a root. Why?

**3.3** Solving $x^2 + 1 = 0$ leads to the solution: $x = \pm i$. Explain this value $i$ and discuss the implications for a real engineering problem searching for a solution.

**3.4** Plot the function $x^2 - 4x + 4 = 0$.

a) Why is it impossible to find a bracketing pair of x values (one positive function value and the other negative) for this equation?
b) Explain why this could be problematic for a numerical root-finding technique.
c) Identify, if possible, three general scenarios for roots of quadratic equations and discuss their implication for numerical searching of the roots.

**3.5** The equation $x^2 - 4x + 4 = 0$ can be re-arranged into the form $x = \sqrt{4x - 4}$. Starting with an initial value of $x = 10$ for the expression on the right of the equals sign, evaluate a new $x$ value. Keep iterating for about 10 steps.

a) Describe what happens to the value of $x$.
b) Try another rearrangement and see what happens.

**3.6** Find where the cubic $y = x^3 - 3x + 2 = 0$ intersects $y = 3x^2$.

a) Plot both curves and estimate their intersections.
b) Rearrange the equation and use the Matlab roots function to check the value of roots from the graphical answer.

**3.7** The movement of a pipeline due to an earthquake is given by the equation $y = 34e^{-kt}\cos(wt)$ where $w = 4$ and $k = 0.4$. Use a graphical method to find the time at which the displacement is reduced from its initial 34 mm to only 10 mm.

**3.8** In the normal depth equation, Equation (3.1), discuss how the initial guesses might be chosen if there is an idea of the solution range. Think about the physical properties of the phenomena being modelled.

**3.9** Consider the equations $ax^{2.1} - bx + c = 0$ and $ax^2 - bx + c = 0$. Which of these equations has an explicit solution? How would the other one be solved?

**3.10** Examine the sketch, shown in Figure 3.8, of the equation:

$$y = \frac{3x - 7}{(x - 3)^2}$$

Explain why this function is going challenge any automatic root-finding technique.

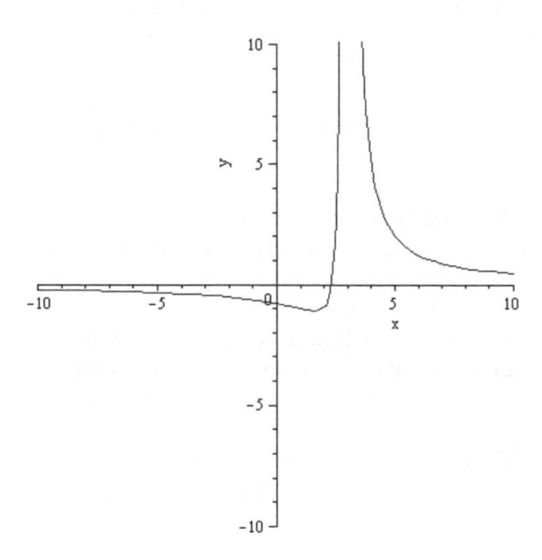


**Figure 3.8** Graph of function where the roots are required to be found.

CHAPTER FOUR

# Roots of Equations (Bracket Methods)

## 4.1 INTRODUCTION

In Chapter 3 it was shown that the task of finding the length of a wave in water of variable depth or normal depth in an open channel could be cast as a particular class of problem: that of finding the roots of an equation. Mathematically this can be expressed as finding $x$ such that $f(x) = 0$. This type of problem often arises in situations where the solution must be determined over and over again and some sort of numerical technique is generally required. The combination of modern computers and numerical techniques provides a powerful engineering tool. One numerical technique that is particularly well suited to this type of problem is based on starting with two guesses for the root, one larger, the other smaller than the actual value, and then gradually refining these until a suitable solution is reached. This is known as a bracket method and is widely applied in engineering.

## 4.2 BRACKET METHODS IN GENERAL

Suppose a solution is required for the equation $x = e^{-x}$. The first step is to re-arrange the function into the standard format such that it is written as a root-finding problem: $f(x) = x - e^{-x} = 0$. The function, once it is in this form, is shown in Figure 4.1. The plot also contains two nominal values, $a$ and $b$, and the value of the function evaluated at those two points, $f(a)$ and $f(b)$.

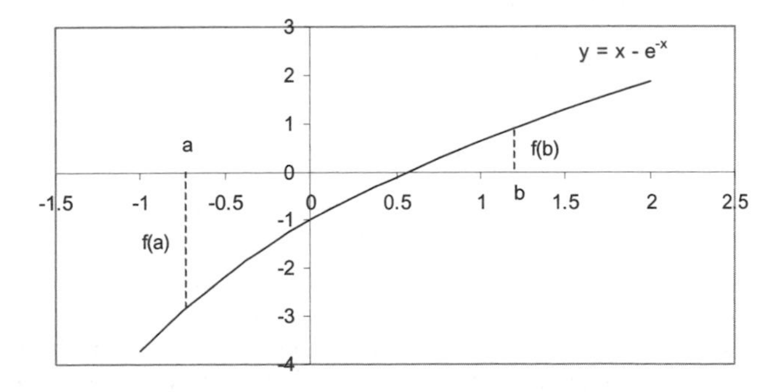


**Figure 4.1** Plot of the function used to solve: $x = e^{-x}$.

The important point to note from Figure 4.1 is that, with $a$ and $b$ chosen appropriately, $f(a)$ and $f(b)$ will be on opposite sides of the $f(x) = 0$ axis and, therefore, the product $f(a)f(b)$ will be negative: this is the key to bracket methods. There is nothing to say that $a$ must be less than $b$, only that the product of their function values is negative. Putting it another way, any two values, $a$ and $b$, that have a negative function product will have at least one root of the equation between them provided the function is continuous; they are said to bracket the root.

This leads to a number of methods where, given initial values $a$ and $b$, a solution can be determined by a search *inside* the interval. The search involves generating a new estimate based on the two start values and then refining the bracketing intervals. With each new estimate of the root that is determined, the search continues with two values, one on each side of the root. It is important to note that there is no need for the initial values for $a$ and $b$ to be anywhere near the solution, just for $f(a)f(b)$ to be negative. A discussion on strategies for determining suitable start guesses, $a$ and $b$, is given later.

## 4.3 BISECTION METHOD

With two values $a$ and $b$ such that $f(a)f(b) < 0$ it is known that a root of the equation lies between these two values. The key to a solution is to gradually refine $a$ and $b$ such that the root can be identified. One way to proceed is to determine a new estimate of the solution by assuming that the root is at the midpoint of $a$ and $b$. This new guess can be labelled $c$. The search then continues using $c$ to replace either $a$ or $b$ depending on which of the two ($a$ and $c$, or $b$ and $c$) bracket the root. Having done that the procedure continues by calculating another new $c$, which is again midway between the updated $a$ and $b$. The method continues until $f(c)$ is sufficiently close to 0 or is at a point where successive guesses of $c$ are changing very little. In the end the value of $c$ is the root (and therefore the solution). The method is set out in Figure 4.2.

```
Select initial a0 and b0 such that f(a0)f(b0) < 0
Then for i = 0,1,2, ... until termination:
    Estimate c = (ai+bi)/2
    If |f(c)| < tolerance then accept c as solution
    If f(ai)f(c) < 0 then
        ai+1 = ai; bi+1 = c
    Else
        ai+1 = c; bi+1 = bi
    Test for termination
```

**Figure 4.2** Pseudocode for the bisection method.

Working with the equation: $f(x) = x - e^{-x}$ and starting with an initial guess of $a = 0$ and $b = 1$ the results, listed in Table 4.1, show that a solution, correct to 6 decimal places, occurs after 12 iterations. It is evident that the method converges reasonably quickly and, because of the bracketing, is a robust approach that is guaranteed to converge. In many situations, this is a significant bonus. A Fortran program to carry out the calculations is shown in Figure 4.3. The program produces the numerical output (but not the column headings) shown in Table 4.1. Notice in the program that results are printed to the screen (unit 6) and to a file (unit 1). The file can then be imported into applications such as Excel or Word (for example).

**Table 4.1** Steps in the solution of the equation $f(x) = x - e^{-x}$.

| $a$ | $f(a)$ | $b$ | $f(b)$ | $c$ | $f(c)$ |
|---|---|---|---|---|---|
| 0.0000 | –1.0000 | 1.0000 | 0.6321 | 0.5000 | –0.10653 |
| 0.5000 | –0.1065 | 1.0000 | 0.6321 | 0.7500 | 0.27763 |
| 0.5000 | –0.1065 | 0.7500 | 0.2776 | 0.6250 | 0.08974 |
| 0.5000 | –0.1065 | 0.6250 | 0.0897 | 0.5625 | –0.00728 |
| 0.5625 | –0.0073 | 0.6250 | 0.0897 | 0.5938 | 0.04150 |
| 0.5625 | –0.0073 | 0.5938 | 0.0415 | 0.5781 | 0.01718 |
| 0.5625 | –0.0073 | 0.5781 | 0.0172 | 0.5703 | 0.00496 |
| 0.5625 | –0.0073 | 0.5703 | 0.0050 | 0.5664 | –0.00116 |
| 0.5664 | –0.0012 | 0.5703 | 0.0050 | 0.5684 | 0.00191 |
| 0.5664 | –0.0012 | 0.5684 | 0.0019 | 0.5674 | 0.00038 |
| 0.5664 | –0.0012 | 0.5674 | 0.0004 | 0.5669 | –0.00039 |
| 0.5669 | –0.0004 | 0.5674 | 0.0004 | 0.5671 | –0.00001 |

```
program Bisection
! root finding problem using bisection
!                       written d walker, nov 2007
implicit none
real:: f,tol,a,b,c
integer:: its

tol = 0.0001
write(*,*)'Type in guesses for a and b'
read(*,*) a,b
do while(f(a)*f(b)>0.0)
 write(*,*)'Poor guesses, try again'
 read(*,*) a,b
end do
its = 0

c=(a+b)/2.0                     ! need this c value to get into loop
do while(abs(f(c)) > tol.and.its < 100)
 c=(a+b)/2.0
 write(*,'(6(f8.4,2x))') a,f(a),b,f(b),c,f(c)       !output to
screen
its=its+1    ! increment counter
 if(f(a)*f(c) < 0.0) then
  b=c                           ! new b, a the same
 else
  a=c                           ! new a, b the same
  endif
end do                          ! at end of loop, result in c
close(1)
stop
end program

real function f(x)
implicit none
real:: x
f = x - exp(-x)
end function
```

**Figure 4.3** Fortran code for the bisection method.

In the program the value *tol* has been set to an arbitrarily small number and, in this case, the method has been limited to 100 iterations. It is not necessary to set $a=a$ or $b=b$ (as shown formally in the algorithm) so these steps have been omitted in the program. Otherwise the code is in close agreement with the algorithm. Note also that although the algorithm seems to keep track of the various values of $a$, $b$ and $c$, there is no need to actually do this in the program. Old guesses arc of little usc and are over-written.

## 4.4 METHOD OF FALSE POSITION (REGULA FALSI)

Although the bisection method is reliable, simply looking at the mid-point of $a$ and $b$ for the new guess does not take into account the observation that, if $f(a)$ is closer to zero than $f(b)$ it would be better to take this into account and look for a new guess $c$ closer to $a$ than $b$. The method of false position does this. By using similar triangles ($a$, $c$, $d$) and ($c$, $b$, $e$) in Figure 4.4, and solving for the intersection point on the $x$ axis, it is possible to get a new estimate. The next guess ($c$) from an original $a$ and $b$ is shown in Figure 4.4.

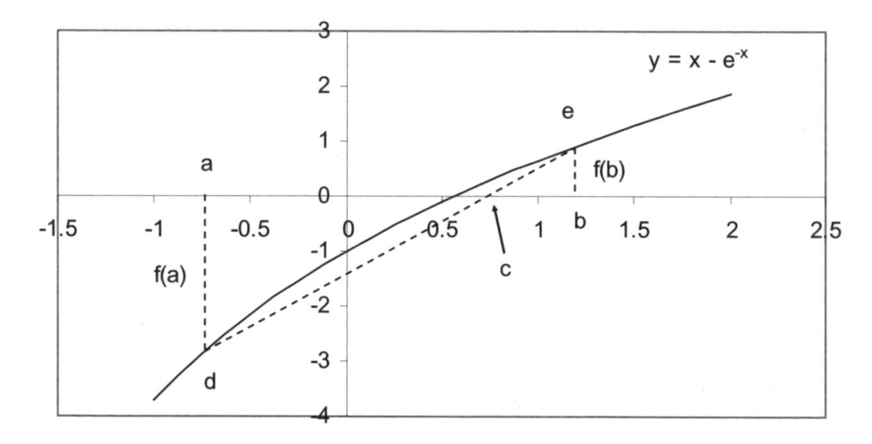


**Figure 4.4** Plot of the function used to solve: $x = e^{-x}$ and the next guess, $c$, using the method of false position.

It is evident that this provides an excellent next guess and one that is closer to the root or solution than would have been selected using the bisection algorithm. The proof, using similar triangles in Figure 4.4, can be listed as:

$$\frac{ad}{ac} = \frac{be}{cb} \tag{4.1}$$

$$\frac{-f(a)}{(c-a)} = \frac{f(b)}{(b-c)} \tag{4.2}$$

$$-f(a)(b-c)=f(b)(c-a) \tag{4.3}$$

$$-f(a)b+f(a)c=f(b)c-f(b)a \tag{4.4}$$

$$af(b)-bf(a)=c(f(b)-f(a) \tag{4.5}$$

$$c=\frac{af(b)-bf(a)}{(f(b)-f(a))} \tag{4.6}$$

Once again the value of $c$ is substituted for either $a$ or $b$ in the next iteration depending on the product of the functions in the same way as the bisection method. This is to make sure that the solution is still properly bracketed. The algorithm is almost identical to that for the bisection method. The only difference is the way that $c$ is estimated in each iteration. The method is set out in Figure 4.5.

Select initial $a_0$ and $b_0$ such that $f(a_0)f(b_0) < 0$
Then for $i = 0,1,2,$ ... until termination:
Estimate $c = a_if(b_i) - b_if(a_i) / f(b_i) - f(a_i)$
    If $|f(c)|$ < tolerance then accept $c$ as solution
    If $f(a_i)f(c) < 0$ then
        $a_{i+1} = a_i; b_{i+1} = c$
    Else
        $a_{i+1} = c; b_{i+1} = b_i$
    Test for termination (solution found or iteration count exceeded).

**Figure 4.5** Pseudocode for the method of false position.

Running the method of false position on the same problem ($x - e^{-x} = 0$) as the bisection method gives the results shown in Table 4.2. A solution is reached in only 6 iterations.

**Table 4.2** Progress using the method of false position.

| $a$ | $f(a)$ | $b$ | $f(b)$ | $c$ | $f(c)$ |
|---|---|---|---|---|---|
| 0.0000 | –1.0000 | 1.0000 | 0.6321 | 0.6127 | 0.07081 |
| 0.0000 | –1.0000 | 0.6127 | 0.0708 | 0.5722 | 0.00789 |
| 0.0000 | –1.0000 | 0.5722 | 0.0079 | 0.5677 | 0.00088 |
| 0.0000 | –1.0000 | 0.5677 | 0.0009 | 0.5672 | 0.00010 |
| 0.0000 | –1.0000 | 0.5672 | 0.0001 | 0.5672 | 0.00001 |
| 0.0000 | –1.0000 | 0.5672 | 0.0000 | 0.5671 | 0.00000 |

Notice that the convergence is much quicker than in the bisection method. That is not to say that the method will always be quicker. It is possible to pose problems where the method of false position converges very slowly and where the bisection method is superior. The way in which a problem is tackled should be based on experience. One way of optimising the approach is to use the method of false position for most of the iterations and to use a bisection calculation occasionally.

## 4.5 CONVERGENCE CRITERIA

Once the methods have been started there has to be a way of stopping. Because the methods are all approximate it is not possible to continue until an exact answer is obtained because there will not generally be an exact answer. So how close is close enough? There are two ways of looking at the problem. It would be possible to keep going until either or both:

- $|f(c)| < \varepsilon$
- the successive estimates of $c$ are changing by so little that there is little point in going on, i.e. $|cnew - cold| < \varepsilon$

where $\varepsilon$ is a small number. The value of $10^{-5}$ is common for single precision and $10^{-10}$ for double precision calculations. The precision is related to the number of digits stored in memory. In engineering problems this termination criterion is important but can be interpreted sensibly in terms of the problem as described below. Both stopping criteria are valid and have their place. The error term, $\varepsilon$, in the first method will relate to the function being solved. In the case of normal depth the function is really written in terms of flow rates, and therefore the error term is an error in the flowrate ($m^3/s$). In the case of wavelength the function is written in terms of the wavelength so the error is in terms of the wavelength itself (m). This allows a sensible tolerance to be determined. For example, if one wanted water wavelengths accurate to 1 cm then the tolerance would be 0.01 m. If one wanted a normal depth that had the flow correct to 1 L/s then a tolerance of 0.001 $m^3/s$ would be appropriate. It is evident that in engineering practice, tolerances much smaller than this might not be appropriate.

The term in the second option listed above, converging until the solution is not changing, is often multiplied by 100 to make it a percentage error. A typical closing error might be 0.1%. The limitations of the number of significant digits stored in a computer memory may limit how small the convergence criterion can be and programmers should be wary of having numbers much smaller than $10^{-6}$.

## 4.6 APPLICATION OF BISECTION METHOD IN EXCEL

Excel Visual Basic macros provide a powerful tool for enhancing the capabilities of Excel spreadsheets. User-defined functions can be used to carry out iterative tasks and to provide a measure of reliability by having code that is well tried and tested. Recalling the expression used to determine normal depth in a trapezoidal channel:

$$Q = \frac{d_n B}{n}\left(\frac{d_n B}{B + 2d_n}\right)^{2/3} S^{1/2} \tag{4.7}$$

where $Q$ is the flow rate, $B$ the width of the channel, $S$ the slope of the channel and $d_n$ the normal depth, and remembering that it must be re-arranged to be in the form of a root-finding problem gives:

$$f(d_n) = 0 = Q - \frac{d_n B}{n}\left(\frac{d_n B}{B + 2d_n}\right)^{2/3} S^{1/2} \tag{4.8}$$

where the product $d_nB$ is the area of the flow and $B + 2d_n$ is the wetted perimeter in a rectangular channel. To make the function more general it can be written taking account of the fact that the cross-sectional area ($A$) and the wetted perimeter ($P$) can be isolated. Equation (4.8) can be reformulated as:

$$f(d_n) = 0 = Q - \frac{A}{n}\left(\frac{A}{P}\right)^{2/3} S^{1/2} \tag{4.9}$$

The function itself is written using the Microsoft Visual Basic Editor which is found under Tools–Macro. A module should then be inserted (Menu Insert–Module) and the code entered into that. It can be seen that the main function calls two other functions, one to calculate the area and the other the wetted perimeter. In the case of a rectangular channel these are of marginal use, but for other shapes and situations the benefits of having a function dedicated to calculating area are well worth the effort of coding the solution in this way. The way the module is called is shown in Figure 4.6 while the module code is listed in Figure 4.7.

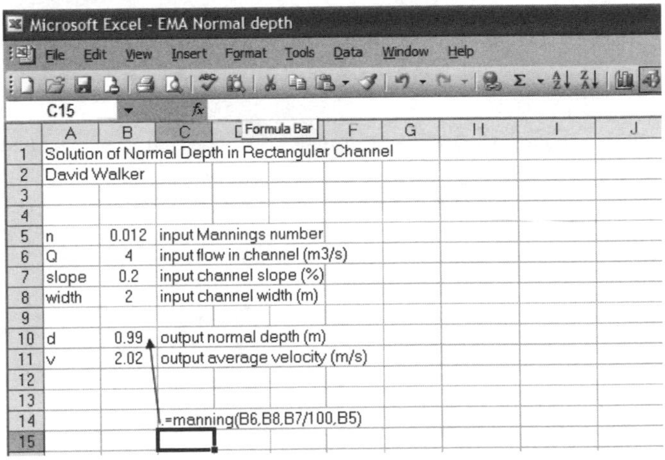

Microsoft Excel - EMA Normal depth

File Edit View Insert Format Tools Data Window Help

C15 *fx* Formula Bar

| | A | B | C |
|---|---|---|---|
| 1 | Solution of Normal Depth in Rectangular Channel | | |
| 2 | David Walker | | |
| 3 | | | |
| 4 | | | |
| 5 | n | 0.012 | input Mannings number |
| 6 | Q | 4 | input flow in channel (m3/s) |
| 7 | slope | 0.2 | input channel slope (%) |
| 8 | width | 2 | input channel width (m) |
| 9 | | | |
| 10 | d | 0.99 | output normal depth (m) |
| 11 | v | 2.02 | output average velocity (m/s) |
| 12 | | | |
| 13 | | | |
| 14 | | | =manning(B6,B8,B7/100,B5) |
| 15 | | | |

**Figure 4.6** Calling a function from a worksheet (in this case a user-defined function MANNING). The contents of cell B10 are indicated on the sheet.

## 4.7 DETERMINING START VALUES FOR BRACKET METHODS

Experience has shown that the starting guesses for the two values that bracket the root need not be particularly close to the final value and, therefore, there is often little point in spending too much time or effort on determining good initial guesses. In many cases a standard set of initial guesses can be used in a range of problems.

For example, in the solution of normal depth where the variable is the discharge, reasonable starting guesses might be 1000 $m^3/s$ and 0.001 $m^3/s$. This means that the method is valid for situations where the flow is between these two values. In water waves, the wavelength is unlikely to be less than 1 mm nor greater than 1000 m so once again start values of 0.001 and 1000 would be appropriate.

In situations where there may be multiple roots a procedure that steps through a range of starting values could be applied. If the product of their functional values ($f(a).f(b)$) is negative it indicates at least one root between them and a search can be made for that root. If the product is not negative then there is either no root or there may be an even number of roots, and without further investigation involving a reduction in the step size it is not possible to differentiate between these two possibilities. For this reason, it is often worth plotting the function to get an understanding of its behaviour and look for reasons why certain brackets may fail.

```
option explicit
Function manning(q, w, s, n)
'function to calculate normal depth in a rectangular channel
'uses bisection method
'q = flow, w = channel width, n = mannings number,
' s = channel slope
' written D.Walker, 1995
Dim a as double
Dim b as double
Dim c as double

a = 0.00001                          ' initial guesses for a and b
b = 1000
c = (a + b) / 2                      ' need a guess for c too

While (Abs(F(c, q, w, s, n)) > 0.0001)      ' bisection method
 c = (a + b) / 2
 If (F(a, q, w, s, n) * F(c, q, w, s, n) < 0) Then
  b = c
 Else
  a = c
 End If
Wend
manning = c
End Function

Function F(x, q, w, s, n)
' solves manning root finding function for a depth x
Dim Area as double
Dim wp as double
Area = w*x
wp = x+2*w
F = q-Area/n*(Area/wp)^(2/3)*(s)^0.5
End Function
```

**Figure 4.7** Visual Basic module for normal depth using the bisection method.

## PROBLEMS

**4.1** Use the bisection method to reproduce the results shown in Table 4.1. Investigate the number of iterations required for a range of tolerances from 0.1 to 0.000 0001. Try this in Excel and Fortran. What are the particular benefits and drawbacks of each approach?

**4.2** Find the roots of the function that formed one of the questions in the previous chapter:

$$y = \frac{3x-7}{(x-3)^2}$$

Use the bisection method in Excel to determine its root or roots. Attempt to find the solution using Excel Goal Seek as an alternative approach.

**4.3** Write and test a Fortran root-finding program using the simple equation $x^2+1=0$. This function will test the initial phase of the procedure, selecting the initial start values. Use the experience to develop a standard warning for the Fortran program that will be given to users of the program.

**4.4** If wavelength in any depth of water can be written as:

$$L = \frac{gT^2}{2\pi}\tanh(\frac{2\pi d}{L})$$

then determine the wavelength for the situation where $T = 10.0$ s and $d = 1.0$ m. Try initial guesses of 0.001 m and 1000 m. Carry out the calculations on an Excel worksheet.

**4.5** Repeat Problem 4.4 using guesses that are closer to the actual value (try, for example, $a$ = 30 m and $b$ = 35 m). How much difference does this make to the number of iterations required to obtain the correct answer to a tolerance of 0.01 m?

**4.6** On an Excel spreadsheet or using Matlab, plot Equation (4.9), which is the function used to solve for normal depth. Assume a rectangular channel with width 1 metre, a Manning's number of 0.020, a flow of 3 $m^3$/s and a slope of 0.001. Identify how many roots the function has for a range of rectangular channels and take note of the general form of the function. Is it possible to anticipate the improvement that would come from using the method of false position rather than the bisection method?

**4.7** Set up an Excel spreadsheet and implement the Visual Basic Module to solve Manning's equation. With the function working correctly, modify the code to use the method of false position.

**4.8** Implement a search method to seek out all the roots in the function:

$$NPV = 200\frac{1-(1+r)^{-5}}{r} - \frac{620}{(1+r)^6} - 400 = 0$$

which solves for the Internal Rate of Return, $r$. [Note: This is Equation (3.7), taken from the previous chapter. It is plotted there as Figure 3.5.]

**4.9** Undertake a search for the root (or roots) of the function: $f(x) = x^3 + 3x^2 - x - 3$.

CHAPTER FIVE

# Roots of Equations (Open Methods)

## 5.1 INTRODUCTION

The bracket methods in the previous chapter are guaranteed to converge; that is, for well behaved functions, if the iterations are repeated often enough, a solution will always be found. There are, however, a number of issues with the methods: they require two starting values and this is not always convenient; they also do not take account of key properties of the function itself as a way to optimise the generation of new estimates.

Open methods on the other hand require only one starting value, and can converge more quickly but also have the possibility of diverging; that is, heading off away from the solution rather than towards it. Once they do diverge they tend to go "off" fairly quickly. Their chief advantage is that they can converge more quickly than the bracket and other methods covered in the earlier chapters. This can be essential for engineering problems which require a large number of evaluations or, more importantly, problems with a large number of equations and variables. Imagine, for example, finding the roots of 10,000 equations with a corresponding 10,000 variables.

## 5.2 SIMPLE ONE-POINT ITERATION

If it is possible to reformulate the root-finding problem $f(x) = 0$ to $x = g(x)$ then it is possible to iterate for $x$ on the LHS using a succession of old values of $x$ on the RHS. A general algorithm for this is shown in Figure 5.1.

```
Select initial x0
Then for i = 0,1,2   , ... until termination:
        Estimate xi+1 = g(xi)
        If |f(xi+1)|  < tolerance then accept c as solution
        Test for number of iterations > maximum.
```

**Figure 5.1** Pseudocode for the one point algorithm.

For example, for the case of $f(x) = x - e^{-x} = 0$; this can be rewritten as $x = e^{-x}$ or $x = g(x)$ where $g(x) = e^{-x}$. The solution can proceed as shown in Table 5.1. It takes 22 iterations to arrive at a solution (for this particular starting value) where the error is less than $10^{-6}$. This compares unfavourably with the bisection and false position methods in the previous chapter, where only 12 and 6 iterations, respectively, were required.

**Table 5.1** One point method algorithm in practice.

| $x$ | $f(x)$ |
|---|---|
| 0.0000 | –1.00000 |
| 1.0000 | 0.63212 |
| 0.3679 | –0.32432 |
| 0.6922 | 0.19173 |
| 0.5005 | –0.10577 |
| 14 iterations not shown | |
| 0.5672 | 0.00002 |
| 0.5671 | –0.00001 |
| 0.5671 | 0.00001 |

The method is simple to implement but in this case slow to converge. Because of the way that the solution proceeds the technique tends to either diverge or converge quickly. The initial guess is critical in this case although for this particular function most start values tend to converge in around 21 to 24 iterations.

## 5.3 NEWTON-RAPHSON METHOD

Just as the method of false position can be superior to the bisection method by taking account of the function so the Newton-Raphson method is superior to the one-point method. The Newton-Raphson method is generally used because of its simplicity and great speed. A potential problem with it is that it requires that the function $f(x)$ be differentiable. This can cause some problems in the case of a complicated function, and a way of overcoming this will be outlined in the next section. A potential complication with the method is that if the initial guess is chosen at a point where the derivative is zero, then the method will fail. The reason for this will become evident when the method is described.

The working of the method is illustrated in Figure 5.2, where a guess, $b$, is improved to $c$. Essentially the process involves using the gradient at the current location, $b$, to drop a tangent to the $x$ axis. Note that in this case the point on the axis, $c$, is much closer to the actual root. The method, listed in Figure 5.3, can be derived using simple geometry:

$$(b-c)f'(b) = f(b)$$
$$\therefore b-c = \frac{f(b)}{f'(b)}, \therefore c = b - \frac{f(b)}{f'(b)} \tag{5.1}$$

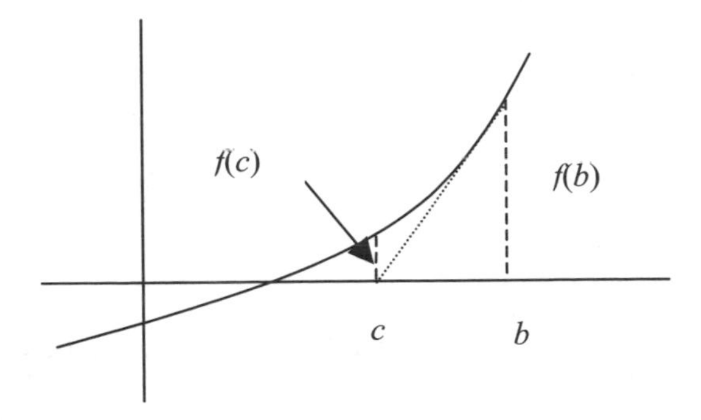


**Figure 5.2** Newton-Raphson method of iteration showing next guess.

```
Select initial guess x0
Then for i = 0,1,2, ... until termination:
    Compute f'(xi)
    If |f'(xi)| < tolerance then stop
    Else
        xi+1 = xi – f(xi)/f'(xi)
    Test for termination
```

**Figure 5.3** Pseudocode for Newton-Raphson algorithm.

For the test case of $f(x) = x - e^{-x}$ it is possible to show that the derivative can be written: $f'(x) = 1 + e^{-x}$ and the solution proceeds as shown in Table 5.2. A solution is obtained in only 4 iterations and this is fairly stable for a range of start values although, in this case, negative start values smaller than –10 slow the process considerably. This is because, values in this range have a very large slope (derivative) and hence, the next guess is very close to the old guess.

**Table 5.2** Newton-Raphson method algorithm in practice.

| $x$ | $f(x)$ | $f'(x)$ |
|---|---|---|
| 1.00000 | 0.63212 | 1.3679 |
| 0.53788 | –0.04610 | 1.5840 |
| 0.56699 | –0.00024 | 1.5672 |
| 0.56714 | –0.00000 | 1.5671 |

## 5.4 SECANT METHOD

One of the difficulties of the Newton-Raphson method is the need for the derivative of the function to be supplied. As already mentioned this can be difficult or impossible. One way of overcoming this is to use an approximation for the derivative based on discrete points on the function. From basic calculus, the derivative can be written:

$$f'(x) = \lim_{h \to 0} \frac{f(x+h) - f(x)}{h} \tag{5.2}$$

When using discrete points it is usual to label them $x_1, x_2, x_3, \ldots x_{i-1}, x_i, x_{i+1}, \ldots x_n$. In this representation the derivative at $x_i$ is:

$$f'(x) \approx \frac{f(x_{i-1}) - f(x_i)}{x_{i-1} - x_i} \tag{5.3}$$

where the spacing between adjacent $x$ values is assumed to be the $h$ in Equation (5.2). This approximation is then substituted into the same algorithm, and the procedure is then referred to as the secant method. Is it listed in Figure 5.4.

Select initial guesses $x_0$, $x_1$
Then for $i$ = 1,2, ... until termination:
Compute $f'(x_i)$
If $|f'(x_i)|$ < tolerance then stop
Else

$$x_{i+1} = x_i - \frac{f(x_i)(x_{i-1} - x_i)}{f(x_{i-1}) - f(x_i)}$$

Test for Termination

**Figure 5.4** Pseudocode for secant method algorithm.

Starting the method requires two initial guesses but since the guesses don't have to bracket the root the method is classified as an open method. With the same function used in the previous examples and using $x_0 = 0.0$ and $x_1 = 1.0$ as the initial guesses, the iteration proceeds as shown in Table 5.3. Solution is found in only 4 iterations.

**Table 5.3** Newton-Raphson method algorithm in practice.

| $x_1$ | $x_2$ | $x_3$ | $f(x_3)$ |
|---|---|---|---|
| 0.0000 | 1.0000 | 0.6127 | 0.07081 |
| 1.0000 | 0.6127 | 0.5638 | –0.00518 |
| 0.6127 | 0.5638 | 0.5672 | 0.00004 |
| 0.5638 | 0.5672 | 0.5671 | 0.00000 |

## 5.5 MULTIPLE ROOTS

In an earlier chapter a solution was sought to the equation to determine Net Present Value for a proposed project. The equation was written as:

$$NPV = 200\frac{1-(1+i)^{-5}}{i} - \frac{620}{(1+i)^6} - 400 = 0 \tag{5.4}$$

and the aim was to find the value $i$ that satisfied the equality. Plotting of the function showed it had two roots in the region between 0 and 0.2.

Although any of the methods discussed so far could be applied to this particular problem, it is evident that while a root might be found it would not be clear if it was the only one, or even if there were more to be sought. To solve this problem completely it is first necessary to know enough about the function to suspect there may be multiple roots, and then to apply an iterative search procedure to find each root. An algorithm to carry out that search is shown in Figure 5.5.

```
Select lower and upper limit for search: x_low, x_high
Select step size for search, step
x = x_low
Do while (x < x_high)
  If f(x)*f(x+step) < 0 then root exists in this interval and should be found
  x = x + step
End do
End
```

**Figure 5.5** Pseudocode for multiple root search.

It is worth noting that in many practical situations in engineering there is only a single root and the function is often well-behaved, making the search relatively straightforward.

## 5.6 OTHER PROBLEMS AND GENERAL CONSIDERATIONS

As a general word of caution it should be remembered that these methods look fine on paper and if calculations are being done by hand then the results and the convergence or otherwise can be monitored. However, in practice these methods tend to be coded into a programming language and buried deep within a program. They help provide a solution rather than being the solution themselves. Therefore, if anything goes wrong with them the reason for the problem often takes some sorting out. It is important to code in checks as the solution proceeds. At the very least a count should be kept as to the number of iterations and if this exceeds some predetermined number (perhaps 100) then some trap should be set and the program should report what is going on.

In many cases it would be wise to be able to vary the starting guess automatically so that if the convergence were found to be slow then a different value could be tried. This is particularly true for the Newton-Raphson method.

If the false position method is being used it might be a good idea to include an occasional bisection method iteration in case the false position method has got stuck down the wrong end of the function as it can do.

It has been suggested that an efficient algorithm is to use the Newton method to finish off the root-finding procedure after one of the bracket methods has been used to get reasonably close to the solution. This helps avoid the potential problems of the open method diverging.

One thing that must be remembered when using these solution techniques is that the number of iterations does not necessarily give an accurate picture of how efficient the technique is. In many cases the calculation of the value of the function will be detailed and take some time, so that methods that require the function to be evaluated frequently may take longer than other methods that make do with fewer calls on the function. The secant method has three function calls per iteration whereas the Newton method has only two. This would make the Newton method more attractive, even if the number of iterations were similar.

It might be argued that with the fast modern computers the efficiency of a method would be of secondary importance. Who cares if it takes a few extra seconds if it is easier to program a simpler method? The problem with that argument is that these methods are, as mentioned before, only a small part of a total solution and may be called on thousands of times rather than once or twice. In this case the speed of the method will be critical and well worth the extra effort in programming.

## 5.7 CASE STUDY: POLLUTANT TRANSPORT IN A RIVER

Swamee et al. (2000) derived empirical relationships to describe the lag time for dye to travel with the flow between two points on a river. If the time was known it was possible to solve the inverse problem and determine the location of the pollutant source, $x_1$. The expression can be written:

$$\Delta t_p = \frac{\lambda (x_1 + \Delta x)^{1.2}}{(gR_2)^{0.2} A_2^{0.1} V_2^{0.6}} - \frac{\lambda x_1^{\ 1.2}}{(gR_1)^{0.2} A_1^{0.1} V_1^{0.6}} \tag{5.5}$$

where all terms, other than $x_1$, are known in terms of the river being studied. According to Swamee et al. (2000) the equation can be solved by trial and error. This would be well suited to a root-finding technique if Equation (5.5) were re-written in the form $f(x) = 0$.

## 5.8 CASE STUDY: WET DETENTION TREATMENT BASINS

The treatment of pollutants in wet detention ponds is receiving increased attention and efforts are being made to enable reliable models of the processes to be developed. Wang et al. (2004) outline such a model, including a module to allow the flow through a pond to be determined. Given the nature of the basic equations it is normal for the solution to require approximate methods such as finite difference modelling, but the authors show that under certain conditions analytical solutions are possible. Taking into account inflow from the catchment ($Q_s$), small perennial inflows ($Q_p$), rainfall intensity ($P$), total evaporation and infiltration ($L$), pond dimensions ($d,b$) and a factor $k$ that is dependent on the pond and outflow structure characteristics, the equation for the rate of change of flow through the pond can be written:

$$\frac{dQ}{dt} = \frac{Q_s + Q_p + P - L}{kQ^{(d/b)-1}} - \frac{Q^{2-(d/b)}}{k} \tag{5.6}$$

For particular values of $d$ and $b$, and assuming $Q$ can be written as $u^3$ (where $u$ is a dummy variable) it is possible (although not obvious) to show that the final equation to be solved can written:

$$\frac{1}{6a}\log\left[\frac{u_t^2 - au_t + a^2}{(u_t + a)^2}\right] + \frac{1}{a\sqrt{3}}arctg\left(\frac{2u_t - a}{a\sqrt{3}}\right) -$$

$$\left[\frac{1}{6a}\log\left[\frac{u_{t0}^2 - au_{t0} + a^2}{(u_{t0} + a)^2}\right] + \frac{1}{a\sqrt{3}}arctg\left(\frac{2u_{t0} - a}{a\sqrt{3}}\right)\right] + \frac{\Delta t}{3k} = 0 \tag{5.7}$$

According to the authors, the equation "is a nonlinear algebraic equation that contains multiple roots. Usually, a Newton iterative method is used to search the roots. In the vicinity of the real root, the rate of change for [the equation] is so steep that the Newton method converges to false roots. To avoid this problem, the approximate value of the real root can be estimated based on the inflow data and initial condition. Therefore, a bisection method can be used to find the solution."

## PROBLEMS

**5.1** Set up an Excel spreadsheet to apply the Newton-Raphson method to find a solution to $x = e^{-x}$. Determine the number of steps required to find the solution as a function of the starting value. What does this say about how robust the method is?

**5.2** Use the secant method to find the wavelength of a 10-second period wave travelling in 1-metre-depth water. Note: the equation to solve for wavelength can be written:

$$L = \frac{gT^2}{2\pi}\tanh\left(\frac{2\pi d}{L}\right)$$

How sensitive is the method to the starting value, and what does this say about the effort that should be put into the initial guess?

**5.3** Compare the number of iterations to find a solution for wavelength with the bracket methods covered in earlier chapters. For example, how does the secant method compare with the method of false position in terms of the number of iterations required to come to a solution and also the relative importance of the start values?

**5.4** Consider the function shown in Figure 5.6 and the rate of progression towards a solution of the root-finding problem for the bisection method and the method of false position (using two guesses, *a* and *b*) as well as the Newton-Raphson method (using guess *a*).

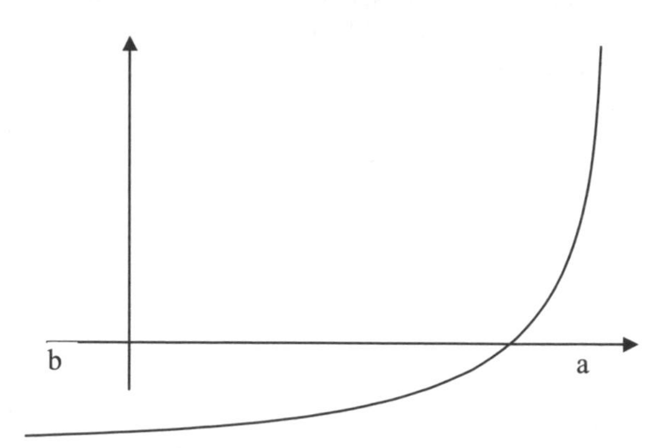


**Figure 5.6** A function $f(x)$ to be solved for roots and two starting guesses, *a* and *b*.

**5.5** Develop a Fortran program to search for multiple roots and test it by finding the roots of the function: $f(x) = x^3 + 3x^2 - x - 3$.

## CHAPTER SIX

# Numerical Integration (Trapezoidal Rule)

### 6.1 INTRODUCTION

Calculus, that powerful mathematical tool first developed by Isaac Newton and Gottfried Leibnitz in the 1600s, has two basic formulations: differential and integral. While both are important in the study of advanced mathematics, both are also central to much of engineering analysis and are put to a range of very practical tasks. In this and the next chapter the focus will be on integral calculus and the task of finding areas under curves or volumes under a surface.

The key to much of the integral calculus taught in mathematics is the determination of analytical expressions for the integral of a range of different functions. For example, for a function $f(x)$ where:

$$f(x) = ax^n \qquad (6.1)$$

the integral can be written:

$$F(x) = ax^{n+1}/(n+1) \qquad (6.2)$$

For many other functions there are a selection of rules that can be applied to determine the required solution and calculus texts generally have an appendix containing a range of standard integrals. In engineering practice, however, there are situations where this approach is simply not possible. The two most common reasons for this are:

- the function is sufficiently complex that it does not have an analytical integral; and
- the quantity being integrated is not in the form of a function at all, but is specified by a set of discrete data points.

In both of these situations some form of numerical approximation may be necessary to evaluate the integral. As an example of integration where the function is too complex to have an analytical solution, consider a problem faced by engineers designing an open-cut mine or assessing the stability of a rock face. As part of the design they have to have some knowledge of the distribution of rock fractures in the area and this can be achieved through the development of a model that considers the surface fractures as the visible part of circular or disc-shaped fractures buried at various depths and angles in the rock. An artist's impression of this is shown in Figure 6.1. It is worth stressing that this may in fact not be the case, but it does constitute a model that has been shown to allow reasonable predictions about behaviour to be made.

**Figure 6.1** Conceptual model of surface fractures based on circular fractures. © N.Oehlers.

Based on surface observations it is possible to determine the three-dimensional spectrum of crack size, $h(l)$, as:

$$h(l) = \int_{l}^{\infty} \left( \frac{l}{m} \right) g(m) dm \tag{6.3}$$

where

$$g(l) = \frac{l f(l)}{\mu_L} \tag{6.4}$$

and

$$f(l) = \left( \frac{l}{\mu_s} \right) \int_{0}^{\pi/2} e(l \sec(s)) \sec(s) ds \tag{6.5}$$

where $e(x)$ is a known function representing the fracture discs (Priest, 2004). In this case, knowing the function $e(x)$ is not sufficient to determine $h(l)$ in anything but the simplest of cases and an approximate method is required if the integrations are to be evaluated.

As an example of integration where the there are only discrete data points rather than a continuous function, consider the cross-sectional area of a stream or river. A series of discrete points would generally be taken at a range of offsets across the stream where the depth of water at that point is determined. Figure 6.2, for example, shows some fictitious depth data plotted as a cross section. The problem is to take this unevenly spaced data set and use it to determine an estimate of the total area of the river channel.

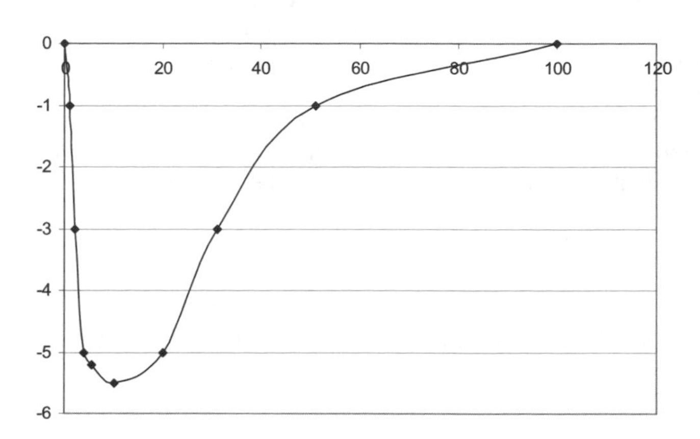


**Figure 6.2** River cross section data used to determine cross-sectional area.

In both of these examples (the rock fractures and the river cross section) analytical expressions for the integrals are unusable or unavailable and this chapter deals with the first of a number of standard methods for carrying out the integration. The basic approach is to split the area into a set of discrete strips and to sum the area over these, as shown in Figure 6.3.

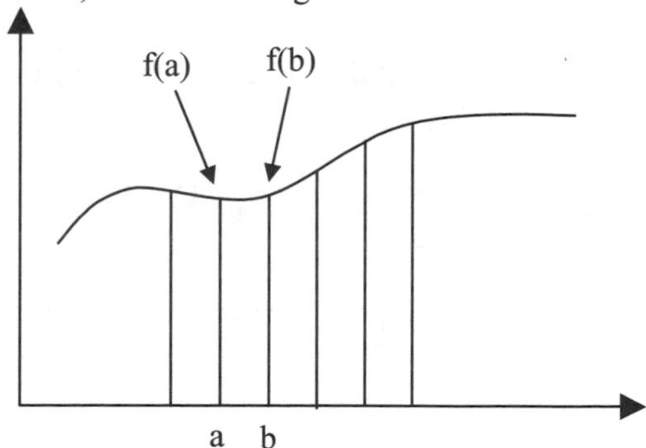


**Figure 6.3** Integration, where an area has been divided into slices. Shown are two typical points $a$ and $b$ and their function values $f(a)$ and $f(b)$.

It is evident that the key to an accurate approximation will be how well the function is approximated between the strips. Straight line segments might be fine with a narrow strip and a well-behaved function (this leads to the trapezoidal rule) but if it were possible to have some sort of curve approximating the function then the results would probably be more accurate. There will be more on this point later.

## 6.2 TRAPEZOIDAL RULE

The trapezoidal rule assumes a series of straight line segments leading to a series of areas of trapezoidal shape to be determined. Although the trapezoidal rule can be derived intuitively it is worth showing a more formal process as this will make the higher-order (and therefore more accurate) methods easier to follow. The basic idea is to approximate the area into a series of strips (as shown in Figure 6.3) with a straight line joining the points along the top of each strip, and then to determine an analytical expression for the integral under this straight line over the width of the strip. In this situation a straight line segment can be written:

$$f_1(x) = f(a) + \frac{f(b) - f(a)}{b - a}(x - a) \tag{6.6}$$

where the subscript indicates a first-order function. Integrating this equation from $a$ to $b$ gives:

$$\int_a^b f_1(x)dx = I = (b - a)\frac{f(a) + f(b)}{2} \tag{6.7}$$

If the distance from $a$ to $b$ is split into $n$ strips the width, $h$, of each strip can be written as:

$$h = \frac{b - a}{n} \tag{6.8}$$

In addition, if the $n$ strips are identified as going from $x_0$ to $x_n$, then the total integral becomes:

$$I = h\frac{f(x_0) + f(x_1)}{2} + h\frac{f(x_1) + f(x_2)}{2} + \ldots + h\frac{f(x_{n-1}) + f(x_n)}{2} \tag{6.9}$$

where it is evident that each internal point appears twice and the end points appear only once each. The situation is illustrated in Figure 6.4. Grouping terms gives:

$$I = \frac{h}{2}\left[ f(x_0) + 2\sum_{i=1}^{n-1}\left(f(x_i) + f(x_n)\right)\right] \tag{6.10}$$

A code segment to calculate the integration in Fortran is listed in Figure 6.5. It assumes that the $x$ data are available in an array defined for $x_0$ to $x_n$ and that $n$ and $h$ are defined with appropriate values. It is evident that the code follows

Equation (6.10) quite closely, although there is an issue with the fact that Fortran is not case-sensitive so the final integral is stored in a variable called `area`. The other aspect worth noting is that in this case the array has elements with indices starting at 0 and continuing to $n$. This is not always the case, and often the first element will have an index of 1, but it is worth knowing how to deal with this and being able to implement algorithms faithfully.

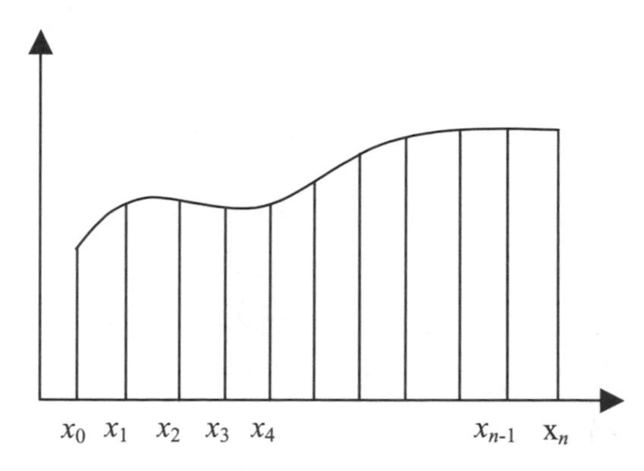


**Figure 6.4** Integration, where an area has been divided into $n$ slices.

```
sum = f(x(0))+f(x(n))      ! start with end values
do i=1,n
 sum = sum + 2*f(x(i))     ! add 2*internal points
end do
area = h/2.0*sum           ! final area
```

**Figure 6.5** Fortran code to apply the trapezoidal rule.

## 6.3 WORKED EXAMPLE: CUMULATIVE NORMAL PROBABILITY

A commonly required value in the field of probability is the cumulative area under the normal probability distribution, shown in Figure 6.6. The cumulative area can be written:

$$F_x(a) = \int_{-\infty}^{a} f_x(x)dx \tag{6.11}$$

where the normal distribution has the formula:

$$f_X(x) = \frac{1}{\sigma\sqrt{2\pi}}\exp\left(-\frac{1}{2}\left(\frac{x-\mu}{\sigma}\right)^2\right) \tag{6.12}$$

Equation 6.12 does not have an easy analytical integral and it is convenient to carry out the integration using an approximate method. For this reason, cumulative values are often listed in tabular form in many probability texts. The trapezoidal rule was programmed to determine the integral from –10 to 2.0 using a variety of step sizes. Note that the integral should be from $-\infty$ but starting at –10 should be sufficient in this case (based on the very small area under the curve at values less than –10). The results are shown in Table 6.1 together with the value returned by the Excel function `NORMDIST`, and the relative error. As a matter of interest, the Excel cell formula to determine the integral is `"=NORMDIST(2,0,1,TRUE)"`.

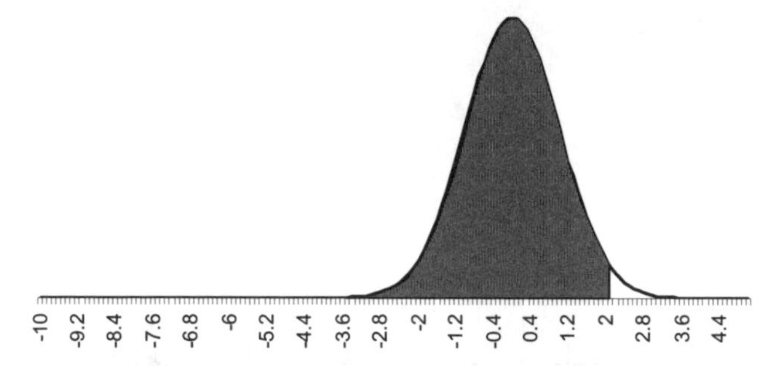


**Figure 6.6** The standard normal distribution showing the area up to z = 2.0.

It is evident from Table 6.1 that a very coarse set of strips leads to an extremely poor estimate for the integral, but that by the time the strip width has reduced to 1.0, giving a total of 12 strips, the error is less than 1%. Further reduction leads to further improvement in the estimates.

**Table 6.1** The cumulative value of the standard normal distribution ($\mu = 0.0$, $\sigma = 1.0$) at z = 2.0. Note: exact value = 0.977249937963813...

| Step Size | Estimate (4 significant figures) | Error (%) |
|---|---|---|
| 6.0 | 0.1628 | –83.3435 |
| 4.0 | 0.8073 | –17.3906 |
| 2.0 | 0.9601 | –1.7523 |
| 1.0 | 0.9684 | –0.9018 |
| 0.5 | 0.9750 | –0.2292 |
| 0.2 | 0.9769 | –0.0368 |
| 0.1 | 0.9772 | –0.0092 |

## 6.4 SOURCES OF ERROR WITH THE TRAPEZOIDAL RULE

The basic approach with the trapezoidal rule is to split the area into a set of discrete strips and to sum the area over these. It might be thought that assuming a straight line between each strip will give a reasonable estimate of the area particularly when the strips become very narrow but the problem is not quite so simple. For example, in the case of an integration over a curve that is convex, the estimate will always give an underestimation of the total area(see Figure 6.7).

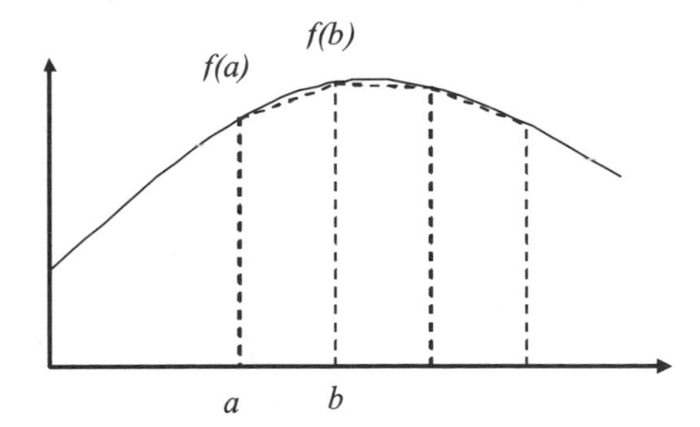


**Figure 6.7** A simple integration that will always be wrong due to each strip 'cutting the corner'.

There is another, potentially more important reason to consider the method being employed. While it might be thought that computers work so fast that there is little problem in splitting any area into minutely small slices and letting the computer do the work, this is often not true. Once again, it is important to put the problem into perspective. The objective of the task will generally not simply be determining the area of a cross section or function. This will be but a small part of the whole process and it may be a part that is required to be calculated literally thousands of times; a smarter algorithm that works twice as fast or ten times as fast will be far superior to a simple but easy algorithm. By all means start with simple but easy, but then work on up to smarter better algorithms. This point will be pursued in the next chapter where more sophisticated algorithms will be set out.

## PROBLEMS

**6.1** Use the trapezoidal rule to integrate $f(x) = 5x$ from $x = 0$ to $x = 20$ taking a strip width of 5.0. Compare the estimate to the exact value determined from integral calculus principles. The estimates should be exactly correct – why?

**6.2** Use the trapezoidal rule to integrate $f(x) = x^2$ from $x = 0$ to $x = 5$ taking a strip width of 1.0. Compare the estimate to the exact value determined from integral calculus principles. The estimates should always be greater than the exact value – why?

**6.3** Write a function to return the area under the standard normal distribution (mean = 0.0, standard deviation = 1.0). Test the program by comparing the results from a range of values given by the Excel function `NORMDIST`. As part of the solution it will be necessary to determine a level of accuracy for the calculation and this may vary as $Z$ varies.

**6.4** Use the trapezoidal rule to integrate the stream survey data given in Table 6.2.

**Table 6.2** Stream offset and depth data.

| Offset (m) | Depth (m) |
|---|---|
| 0 | 0.0 |
| 10 | –1.0 |
| 20 | –1.5 |
| 25 | –3.7 |
| 30 | –2.0 |
| 40 | –0.5 |
| 50 | –0.2 |
| 90 | 0.0 |

**6.5** Set up an Excel spreadsheet to determine the integral of the normal distribution. If possible, set up the sheet so that there is a variable number of strips used in the calculation and compare the results with the exact value using the Excel function `NORMDIST`. (This is applied as a formula in a cell in the form "`=NORMDIST(`*`z`*`,0.0,1.0,`*`true`*`)`" where the function is integrated up to the value *z*, and the 0.0 and 1.0 refer to the mean and standard deviation of the distribution which for the standard normal distribution is 0.0 and 1.0 respectively. The "*true*" causes the cumulative distribution to be calculated.)

**6.6** A pond has been surveyed to determine the water depth over a regular grid (10 metres by 10 metres) of points. The results are listed in Table 6.3. Determine the volume of the pond. (Hint: This is a two-step process. First, determine the area under each of the sections, working either across the rows or down the columns. Then, if they were plotted the "area" under the curve would be the volume of the pond.)

**Table 6.3** Depths recorded over a two-dimensional grid of points.

| Depth (metres) | | | | | | | |
|---|---|---|---|---|---|---|---|
| 0 | 0 | 0 | 0.5 | 0 | 0 | 0 | 0 |
| 0 | 0.2 | 0.5 | 0.8 | 0.6 | 0.4 | 0.2 | 0 |
| 0 | 0.3 | 0.9 | 1.5 | 1.2 | 0.8 | 0.3 | 0 |
| 0.1 | 0.5 | 1.9 | 2.5 | 1.9 | 1.5 | 1.1 | 0.1 |
| 0.1 | 0.6 | 2.9 | 4.1 | 2.8 | 2.0 | 0.4 | 0.1 |
| 0.1 | 0.4 | 3.0 | 3.9 | 2.7 | 2.0 | 0.5 | 0 |
| 0 | 0.2 | 2.5 | 3.5 | 2.0 | 1.0 | 0.2 | 0 |
| 0 | 0.2 | 1.6 | 1.9 | 1.1 | 0.5 | 0.1 | 0 |
| 0 | 0 | 0.1 | 0.1 | 0.2 | 0.1 | 0 | 0 |
| 0 | 0 | 0 | 0 | 0 | 0 | 0 | 0 |

## CHAPTER SEVEN

# Numerical Integration (Simpson's Rule)

### 7.1 INTRODUCTION

In many situations the assumption of a straight line segment between points on a function or between data values may not be suitable and a better approximation might be formed by fitting some sort of curve – for example, a parabola. Since any three non-collinear points can be joined by a second-order parabola it is possible to derive an expression for this function (in this case in the form of a second-order Lagrange polynomial as set out in Chapter 8). It can be written:

$$f_2(x) = \frac{(x-x_1)(x-x_2)}{(x_0-x_1)(x_0-x_2)} f(x_0) + \frac{(x-x_1)(x-x_2)}{(x_1-x_0)(x_1-x_2)} f(x_1) + \frac{(x-x_0)(x-x_1)}{(x_2-x_0)(x_2-x_1)} f(x_2) \tag{7.1}$$

where $x_0$, $x_1$ and $x_2$ are the $x$ coordinates of the three points. The situation is shown in Figure 7.1. If this is done the area under the three points is then the integral of the function, that is:

$$I = \int f_2(x)dx \tag{7.2}$$

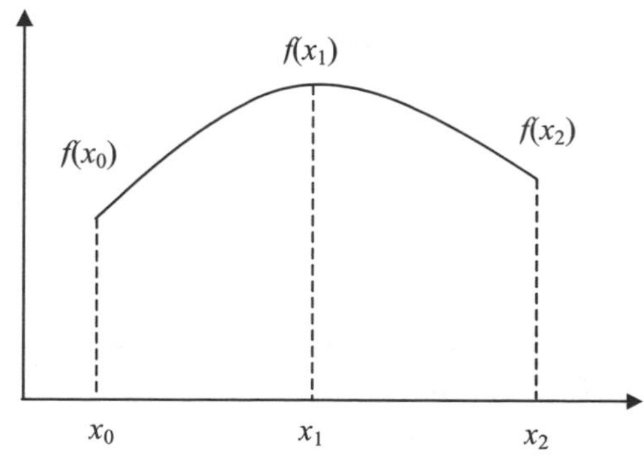


**Figure 7.1** The basic layout of the integration over two adjacent segments.

If the integration is carried out between the two end points of this segment, $x_0$ and $x_2$, then (after some manipulation) the integral can be determined as:

$$I = \frac{h}{3}\left[f(x_0) + 4f(x_1) + f(x_2)\right] \tag{7.3}$$

where

$$h = \frac{b-a}{2} \tag{7.4}$$

The method can be programmed to work on an equal strip width, dividing the area into an even increasing number of strips and this is referred to as Simpson's rule. In this case the formula for the area can be written:

$$I = \frac{(b-a)}{3n}\left[f(x_0) + 4\sum_{i=1,3,5,\ldots}^{n-1} f(x_i) + 2\sum_{j=2,4,6,\ldots}^{n-2} f(x_j) + f(x_n)\right] \tag{7.5}$$

Running the method on the same problem used to illustrate the trapezoidal rule in the previous chapter demonstrates the benefits that come from using Simpson's rule. The results are shown in Table 7.1. Notice that the number of segments must be a multiple of 2 for the method to work. This comes from the fact that three points are required to define a parabola and three points enclose two sub-segments of area. Notice also the impressive improvement in accuracy over the trapezoidal rule. It is worth the additional effort.

Figure 7.2 shows the area to be integrated. It is worth considering the issues involved with specifying the lower limit of the integral and particularly how this will affect the integration using the Simpson rule. With a lower limit of –10 and a step size of 6.0, then two of the three points fall on the section of the curve where the value is essentially 0.0 and this leads to a serious underestimation of the total area.

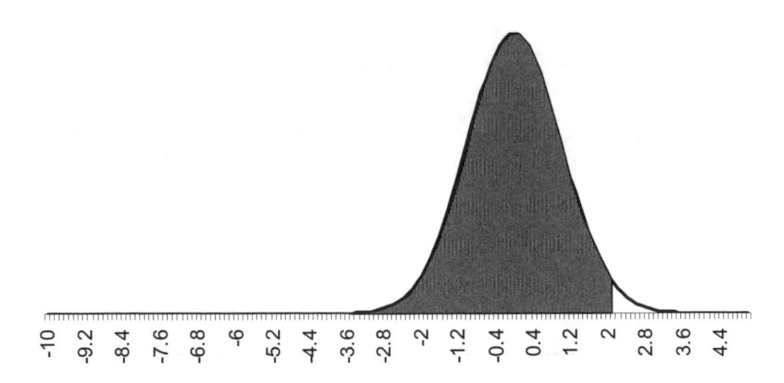


**Figure 7.2** The standard normal distribution showing the area up to z = 2.0.

**Table 7.1** The cumulative value of the standard normal distribution (m = 0.0, s = 1.0) at z = 2.0. Note: exact value = 0.977249937963813...

| Step size (s) | Estimate (4 significant figures) | Error (%) | Trapezoidal rule error (%) |
|---|---|---|---|
| 6.0 | 0.1091 | –88.8409 | –83.3435 |
| 4.0 | 1.0221 | 4.5937 | –17.3906 |
| 2.0 | 1.1722 | 19.9473 | –1.7523 |
| 1.0 | 0.9712 | –0.6182 | –0.9018 |
| 0.5 | 0.9772 | –0.0050 | –0.2292 |
| 0.2 | 0.9772 | –0.0001 | –0.0368 |
| 0.1 | 0.9772 | 0.0000 | –0.0092 |

## 7.2 SIMPSON'S 3/8 RULE

By taking four points at a time it is possible to fit a third-order polynomial to the data. In this case (skipping intermediate steps) it is possible to write the integral as:

$$I = \frac{3h}{8}[f(x_0) + 3f(x_1) + 3f(x_2) + f(x_3)] \tag{7.6}$$

where

$$h = \frac{b-a}{3} \tag{7.7}$$

The 1/3 rule is generally preferred as the additional effort for the 3/8 rule is not rewarded with extra accuracy.

**Estimating Sediment Volumes**

Understanding and managing tidal inlets requires a knowledge of the main sediment paths taken during normal conditions and part of this can be derived by surveying large areas of coastal land and calculating sediment volumes and comparing them to previous estimates. In a study reported by Pacheco et al. (2008) the authors used the trapezoidal rule, Simpson's 1/3 rule and Simpson's 3/8 rule and simply averaged the three results to determine what they said was a best estimate of the true volume. It should not be assumed that this is good practice!

## 7.3 AN EFFICIENT SIMPSON'S RULE

The methods illustrated so far have shown an increase in accuracy with increasing segments used in the integration, but no guidance has been given on how to determine this *a priori*. Listed in Figure 7.3 is an intelligent integration technique based on Simpson's Rule where the number of segments is doubled until the increase in accuracy is sufficiently small that the process can be halted. An advantage of the method is that calculations at any particular point are not repeated when the number of segments is doubled. In this case the sum of the odd and even points from the previous estimate becomes the new estimate of the sum of the even points and need not be re-calculated. The essential elements of a function to calculate the area from lower to upper and defined by function $f(x)$ are shown.

```
real function simpson(lower,upper)
!      function to iterate simpsons rule to convergence
!      Note: function f(x) must be defined
implicit none
real:: tolerance,sum,temp,lower,upper,oddsum,evensum,endsum,oldsum,h
real:: f
integer:: n, i

tolerance = 0.001          ! user to set value
n=2           !start with 2 segments
h=(upper - lower)/n
oddsum= f(lower + h)       ! sum of odd points is middle
evensum= 0 ! there are no even points yet
endsum=f(lower)+f(upper) ! sum up the two end points
sum=(endsum+4.0*oddsum)*h/3.0 ! initial sum
                           ! main iteration loop
oldsum=0.0  ! force do while to work for first time
do while(abs((sum-oldsum)/sum) >= tolerance)
 n=n*2                     ! now double segments
 oldsum = sum              !this is last estimate
 h= (upper - lower) / n
                           ! the old odd points are now even
 evensum= evensum + oddsum
 oddsum= 0
                           ! now do odd points
 do i= 1, n/2
  temp= lower + h * (2.0 * i - 1.0)
  oddsum= oddsum + f(temp)
 end do
 sum=(endsum+4*oddsum+2*evensum)*h/3.0
end do
simpson=sum                ! converged, return value
return
end
```

**Figure 7.3** Fortran code for Simpson's 1/3 rule.

## 7.4 INTEGRATION IN MATLAB

The adaptive integration routine described in the previous section can be implemented in Matlab using the quad function. Information about this particular function can be found by typing "help quad" at the Matlab prompt. This will provide information about the function and how to implement it. It is relatively simple, but quickly gets complex once it starts dealing with vector arguments.

*Example*
Integrate the standard normal distribution using Matlab.

**Groundwater Flow and the Two-Well Problem**

The problem of describing groundwater flow through a two-dimensional aquifer between a well that is being drawn down and a well that is being recharged (see Figure 7.4) was tackled by Maloof and Protopapas (2001). Two key integrals were necessary for the solution:

$$a = \int_{-\infty}^{\eta} \frac{d\eta}{(\cosh(\eta) + \cos(\xi))^2}$$

$$b = \int_{-\infty}^{\eta} \left[ \frac{\beta}{(\cosh(\eta) + \cos(\xi))^3} + \frac{\gamma}{(\cosh(\eta) + \cos(\xi))^4} \right] d\eta$$

where there were a range of $\eta$ and $\xi$ values (9 combinations of either positive, negative or equal to zero). They developed a new analytical solution that was valid for all combinations and used a numerical integration technique employing Simpson's rule to verify the results. They used the close agreement between the numerical results and their analytical results to argue for the accuracy of the analytical method: an opposite approach to that normally taken where analytical results are generally used to confirm the numerical approximation.

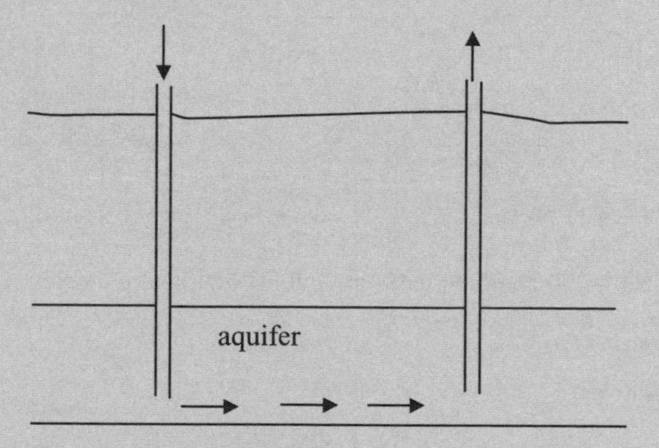


**Figure 7.4** Adjacent aquifers, one being drawn down, the other being recharged.

*Solution*

For this there are a number of steps that go from creating a Matlab file to implementing the solution:

1. First create a new "m-file" from the main Matlab environment File/New/M file.
2. Save this file in the current directory. In this case it is C:\Program Files\MATLAB704\work as shown in Figure 7.5.

**Figure 7.5** Matlab screen shot showing the saving of the m-file.

3. Enter the text shown in Figure 7.6 into the m-file that will define the function to be integrated. Save the file. Be careful with the use of capitalisation. Use array operators .*, ./ and .^ in the definition of the function so that it can be evaluated with a vector argument.

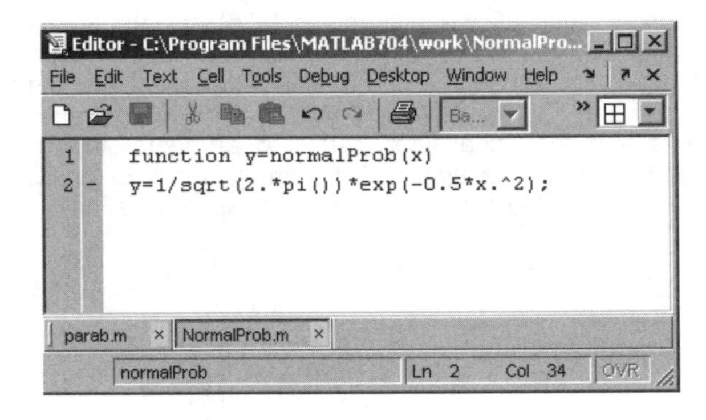


**Figure 7.6** Matlab screen shot showing the function definition.

4. In the main Matlab environment type the following at the prompt:

```
>> quad(@normalProb,-10,2)
ans =
   0.97725019699048
```

The display may only show 0.9773. If this is case type the command "`format long`" to increase the number of significant figures shown in the output.

## PROBLEMS

**7.1** Use Simpson's 1/3 Rule to integrate $f(x) = 5x$ from $x = 0$ to $x = 20$. Compare the estimates to the exact value determined from integral calculus principles. The estimates should be exactly correct – why?

**7.2** Use Simpson's 1/3 rule to integrate $f(x) = x^2$ from $x = 0$ to $x = 5$. Compare the estimates to the exact value determined from integral calculus principles. The estimates should be exactly correct. Again, why?

**7.3** Write a function to return the area under the standard normal distribution (mean = 0.0, standard deviation = 1.0) using Simpson's 1/3 Rule. Test the program by comparing the results from a range of values given by the Excel function `NORMDIST`. As part of the solution it will be necessary to determine a level of accuracy for the calculation and this may vary as $Z$ varies.

**7.4** Use Simpson's 1/3 Rule to integrate the stream survey data given in Table 7.2.

**Table 7.2** Stream offset and depth data.

| Offset (m) | Depth (m) |
|---|---|
| 0 | 0.0 |
| 10 | –1.0 |
| 20 | –1.5 |
| 25 | –3.7 |
| 30 | –2.0 |
| 40 | –0.5 |
| 50 | –0.2 |
| 90 | 0.0 |

**7.5** A pond has been surveyed to determine the water depth over a regular grid (10 metres by 10 metres) of points. The results are listed in Table 7.3. Determine the volume of the pond and compare it to the answer obtained using the trapezoidal rule in the previous chapter. (Hint: This is a two-step process. First, determine the area under each of the sections, working either across the rows or down the columns. Then, if they were plotted the "area" under the curve would be the volume of the pond.)

**Table 7.3** Depths recorded over a two-dimensional grid of points.

| Depth (metres) | | | | | | | |
|---|---|---|---|---|---|---|---|
| 0 | 0 | 0 | 0.5 | 0 | 0 | 0 | 0 |
| 0 | 0.2 | 0.5 | 0.8 | 0.6 | 0.4 | 0.2 | 0 |
| 0 | 0.3 | 0.9 | 1.5 | 1.2 | 0.8 | 0.3 | 0 |
| 0.1 | 0.5 | 1.9 | 2.5 | 1.9 | 1.5 | 1.1 | 0.1 |
| 0.1 | 0.6 | 2.9 | 4.1 | 2.8 | 2.0 | 0.4 | 0.1 |
| 0.1 | 0.4 | 3.0 | 3.9 | 2.7 | 2.0 | 0.5 | 0 |
| 0 | 0.2 | 2.5 | 3.5 | 2.0 | 1.0 | 0.2 | 0 |
| 0 | 0.2 | 1.6 | 1.9 | 1.1 | 0.5 | 0.1 | 0 |
| 0 | 0 | 0.1 | 0.1 | 0.2 | 0.1 | 0 | 0 |
| 0 | 0 | 0 | 0 | 0 | 0 | 0 | 0 |

**7.6** Set up an Excel spreadsheet to determine the integral of the normal distribution using Simpson's 1/3 Rule. If possible, set up the sheet so that there is a variable number of strips used in the calculation and compare the results with the exact value using the Excel function `NORMDIST`. (This is applied as a formula in a cell in the form "`=NORMDIST(z,0.0,1.0,true)`" where the function is integrated up to the value $z$, and the 0.0 and 1.0 refer to the mean and standard deviation of the distribution which for the standard normal distribution is 0.0 and 1.0 respectively. The "*true*" causes the cumulative distribution to be calculated.)

**7.7** Write a program to determine the integral to the equation that John Bernoulli was working on in 1697 (Nahin, 1998):

$$I = \int_0^1 x^x dx$$

which can also be determined as the sum of the series:

$$1 - \frac{1}{2^2} + \frac{1}{3^3} - \frac{1}{4^4} + \frac{1}{5^5} - \ldots = 0.78343$$

CHAPTER EIGHT

# Numerical Interpolation (Newton's Method)

## 8.1 INTRODUCTION

As an introduction to the use of numerical interpolation consider the calibration of an orifice plate, a common device for measuring flow in pipes. It fits into a pipe between a pair of flanges and, by constricting the flow and generating turbulence, leads to a drop in pressure that can be measured by reading the pressure head upstream and downstream of the plate. A plate, and a pipe with a plate fitted are shown in Figure 8.1.

**Figure 8.1** A pipeline with orifice plate and associated pressure tappings. Insert shows orifice plate.

The plates can be calibrated using a volumetric method where the flow at a particular change in pressure (head loss) is accumulated over a chosen duration and measured. Results for a particular orifice plate at the University of Adelaide hydraulics laboratory are listed in Table 8.1 and plotted in Figure 8.2. Since it is not realistic to take more than a few key readings, the problem is how to determine intermediate values, especially if a direct numerical output is required.

The aim of this chapter is to introduce a technique that can be used in situations such as this.

**Table 8.1** Data used to develop a calibration for an orifice plate flow measurement device.

| Head (mm) | Flow (L/s) |
|---|---|
| 18 | 1.44 |
| 77 | 2.97 |
| 134 | 3.92 |
| 217 | 4.98 |
| 873 | 10.05 |

It is worth noting that what is required is not a line of best fit as might be determined using some sort of regression analysis. The equation or equations in this case are expected to pass exactly through the given points rather than fitting a line that best describes the points in general without necessarily going through any of them.

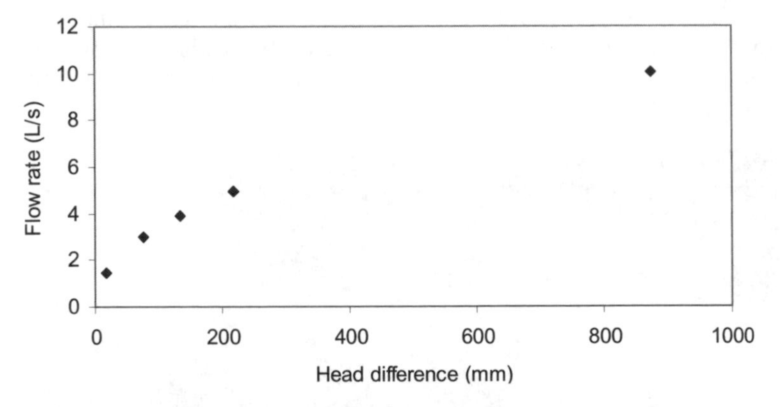


**Figure 8.2** Plot of output from orifice plate calibration.

**Missing Meteorological Data**

Masselink and Pattiaratchi (2001), in a study of the characteristics of the sea breezes occurring at Perth, Western Australia, found that in the data from 1949 to 1964 did not include a reading at 9pm each day (21:00 hours). To overcome this the authors used linear interpolation to "fill in" the missing values. They argued that this had no effect on the time-domain analysis and an insignificant effect on the frequency-domain analysis. Other missing data were also filled in using linear interpolation.

## 8.2 POLYNOMIAL EQUATIONS

As a first approach it is worth remembering that it is possible to write the equation of a general $n^{th}$ order polynomial as:

$$f(x) = a_0 + a_1 x + a_2 x^2 + ... + a_n x^n \tag{8.1}$$

and that for any $n+1$ non-collinear data points there is one and only one polynomial of order $n$ or less that passes through all points. For example, there is only one straight line (polynomial of order 1) which passes through 2 points, only onc quadratic (order 2) which passes through any 3 given points, and so on.

This then provides one possible way of dealing with the problem of determining an equation to fit the $n$ data points: simply fit an order $n$–1 polynomial through them. Having determined this polynomial it is then possible to calculate values of the function at any intermediate point. This approach will be set out in the following sections before some problems are identified and more robust methods are introduced.

## 8.3 NEWTON'S LINEAR INTERPOLATION

Although Newton's method is for a general $n^{\text{th}}$ order polynomial it is convenient to illustrate its performance on a simple first-order case. The simplest form of interpolation is to join each adjacent point by a straight line. This can be written as:

$$f_1(x) = f(x_0) + \frac{f(x_1) - f(x_0)}{x_1 - x_0}(x - x_0) \tag{8.2}$$

which is based on a consideration of similar triangles as shown in Figure 8.3, where it is evident that $(f(x_1) - f(x_0)) / (x_1 - x_0) = (f(x) - f(x_0)) / (x - x_0)$. Equation (8.2) follows from a simple re-arrangement of this equality. In the expression $f_1$ the subscript indicates that it is a first order function. In general the smaller thc interval between the points the better the approximation.

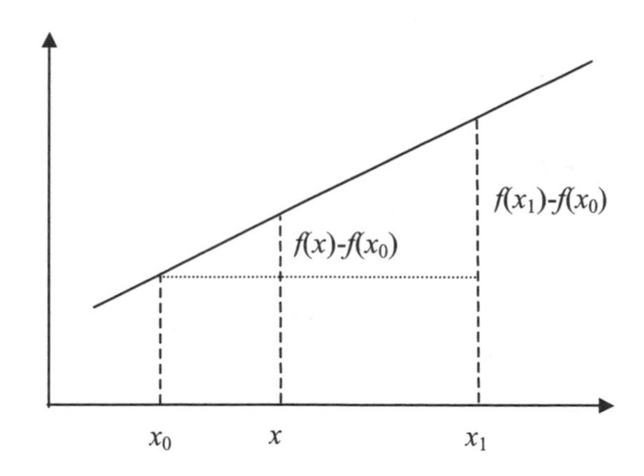


**Figure 8.3** Similar triangles used to derive the formula for linear interpolation.

*Example*

Estimate a flow rate for a head difference of 100 mm using the data in Table 8.1.

*Solution*
Applying Equation (8.2) with $x_0$ = 77 mm and $x_1$ = 134 mm (so that the required value is an intermediate value) gives:

$$f_1(100) = 2.97 + \frac{3.92 - 2.97}{134 - 77}(100 - 77) = 3.35 \text{ L/s}.$$

This equation could be used for any points between 77 mm and 134 mm. To apply the method to the whole range of values from 18 mm to 873 mm, therefore, requires a total of 4 linear equations. This might be inconvenient, but a more serious problem can be appreciated by viewing the data points in Figure 8.3: the data appear to follow a smooth curve rather than a series of straight lines. The estimate in the example might then be expected to be less than the "correct" value.

## 8.4 NEWTON'S QUADRATIC INTERPOLATION

For data that appear to follow a curve a quadratic between three points may be superior. A convenient form of the quadratic can be written as:

$$f_2(\mathrm{x}) = \mathrm{b}_0 + b_1(x - x_0) + b_2(x - x_0)(x - x_1) \tag{8.3}$$

To determine the coefficients first consider the point $x = x_0$. Here $f(x_0) = \mathrm{b}_0$ which gives the first value required. At $x = x_1$ it can be shown that:

$$b_1 = \frac{f(x_1) - f(x_0)}{x_1 - x_0} \tag{8.4}$$

At $x = x_2$, after some manipulation:

$$b_2 = \frac{\dfrac{f(x_2) - f(x_1)}{x_2 - x_1} - \dfrac{f(x_1) - f(x_0)}{x_1 - x_0}}{x_2 - x_0} \tag{8.5}$$

Notice that the first two terms of the expansion associated with $b_0$ and $b_1$ are similar to the first-order solution. It is only the last term which includes some curvature into the estimate.

*Example*
Use a quadratic interpolation to estimate the flow rate for a head difference of 100 mm using the data in Table 8.1.

*Solution*
Applying Equation (8.3) with $x_0$ = 77 mm, $x_1$ = 134 mm and $\mathrm{x}_2$ = 217 mm gives:

$$f_2(100) = 2.97 + 0.0166(100 - 77) + (-2.78.10^{-5})(100 - 77)(100 - 134) = 3.375 \text{ L/s}$$

As expected, the estimate is slightly higher than that from the linear interpolation. It is worth noting that it is also possible to use other values for the $x_0$, $x_1$ and $x_2$. For example, there is nothing to stop the use of $x_0 = 18$ mm, $x_1 = 217$ mm and $x_2 = 873$ mm; however, it is considered good practice to use the points closest to the point being estimated. As a matter of interest, using these points gives an estimate of the flow rate as 3.01 L/s, which looks a poor value based on the behaviour of the data in the vicinity of a head difference of 100 mm. The behaviour of the equation with these ill-chosen data values is shown in Figure 8.4.

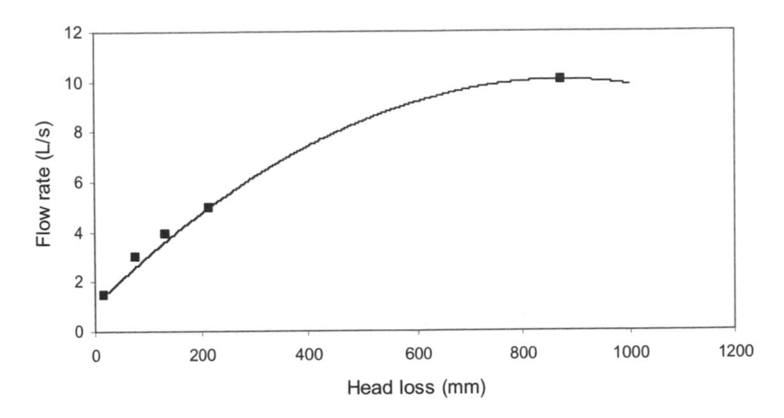


**Figure 8.4** Plot of output from quadratic interpolation where inappropriate points have been used to define the polynomial.

Two key problems become evident. Firstly, the reason for the under-prediction based on an inappropriate choice of points is evident with the function passing well below the points at $x = 77$ mm and $x = 134$ mm which were ignored for the calculation. The other point of interest is to observe the behaviour of the polynomial in the region to the right of $x = 873$ mm: it starts to fall! In reality this is not possible; an increasing head loss across the orifice plate is due to an increasing discharge, and what is seen here is a particular problem of using polynomials to fit data: there is the possibility of odd behaviour both when interpolating and especially when extrapolating. There will be further discussion on this in a later section.

## 8.5 GENERAL FORM OF NEWTON'S POLYNOMIALS

A general equation for a $n^{\text{th}}$ order polynomial can be written:

$$f_n(x) = b_0 + b_1(x - x_0) + \ldots + b_n(x - x_0)(x - x_1)..(x - x_{n-1}) \tag{8.6}$$

As before the $b$ values can be written in terms of the function at various points on the curves:

$$\begin{aligned} b_0 &= f(x_0) \\ b_1 &= f[x_1, x_0] \\ b_2 &= f[x_2, x_1, x_0] \\ &. \\ &. \\ b_n &= f[x_n, x_{n-1}, \ldots, x_1, x_0] \end{aligned} \tag{8.7}$$

where the square-bracketed evaluations are referred to as finite divided differences. The first finite divided difference can be written:

$$f[x_i, x_j] = \frac{f(x_i) - f(x_j)}{x_i - x_j} \tag{8.8}$$

The second finite difference divided is:

$$f[x_i, x_j, x_k] = \frac{f[x_i, x_j] - f[x_j, x_k]}{x_i - x_k} \tag{8.9}$$

and so on until the $n^{\text{th}}$ divided difference:

$$f[x_n, x_{n-1}, \ldots, x_1, x_0] = \frac{f[x_n, x_{n-1}, \ldots, x_1] - f[x_{n-1}, x_{n-2}, \ldots, x_0]}{x_n - x_0} \tag{8.10}$$

These can be used to calculate the coefficients $b_0$ to $b_n$ and substituted back into the equation to give:

$$f_n(x) = f(x_0) + (x - x_0) f[x_1, x_0] + (x - x_0)(x - x_1) f[x_2, x_1, x_0] + \ldots \tag{8.11}$$

It is important to note that there is no need for the data points to be equally spaced or even for the order of the points to be correct (in some ascending or descending sense).

Notice also that the calculation of the divided differences is recursive, that is one term relies on terms of lower order. An algorithm to calculate the $n^{\text{th}}$ order polynomial can be developed to evaluate the terms in an efficient manner. It begins with the decision on the order of the polynomial to be determined and then proceeds by calculating the first order divided differences, followed by the second order divided differences, and so on until the appropriate level has been reached.

**Newton Interpolating Polynomial Algorithm**

1. Decide on Order of Scheme, *n*.
2. Calculate first divided differences:

$$f[x_{i+1},x_i]=\frac{f(x_{i+1})-f(x_i)}{x_{i+1}-x_i} \text{ for } i = 0, \ldots n{-}1$$

3. Calculate subsequent divided differences:

for $j$ = 2, ... $n$
for $i$ = $j$, ... $n$

$$f[x_i,\ldots,x_{i-j}]=\frac{f[x_i,\ldots x_{i-j+1}]-f[x_{i-1},\ldots,x_{i-j}]}{x_i-x_{i-j}}$$

4. Evaluate function:

$$f_n(x)=f(x_0)+\sum_{i=1}^{n}\left(\prod_{j=0}^{i-1}(x-x_j)\right)f[x_i,\ldots,x_0]$$

where $\Pi$ designates the product of.

## 8.6 LAGRANGE'S INTERPOLATING POLYNOMIALS

Lagrange's approach to the derivation of interpolating polynomials is simply a reformulation of the Newton method that avoids the computation of the divided differences. This makes the method much easier to program. The final results are identical. It can be represented as:

$$f_n(x)=\sum_{i=0}^{n}L_i(x)f(x_i) \tag{8.12}$$

where

$$L_i(x)=\prod_{j=0,j\neq i}^{n}\frac{x-x_j}{x_i-x_j} \tag{8.13}$$

where $\Pi$ designates the product of. The first-order (linear) version becomes:

$$f_1(x)=\frac{x-x_1}{x_0-x_1}f(x_0)+\frac{x-x_0}{x_1-x_0}f(x_1) \tag{8.14}$$

and the second-order version is:

$$f_2(x) = \frac{(x-x_1)(x-x_2)}{(x_0-x_1)(x_0-x_2)} f(x_0) + \frac{(x-x_1)(x-x_2)}{(x_1-x_0)(x_1-x_2)} f(x_1) + \frac{(x-x_0)(x-x_1)}{(x_2-x_0)(x_2-x_1)} f(x_2) \tag{8.15}$$

A code segment that would carry out these steps in Fortran is listed in Figure 8.5. Note that the data points are numbered from 0 to $n$, giving the $n$+1 points required. For a given $x$ (included as a variable called *xvalue*) the code will determine the $y$ value that would fit on the polynomial.

```
y=0
do i=0,n
 product=f(i)
 do j = 0,n
  if(i /= j) then
   product=product*(xvalue-x(j))/(x(i)-x(j))
  endif
 enddo
 y = y+product
enddo
```

**Figure 8.5** Fortran code segment to carry out a Lagrange interpolation.

## 8.7 POLYNOMIAL INTERPOLATION IN PRACTICE

Although the idea of fitting a polynomial to a set of data may sound appealing it is worth being aware of some of the issues that may arise. The higher the order of polynomial the more the chance for wild fluctuations. Consider, for example, the situation where seven data points from the error function have been used to test the polynomial interpolating procedure. The points and the sixth-order polynomial that results are shown in Figure 8.6.

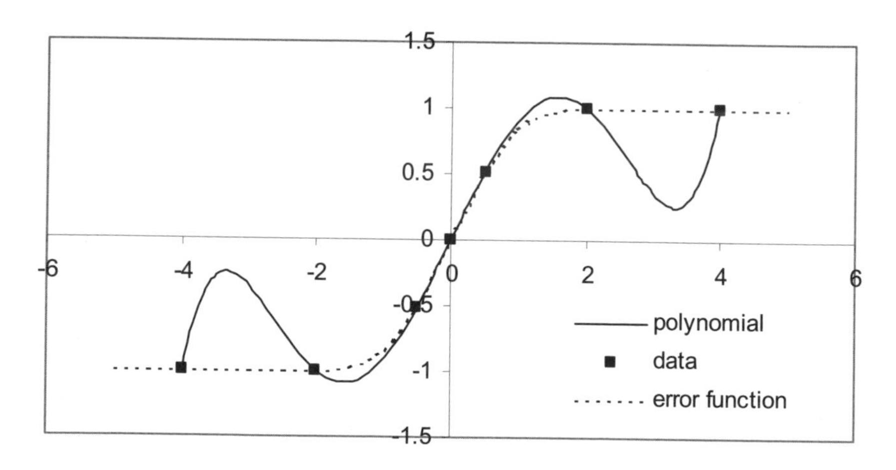


**Figure 8.6** Sixth-order Lagrange polynomial fit to selected data from error function.

It is evident from the plot that, although the function goes through the points exactly, in between those particular places it oscillates wildly. This is an important consideration, particularly where it may be thought more convenient to have a single high-order polynomial that represents all the points rather than a series of lower-order polynomials that apply only to a specified sub-range of values.

## 8.8 POLYNOMIAL INTERPOLATION IN EXCEL

In Excel it is possible to fit polynomials to data, although the package is a mixture of numerical interpolation and regression analysis. The data are plotted and the line on the plot is selected by clicking on it with the right mouse button to bring up a range of options, one of which is "Add Trendline". This is selected and the user is offered a range of options including a polynomial of an order that can be specified. The procedure is shown in Figure 8.7. It is also possible to have the equation of the line shown adjacent to the plot as well as an $R^2$ value (appropriate to the regression aspect of the procedure).

It should be noted that if the order of polynomial selected is less than $n$–1 (where $n$ is the number of data) the package will fit a line of best fit; otherwise it will be the full polynomial interpolation.

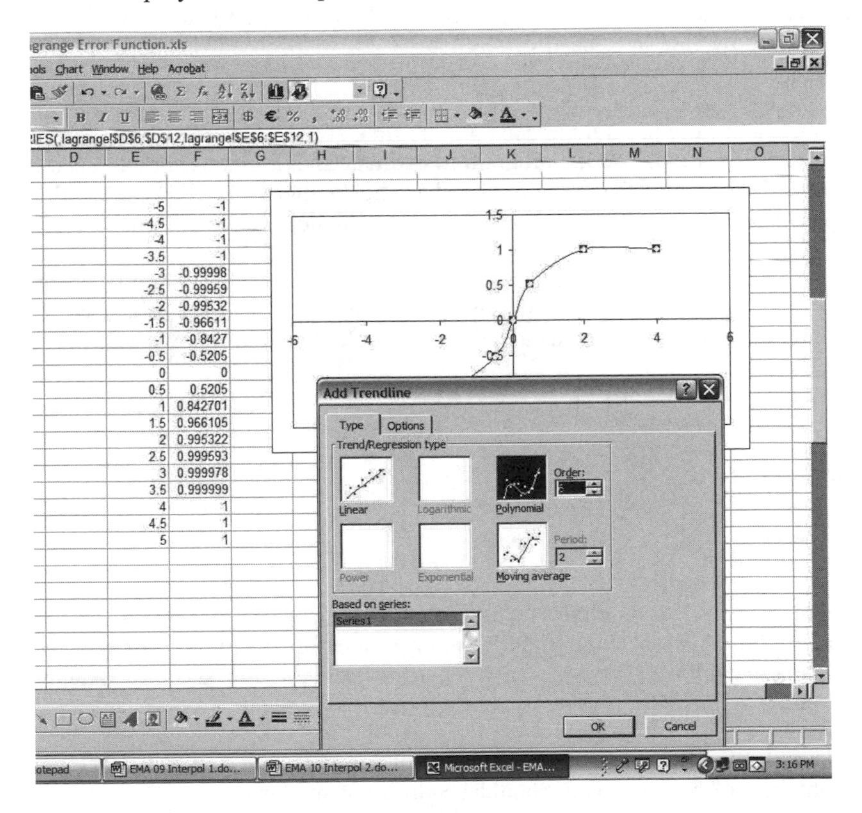


**Figure 8.7** Application of polynomial in Excel using "Add Trendline" option.

## 8.9 POLYNOMIAL INTERPOLATION IN MATLAB

While a program can be written in the Matlab scripting language, Matlab also has many in-built functions for this purpose. One of these functions is `interp`. To get more details type "help interp" at the command prompt. This function views the problem slightly differently. It views the interpolation problem as one of increasing the sampling rate of the available data so the function call:

```
>>y=interp(x,10)
```

increases the number of values in $x$ by a factor of 10, so 10 $x$ values become 100 $y$ values. To do this it uses a simple linear interpolation which may not be appropriate in many situations.

## PROBLEMS

**8.1** Use linear interpolation (hand calculation) to estimate a value for the discharge flow rate at a head difference of 200 mm based on the data given in Table 8.1.

**8.2** Use quadratic interpolation (hand calculation) to estimate a value for the discharge flow rate at a head difference of 200 mm based on the data given in Table 8.1.

**8.3** Write a Fortran program to generate the first, second and $n^{th}$ divided differences in Newton's method. Use the program to estimate the discharge for any head given in Table 8.1. Ensure that the program has the facility to select heads within the range covered by the data and that it provides guidance for selecting such values.

**8.4** Write Fortran code to interpolate between a general number of data points using the Lagrange interpolating polynomial method. The input should be from a file and (as a suggestion) the data in the file should be in the following format:

```
Line 1: A general header (text)
Line 2: Number of data to follow (integer)
Line 3: x1 y1
Line 4: x2 y2
.
.
Line n+2: xn yn
```

**8.5** Use the Excel "Trendline" feature to fit the data given in Table 8.1 where the order of the equation to be fitted is appropriate for the number of data points. Compare the solution with the answers from Problems 8.1 and 8.2. Explain any differences.

**8.6** Use the Excel "Trendline" feature to fit the data given in Table 8.1 using a straight line of best fit. This should be different to the result from the previous question – why?

**8.7** Table 8.2 lists three-hourly wind speed and direction data collected at Adelaide Airport in 1955 by the Bureau of Meteorology. Develop a number of interpolation schemes to estimate the value at 21:00 each day assuming the actual value was not available and compare the estimate with the actual value.

**Table 8.2** Wind speed and direction data from Adelaide Airport. Provided by the Bureau of Meteorology, Adelaide, Australia.

| Date | Time | Direction (degrees) | Speed (m/s) |
|---|---|---|---|
| 2/08/1955 | 0:00 | 22.5 | 6.2 |
| 2/08/1955 | 3:00 | 337.5 | 8.8 |
| 2/08/1955 | 6:00 | 337.5 | 7.2 |
| 2/08/1955 | 9:00 | 337.5 | 6.2 |
| 2/08/1955 | 12:00 | 292.5 | 10.3 |
| 2/08/1955 | 15:00 | 292.5 | 6.2 |
| 2/08/1955 | 18:00 | 315 | 3.1 |
| 2/08/1955 | 21:00 | 270 | 7.7 |
| 3/08/1955 | 0:00 | 292.5 | 7.2 |
| 3/08/1955 | 3:00 | 292.5 | 7.2 |
| 3/08/1955 | 6:00 | 270 | 4.6 |
| 3/08/1955 | 9:00 | 0 | 0 |
| 3/08/1955 | 12:00 | 270 | 3.6 |
| 3/08/1955 | 15:00 | 292.5 | 2.1 |
| 3/08/1955 | 18:00 | 270 | 2.6 |
| 3/08/1955 | 21:00 | 0 | 0 |
| 4/08/1955 | 0:00 | 45 | 1 |
| 4/08/1955 | 3:00 | 67.5 | 1 |
| 4/08/1955 | 6:00 | 45 | 1 |
| 4/08/1955 | 9:00 | 0 | 0 |
| 4/08/1955 | 12:00 | 360 | 6.2 |
| 4/08/1955 | 15:00 | 22.5 | 4.1 |
| 4/08/1955 | 18:00 | 45 | 2.1 |
| 4/08/1955 | 21:00 | 22.5 | 2.6 |
| 5/08/1955 | 0:00 | 45 | 1.5 |
| 5/08/1955 | 3:00 | 67.5 | 1.5 |
| 5/08/1955 | 6:00 | 90 | 1 |
| 5/08/1955 | 9:00 | 0 | 0 |
| 5/08/1955 | 12:00 | 45 | 6.2 |
| 5/08/1955 | 15:00 | 292.5 | 6.2 |
| 5/08/1955 | 18:00 | 0 | 0 |
| 5/08/1955 | 21:00 | 22.5 | 2.6 |
| 6/08/1955 | 0:00 | 45 | 4.1 |

**8.8** In a site investigation for geotechnical design the cone penetrometer data in Table 8.3 were collected. Assuming there had been a malfunction in the instrument and the data between 200 mm and 250 mm were not recorded, develop a numerical interpolation program that could be used to estimate the missing data. Use a variety of techniques and compare the predictive ability of each.

**Table 8.3** Cone penetrometer test (CPT) data for a site in Adelaide, South Australia.

| Depth (mm) | Cone Tip Resistance (MPa) | Sleeve Friction (kPa) | Porewater Pressure (kPa) |
|---|---|---|---|
| 10 | 0.58 | 2 | 51 |
| 20 | 0.7 | 2 | 51 |
| 30 | 1.15 | 3 | 52 |
| 40 | 1.7 | 18 | 53 |
| 50 | 2.28 | 35 | 54 |
| 60 | 3.88 | 47 | 54 |
| 70 | 6.1 | 50 | 54 |
| 80 | 9.73 | 60 | 54 |
| 90 | 14.75 | 80 | 54 |
| 100 | 18.1 | 85 | 54 |
| 110 | 19.55 | 92 | 54 |
| 120 | 20.78 | 103 | 54 |
| 130 | 21.88 | 123 | 55 |
| 140 | 22.03 | 165 | 54 |
| 150 | 22.15 | 265 | 53 |
| 160 | 21.7 | 293 | 52 |
| 170 | 21.03 | 293 | 52 |
| 180 | 19.55 | 321 | 52 |
| 190 | 19 | 340 | 52 |
| 200 | 18.28 | 336 | 51 |
| 210 | 17.98 | 333 | 51 |
| 220 | 16.18 | 347 | 50 |
| 230 | 14.45 | 336 | 50 |
| 240 | 13.03 | 344 | 50 |
| 250 | 12.23 | 352 | 50 |
| 260 | 11.43 | 332 | 48 |
| 270 | 10.85 | 317 | 47 |
| 280 | 10.48 | 276 | 48 |
| 290 | 9.95 | 236 | 47 |
| 300 | 9.5 | 215 | 46 |

# CHAPTER NINE

# Numerical Interpolation (Cubic Splines and Other Methods)

## 9.1 INTRODUCTION

In the previous chapter the first illustrative problem was to derive a way of interpolating values between a number of discrete points: specific head and discharge readings from an orifice plate flow measurement device. Using the methods covered so far there is a range of options for this, some of which are:

1. linearly interpolate between each pair of points, leading to four equations to cover the range of values;
2. fit a quadratic between the first three points and a second over the last three points (using the middle point twice, once for each curve);
3. fit a cubic over the first four points and then a linear interpolation over the other interval;
4. fit a fourth-order polynomial over the five points.

Each of these options has a problem. For Option 1 the linear fit is not particularly good and there is a break in the slope of the curves at each interior point. For Option 2 there is a change in equations and therefore slope where the two curves meet. Option 3 is an odd mixture of curve and straight with, once again, a break in the slope where the two meet. Option 4 has the potential to produce unrealistic (and unwanted) oscillations and variations from what might be expected.

A method that overcomes some of these problems is to fit a set of third-order polynomials with one for each segment while making sure that the gradient matches on each side of each internal point. This has the potential to give a smooth curve over all points and one that does not have the discontinuity associated with changing polynomials. This solution is referred to as a cubic spline: a series of third-order polynomials for each segment that transition smoothly from one to the next.

## 9.2 CUBIC SPLINES

A cubic spline has the general equation:

$$f_i(x) = a_i x^3 + b_i x^2 + c_i x + d_i \tag{9.1}$$

and so for $n$+1 data points ($i$ = 0, 1, 2,..., $n$) there are $n$ intervals and four unknowns ($a_i$, $b_i$, $c_i$, $d_i$) per interval giving 4$n$ unknowns in total. This is shown in Figure 9.1.

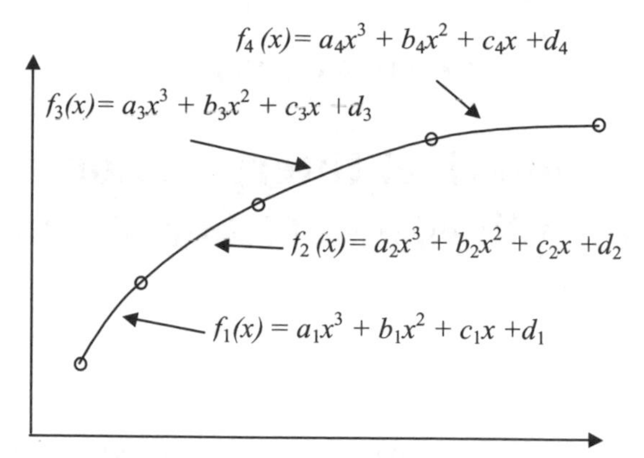


**Figure 9.1** Cubic splines for a set of 5 data points.

Therefore a solution requires 4$n$ equations to solve for the 4$n$ unknowns. These equations and conditions come from:

- the functions on each side must be equal at the interior points (2$n$ – 2);
- the first and last functions must pass through the end points (2);
- the first derivatives must be equal at the interior points ($n$–1);
- the second derivatives must be equal at the interior points ($n$–1);
- the second derivative at the end points are zero (2) and this condition gives rise to what is known as a natural spline.

Although this set of equations would allow a viable solution technique a method has been developed which reduces the number of unknowns to $n$–1 which requires only the solution of $n$–1 equations. The method has the additional advantage that the matrix that is formed for the solution is tri-diagonal and can therefore be solved using the Thomas algorithm which is described in Chapter 12.

The method uses the fact that a cubic function between two points ($x_i$, $x_{i-1}$) will have a second derivative which varies linearly with $x$ (a straight line). This can be represented by first-order Lagrange polynomial which can be written:

$$f_i''(x) = f''(x_{i-1})\frac{x - x_i}{x_{i-1} - x_i} + f''(x_i)\frac{x - x_{i-1}}{x_i - x_{i-1}} \tag{9.2}$$

where $f''(x_{i-1})$ and $f''(x_i)$ are the values of the second derivatives at the two end-of-segment points. If this is integrated twice an expression for the function is returned. There are two constants of integration, and these can be determined by setting adjacent functions equal at the internal points. The equation of the line can then be written:

$$f_i(x) = \frac{f''(x_{i-1})}{6(x_i - x_{i-1})}(x_i - x)^3 + \frac{f''(x_i)}{6(x_i - x_{i-1})}(x - x_{i-1})^3$$

$$+\left(\frac{f(x_{i-1})}{x_i - x_{i-1}} - \frac{f''(x_{i-1})(x_i - x_{i-1})}{6}\right)(x_i - x) + \left(\frac{f(x_i)}{x_i - x_{i-1}} - \frac{f''(x_i)(x_i - x_{i-1})}{6}\right)(x - x_{i-1}) \quad (9.3)$$

Notice that the line is a cubic (highest power of $x$ is 3) and that it applies to each line segment between internal points. This equation involves only two unknowns: the second derivatives at $x_i$ and $x_{i-1}$. These can be evaluated by taking the first derivative of the expression and using the property that the first derivative will be equal at internal points. With this it is possible to write:

$$(x_i - x_{i-1})f''(x_{i-1}) + 2(x_{i+1} - x_{i-1})f''(x_i) + (x_{i+1} - x_i)f''(x_{i+1}) =$$

$$\frac{6}{x_{i+1} - x_i}[f(x_{i+1}) - f(x_i)] + \frac{6}{x_i - x_{i-1}}[f(x_{i-1}) - f(x_i)] \quad (9.4)$$

If this equation is written for all interior points ($i = 1, n-1$) then there are $n-1$ equations in $n-1$ unknowns (the second derivatives). The boundary conditions are that the second derivatives at the end points are zero. It will be seen that the equations form a tri-diagonal matrix which can be solved using the Thomas algorithm.

*Example*
Estimate a flow rate for a head difference of 100 mm by developing a set of cubic splines to describe the data from the previous chapter (repeated here as Table 9.1).

**Table 9.1** Raw data used to develop a calibration for an orifice plate flow measurement device.

| Head (mm) | Flow (L/s) |
|---|---|
| 18 | 1.44 |
| 77 | 2.97 |
| 134 | 3.92 |
| 217 | 4.98 |
| 873 | 10.05 |

*Solution*
The data are allocated subscripts in line with the requirement of the method. Therefore $x_0 = 18.0, f(x_0) = 1.44$, and so on up to $x_4 = 873.0, f(x_4) = 10.05$.

For $i = 2$, for example, Equation (9.4) can be written:

$$(x_2 - x_1)f''(x_1) + 2(x_3 - x_1)f''(x_2) + (x_3 - x_2)f''(x_3) = RHS_2$$

where

$$RHS_2 = \frac{6}{x_3 - x_2}[f(x_3) - f(x_2)] + \frac{6}{x_2 - x_1}[f(x_1) - f(x_2)]$$

Similar expressions can be written for the case where $i$=1 and $i$=3 although it will be necessary to make use of the fact that at the end points the second derivative is assumed to be zero (representing a straight line). Given the pattern that emerges, it is convenient to write the three equations in matrix form. This is given as:

$$\begin{bmatrix} 2(x_2 - x_0) & (x_2 - x_1) & 0 \\ (x_2 - x_1) & 2(x_3 - x_1) & (x_3 - x_2) \\ 0 & (x_3 - x_2) & 2(x_4 - x_2) \end{bmatrix} \begin{Bmatrix} f''(x_1) \\ f''(x_2) \\ f''(x_3) \end{Bmatrix} = \begin{Bmatrix} RHS_1 \\ RHS_2 \\ RHS_3 \end{Bmatrix}$$

where

$$\begin{Bmatrix} RHS_1 \\ RHS_2 \\ RHS_3 \end{Bmatrix} = \begin{Bmatrix} \frac{6}{x_2 - x_1}[f(x_2) - f(x_1)] + \frac{6}{x_1 - x_0}[f(x_0) - f(x_1)] \\ \frac{6}{x_3 - x_2}[f(x_3) - f(x_2)] + \frac{6}{x_2 - x_1}[f(x_1) - f(x_2)] \\ \frac{6}{x_4 - x_3}[f(x_4) - f(x_3)] + \frac{6}{x_3 - x_2}[f(x_2) - f(x_3)] \end{Bmatrix}$$

Although this may appear complicated, it actually becomes quite simple when programmed. When using the Thomas algorithm the left-hand element in each row of the matrix is stored in the *A* array, the middle (diagonal) element in the *B* array and the right-hand element in the *C* array. The instructions to populate the matrix and solve it are shown in a code segment in Figure 9.2. Having solved the matrix it is convenient to transfer the solution to a new array named `f2` (so that it is similar to the algorithm for the cubic). The final plot for the orifice data is shown in Figure 9.3.

```
do i=1,n-1
 a(i)=x(i)-x(i-1)             !  set up a, b, c and d for the matrix
 b(i)=2*(x(i+1)-x(i-1))
 c(i)=x(i+1)-x(i)
 d(i)=6.0/(x(i+1)-x(i))*(f(i+1)-f(i))+6.0/(x(i)-x(i-1))* &
      (f(i-1)-f(i))
end do
call thomas(a,b,c,d,n-1)       ! solve using thomas algorithm
do i=1,n-1
 f2(i) = d(i)                  ! solution matrix back from thomas
end do
f2(0)=0.0                  ! first and last f'' = 0 (straight lines)
f2(n)=0.0
```

**Figure 9.2** Fortran code segment to populate and solve matrix for cubic spline.

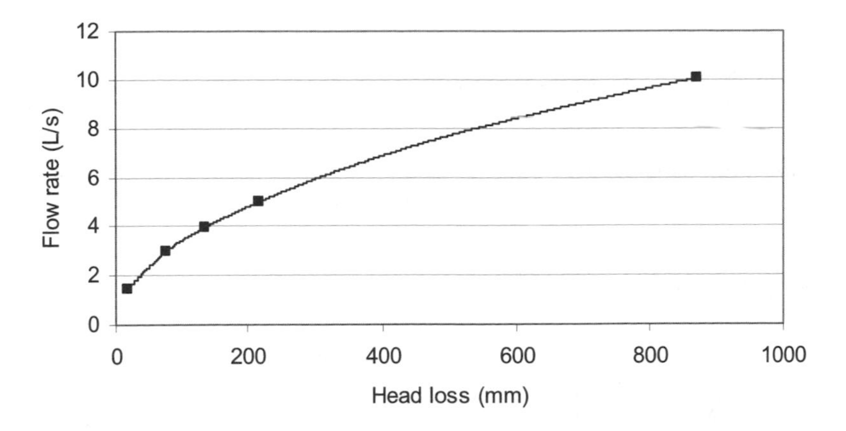


**Figure 9.3** Plot of output from the cubic spline interpolation.

It should be noted that the solution using a cubic spline will be different to that if a cubic polynomial was used. This is because the spline requires zero second derivatives at the end points whereas the cubic polynomial does not. As a check, the cubic spline approach was used on the error function data plotted earlier in Figure 8.6. It is shown in Figure 9.4. It is clear that the splines give a much better fit compared to the high-order polynomials used in the previous chapter.

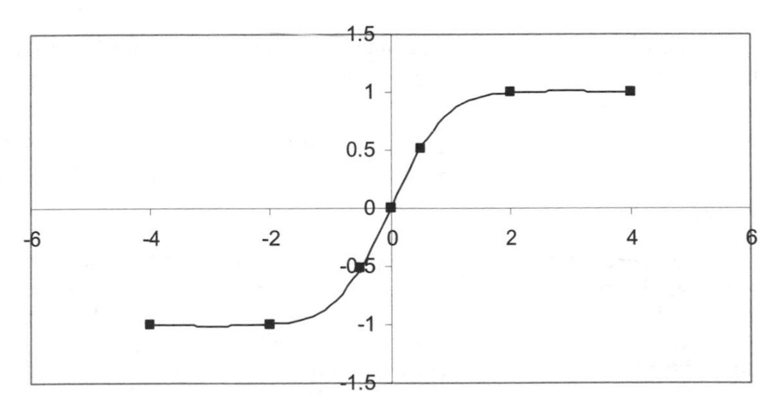


**Figure 9.4** Cubic spline fit to selected data from error function.

## 9.3 TRIGONOMETRIC INTERPOLATION

Fourier was able to demonstrate that any data (and particularly data that is periodic in nature) can be fitted by the sum of sine and cosine waves in the form:

$$f(t) = a_0 + a_1 \cos(\omega_0 t) + b_1 \sin(\omega_0 t) + a_2 \cos(2\omega_0 t) + .. \tag{9.5}$$

or more concisely

$$f(t) = a_0 + \sum_{k=1}^{\infty}[a_k \cos(k\omega_0 t) + b_k \sin(k\omega_0 t)] \tag{9.6}$$

where $\omega_0$ is the fundamental frequency and the other frequencies, which are whole multiples of it, are called harmonics. The form of the constants can be expressed:

$$a_n = \frac{2}{N}\sum_{k=1}^{N} f_k \cos\left(\frac{2\pi k}{T} t_k\right) \tag{9.7}$$

$$b_n = \frac{2}{N}\sum_{k=1}^{N} f_k \sin\left(\frac{2\pi k}{T} t_k\right) \tag{9.8}$$

$$a_0 = \frac{1}{N}\sum_{k=1}^{N} f_k \tag{9.9}$$

where $N$ is the number of data points, $f_k$ is the $k^{\text{th}}$ value of the raw data, $t_k$ is $k^{\text{th}}$ value of the time increment and $T$ is the fundamental period of the data set which is taken equal to its length.

This method is the basis of spectral analysis, which will be dealt with in some detail later in the book. One of the features of the spectral approach is the assumption that the amplitude and frequency of the components necessary to fit the data exactly actually have something to tell about the fundamental processes that generated the data in the first place, so the method becomes much more than simply curve fitting.

**Numerical Interpolation and the Generation of GPS Data**

Yousif and El-Rabbany (2007) provide a current application of numerical interpolation in the quest to provide highly accurate GPS data available at a higher rate than the 15 minutes that they quote as the typical rate. In this case they had as many data as they wanted at 15-minute intervals and required a numerical interpolation to reduce the effective time interval. The authors tested a number of methods including Newton divided differences, Lagrange polynomials, cubic splines and a trigonometric interpolation technique.

For the data they were working with, accuracy was best towards the centre of the interpolation interval and degraded near the end points, the best accuracy was achieved using $n$ = 9 data points, the Newton interpolation gave identical results to the Lagrange (as might be expected) and for this particular application the cubic splines gave low accuracy.

## 9.4 OTHER METHODS OF INTERPOLATION

Consider a thin strip of spring steel that has been bent around a series of fixed points on a board (see Figure 9.5). The question naturally arises: what shape will it take, and how might that shape vary if the steel is put into tension with a force, *T*? Horn (1983) investigated this and found that the shape will be one that minimises the elastic energy in the steel, where the elastic energy is a function of the curvature of the steel integrated over its length. Qu and Ye (2000) state that this is a "commonly used and recognized criterion for curve modelling" in engineering while Horn (1983) argues that the human vision system tends to use a curve of low energy when completing a contour visually.

It is tempting to assume that the minimum energy curve should be easy to develop but in fact this is not the case and the tendency is to use other forms, such as cubic splines, in an effort to produce forms that minimise the elastic energy. Qu and Ye (2000) tested a number of forms and found that, while the cubic spline was not the best in many cases, it did produce results that were close to the minimum energy and typically produced satisfactory data-fitting curves. The standard polynomial functions performed poorly, mainly because of the potential for the oscillations that can occur when high-order polynomials are used to fit a large number of data points. Some of the forms they tested included a tension term (*T*) and although these were efficient in some situations the authors argued that "curves with less strained energy and normal length look *nice*".

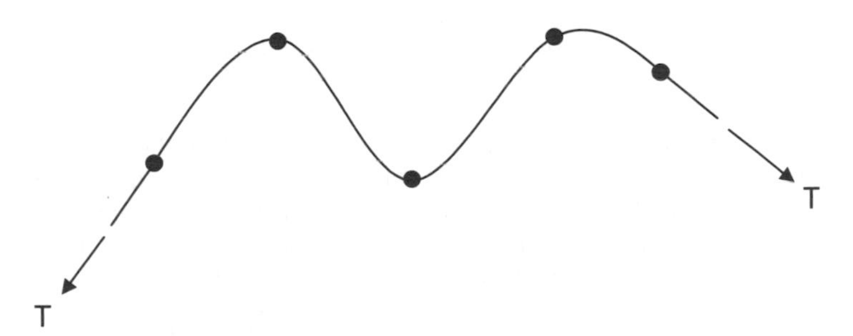


**Figure 9.5** A thin spring steel bent around some fixed points with an optional tension force applied.

**Numerical Interpolation and the Automotive Industry**

One of the early applications of numerical interpolation occurred in the 1950s in the car industry (Birkhoff, 1990). At that time managers at General Motors were keen to investigate methods to improve the process of designing cars which, at the time, was a long and laborious process involving "acres of drafting boards". One key problem was designing and then implementing the curves and bends in sheet metal that make up much of the car body. Cubic splines were one option that was tried but a superior solution in the form of bicubic splines was soon implemented. These required a computer to put them into effect and it was fortunate that, around this time, Fortran was developed and saved the programmers from working in machine language.

## 9.5 EXTRAPOLATION

The discussion to date has centred on interpolation, that is estimating points inside the bounds of known points. There might be an inclination to use these methods to extrapolate, that is estimate values outside the bounds of the known points. The methods may be satisfactory in some situations but be warned: in some cases large errors may follow.

There are two basic reasons for this. Firstly, the form of the polynomials means that although they fit the points exactly there is no guarantee that they behave properly, or that they continue to provide a sensible fit once they are past the range of the data. Secondly, extrapolation has a history of getting things spectacularly wrong. For example, at the start of the 19th century a French naturalist measured ocean temperatures and the variation with depth. On the basis of his measurements (which showed a decrease in temperature with depth) he concluded that the bottom of the ocean was covered with a layer of ice. This unfortunately ignored the fact that the density of water below 4° C reduces and therefore ice floats (Kunzig, 2000).

## PROBLEMS

**9.1** Develop a Fortran cubic spline program to verify the results shown in the chapter for the head loss data.

**9.2** Modify the cubic spline program to verify the results for the error function plot in the chapter.

**9.3** Given a set of discrete points and the desire to determine the area under them (see, for example, Figure 9.6), one way forward would be to use the Lagrange interpolating polynomials to determine a function that links the points and then to integrate the polynomial using standard calculus. Attempt this for the case shown in Figure 9.6 where the raw data points are: (0,10), (2,10), (4,2).

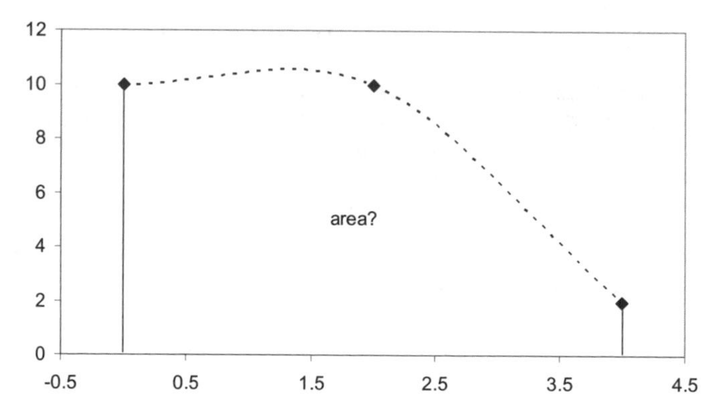


**Figure 9.6** Three data points and an area under the curve to be determined.

**9.4** Modify the cubic spline program to produce a general curve-fitting program that reads data from an appropriately formatted file and produces a general output over a user-specified range of values.

**9.5** Use the data listed in Table 9.2 to compare the predictions of the cubic spline program to a third-order polynomial interpolator.

**9.6** Use one of the polynomial interpolation programs developed previously to investigate the range of reliable predictions outside the limits of the input data for the error function. That is, use the program to estimate values for ERF(5) to ERF(20).

**9.7** Use the cubic spline program to generate a set of splines to fit the cone tip resistance data listed in Table 9.2. Investigate the performance of the splines in the region of the local peak in the data around a depth of 1720 mm.

**Table 9.2** Cone Tip Resistance data recorded at an Adelaide, South Australia, site.

| Depth (mm) | Cone Tip Resistance (MPa) |
|---|---|
| 1650 | 9.75 |
| 1660 | 10.75 |
| 1670 | 12.48 |
| 1680 | 13.28 |
| 1690 | 14.78 |
| 1700 | 15.63 |
| 1710 | 16.20 |
| 1720 | 16.33 |
| 1730 | 16.00 |
| 1740 | 15.35 |
| 1750 | 14.10 |
| 1760 | 12.95 |
| 1770 | 11.90 |
| 1780 | 10.90 |
| 1790 | 9.33 |
| 1800 | 9.03 |
| 1810 | 8.73 |
| 1820 | 8.30 |
| 1830 | 7.90 |

CHAPTER TEN

# Systems of Linear Equations (Introduction)

## 10.1 INTRODUCTION

In engineering, numerical models provide a significant tool in the analysis of structures, soils, fluid flow and environmental modelling. In many cases the problem is tackled by developing the basic equations that describe the phenomenon, then applying these over the area of interest. To provide a solution the area is often split into a grid of points and the equations applied at each point. An example of a flow problem is shown in Figure 10.1, where a model of the River Murray mouth region was developed to investigate the effect of the barrages on the flow patterns in the area.

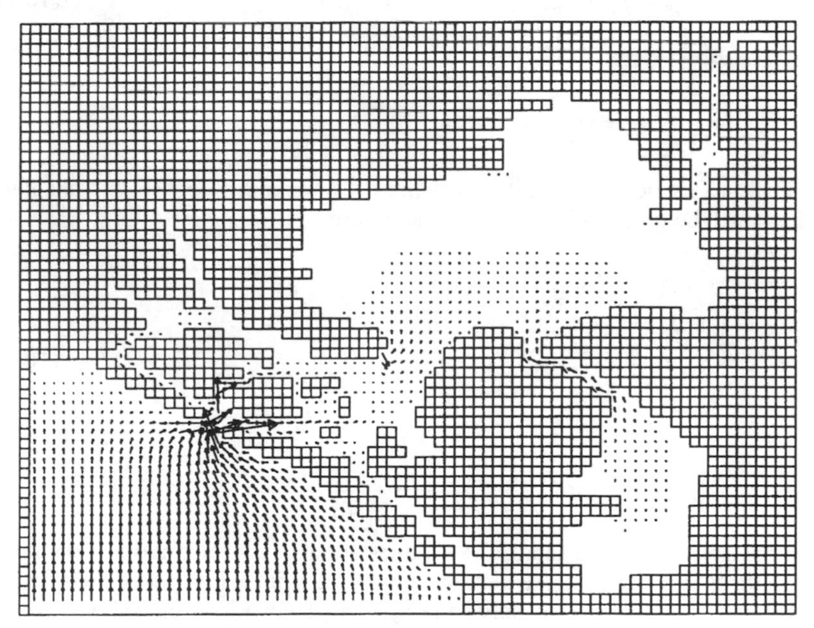

**Figure 10.1** River Murray mouth tidal flow simulation model.

This approach leads to a large set of equations, involving a large number of unknowns that need to be solved. It would not be uncommon to have a grid measuring hundreds of points by hundreds of points leading to tens of thousands of equations in tens of thousands of unknowns. Solution of this set of equations must often be carried out in a series of discrete time-steps leading to the need for a fast and efficient method for dealing with such large sets of equations.

As an example, O'Hanlon (2000) reports on a model of Yucca Mountain in the Nevada Desert (the proposed site for a nuclear dump) which has been developed to simulate long-term performance of the area. In the model the mountain is split into 40 million cells with 120 geological parameters specified at each cell. Simulations are run over millions of years at a reasonably small time-step and have turned up some interesting predictions for those who thought the area was stable and safe.

## 10.2 THE BASIC EQUATIONS

In general terms the equations that are developed to solve a wide range of problems can be written in a quite general form:

$$\begin{aligned} &f_1(x_1, x_2, ..., x_n) = 0 \\ &f_2(x_1, x_2, ..., x_n) = 0 \\ &f_3(x_1, x_2, ..., x_n) = 0 \\ &. \\ &. \\ &f_n(x_1, x_2, ..., x_n) = 0 \end{aligned} \qquad (10.1)$$

This in fact describes quite a general set of possibilities. In this chapter the emphasis will be on equations of the general form:

$$\begin{aligned} &a_{11}x_1 + a_{12}x_2 + ... + a_{1n}x_n = c_1 \\ &a_{21}x_1 + a_{22}x_2 + ... + a_{2n}x_n = c_2 \\ &a_{31}x_1 + a_{32}x_2 + ... + a_{3n}x_n = c_3 \\ &. \\ &. \\ &a_{n1}x_1 + a_{n2}x_2 + ... + a_{nn}x_n = c_n \end{aligned} \qquad (10.2)$$

where the $a$ values are constant, the $c$ values are also constant and $n$ is the number of equations. In this case the equations are linear. With $n$ independent equations in $n$ unknowns it is feasible to solve the system of equations providing a solution exists.

Before looking at computer methods it is worth reviewing a couple of ways of dealing with sets of equations that predate computers: graphical methods and Cramer's rule.

## 10.3 GRAPHICAL METHODS

Graphical methods allow the solution of simple sets of equations (up to 3 unknowns). For the case of 2 equations in 2 unknowns each of the equations represents a straight line. These can be plotted and the solution is the intersection of the two lines. This method is useful since it would also be applicable to non-linear sets of equations although in this case the lines would not be straight.

It is also possible to illustrate graphically a number of situations where there are likely to be problems. These problems will also become apparent in later developments working with many unknowns but are easier to illustrate for two. For two equations there can be either (a) one solution, (b) no solution, (c) infinite solutions or (d) a solution although one that is difficult to see due to the system being ill-conditioned. These are shown in Figure 10.2. Both (b) and (c) are said to be singular. Singularity and ill-conditioned systems of equations will be mentioned later in the chapter.

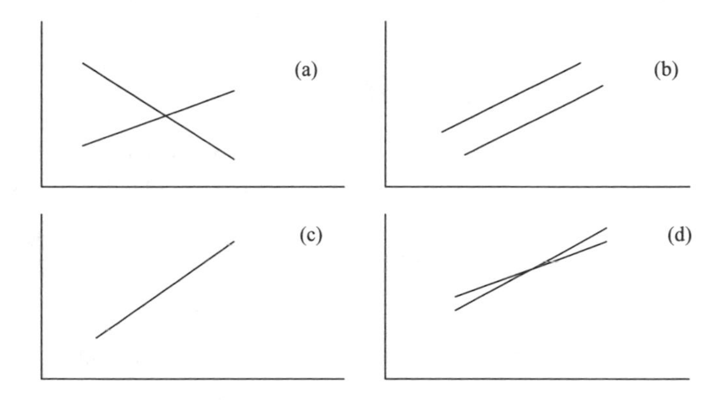


**Figure 10.2** Four cases of possible solutions to two equations in two unknowns. In (a) there is a well-defined solution, in (b) the lines are parallel and there is no solution, in (c) the lines are identical and there are infinite solutions, and in (d) there is a solution but there are many points where there is an approximate solution.

## 10.4 CRAMER'S RULE

If the equations to be solved are expressed in matrix form it is possible to write:

$$[A]\{X\}=\{C\} \tag{10.3}$$

where $A$ is a 2D matrix, $X$ is a column matrix containing the unknown $x$ values and $C$ is a column matrix. Cramer's rule states that each unknown can be expressed as a fraction of two determinants. The denominator is the determinant of $A$ (which will be called $D$) and the numerator obtained from $D$ by replacing the column of coefficients of the unknown in question by the constants $c_1, c_2 \ldots c_n$.

In the case of a 3 × 3 matrix problem representing 3 equations in 3 unknowns, the unknowns $x_1$ and $x_2$ (for example) can be determined as:

$$x_1 = \frac{\begin{vmatrix} c_1 & a_{12} & a_{13} \\ c_2 & a_{22} & a_{23} \\ c_3 & a_{32} & a_{33} \end{vmatrix}}{D} \qquad x_2 = \frac{\begin{vmatrix} a_{11} & c_1 & a_{13} \\ a_{21} & c_2 & a_{23} \\ a_{31} & c_3 & a_{33} \end{vmatrix}}{D} \tag{10.4}$$

where $D = \text{Det}\,A$. The determinant of the 3 × 3 determinant of $A$ is written as:

$$D = a_{11}(a_{22}a_{33} - a_{23}a_{32}) - a_{12}(a_{21}a_{33} - a_{23}a_{31}) + a_{13}(a_{21}a_{32} - a_{22}a_{31}) \tag{10.5}$$

Cramer's rule works well for two or three equations but, as the number increases, the determinant becomes more time-consuming to calculate and there are better alternatives.

## 10.5 ELIMINATION OF UNKNOWNS

Another simple method is based on hand calculation methods for solving simultaneous equations. Given a general problem of 2 equations in 2 unknowns:

$$\begin{aligned} a_{11}x_1 + a_{12}x_2 &= c_1 \\ a_{21}x_1 + a_{22}x_2 &= c_2 \end{aligned} \tag{10.6}$$

the solution can be obtained by multiplying the first equation by $a_{21}$ and the second by $a_{11}$. This gives:

$$\begin{aligned} a_{11}a_{21}x_1 + a_{12}a_{21}x_2 &= a_{21}c_1 \\ a_{11}a_{21}x_1 + a_{11}a_{22}x_2 &= a_{11}c_2 \end{aligned} \tag{10.7}$$

and subtracting the second equation from the first and rearranging gives:

$$x_2 = \frac{a_{11}c_2 - a_{21}c_1}{a_{11}a_{22} - a_{12}a_{21}} \tag{10.8}$$

This can then be substituted into one of the original equations to obtain an expression for $x_1$. Note that this expression for $x_1$ is exactly what would have been obtained from Cramer's rule.

$$x_1 = \frac{c_1a_{22} - a_{12}c_2}{a_{11}a_{22} - a_{12}a_{21}} \tag{10.9}$$

$$x_1 = \frac{\begin{vmatrix} c_1 & a_{12} \\ c_2 & a_{22} \end{vmatrix}}{\begin{vmatrix} a_{11} & a_{12} \\ a_{21} & a_{22} \end{vmatrix}} = \frac{c_1 a_{22} - a_{12} c_2}{a_{11} a_{22} - a_{12} a_{21}} \tag{10.10}$$

The elimination of unknowns is a reliable method of solving a set of equations, although for more than a $3 \times 3$ set the method becomes unmanageable by hand. However, it does lend itself to computers as the rules for its implementation can be easily formulated. It is the basis of the Gauss elimination method.

## 10.6 GAUSS ELIMINATION

Gauss elimination is a procedure to solve sets of equations using the elimination of unknowns. The method is efficient up to about 100 equations in 100 unknowns (for reasons that will be outlined later). The method works by gradually eliminating variables from the equations until the stage is reached where there is a single equation with only one unknown. This is then back-substituted to get the solution for that unknown. This is then put back into the next most recent equation and so on until all unknowns have been determined.

In terms of matrix manipulation the aim is to get a matrix that is upper triangular. For the example in Equation (10.11) it will be of the form, (for a 4 x 4):

$$\begin{bmatrix} a_{11} & a_{12} & a_{13} & a_{14} \\ 0 & a_{22} & a_{23} & a_{24} \\ 0 & 0 & a_{33} & a_{34} \\ 0 & 0 & 0 & a_{44} \end{bmatrix} \begin{Bmatrix} x_1 \\ x_2 \\ x_3 \\ x_4 \end{Bmatrix} = \begin{Bmatrix} c_1 \\ c_2 \\ c_3 \\ c_4 \end{Bmatrix} \tag{10.11}$$

In this case $x_4$ can be determined from the fourth equation directly:

$$a_{44} x_4 = c_4 \quad \text{therefore} \quad x_4 = c_4 / a_{44} \tag{10.12}$$

Then knowing $x_4$ the third equation has only one unknown ($x_3$) which can be solved. This process continues until $x_1$, $x_2$, $x_3$ and $x_4$ have all been solved.

The Gauss elimination method has two stages. In the first the matrix is manipulated to get it into upper triangular form (forward elimination) and in the second the back substitution is carried out.

*Forward elimination*

In the following, the equations will be numbered to make the manipulations easier to follow. Starting with the given set of equations (again taking a $4 \times 4$ ) it is possible to write:

$$\begin{bmatrix} a_{11} & a_{12} & a_{13} & a_{14} \\ a_{21} & a_{22} & a_{23} & a_{24} \\ a_{31} & a_{32} & a_{33} & a_{34} \\ a_{41} & a_{42} & a_{43} & a_{44} \end{bmatrix} \begin{Bmatrix} x_1 \\ x_2 \\ x_3 \\ x_4 \end{Bmatrix} = \begin{Bmatrix} c_1 \\ c_2 \\ c_3 \\ c_4 \end{Bmatrix} \begin{matrix} (1) \\ (2) \\ (3) \\ (4) \end{matrix} \tag{10.13}$$

*Step 1*

Produce an equation by multiplying (1) by $a_{21}/a_{11}$ and subtract the result from (2). Leave (1) as it is. Similarly produce an equation by multiplying (1) by $a_{31}/a_{11}$ and subtracting that from (3). Do the same for (4). The result of this is:

$$\begin{bmatrix} a_{11} & a_{12} & a_{13} & a_{14} \\ 0 & a'_{22} & a'_{23} & a'_{24} \\ 0 & a'_{32} & a'_{33} & a'_{34} \\ 0 & a'_{42} & a'_{43} & a'_{44} \end{bmatrix} \begin{Bmatrix} x_1 \\ x_2 \\ x_3 \\ x_4 \end{Bmatrix} = \begin{Bmatrix} c_1 \\ c'_2 \\ c'_3 \\ c'_4 \end{Bmatrix} \begin{matrix} (1) \\ (2) \\ (3) \\ (4) \end{matrix} \tag{10.14}$$

The primes (') indicate that the coefficients have changed. In this manipulation $a_{11}$ is known as the pivot element. It will be seen later that the magnitude of this value has important ramifications for the operation of this method.

*Step 2*

Now repeat the procedure, but leave (1) and (2) as is and work on (3) and (4). Produce an equation by multiplying (2) by $a'_{32}/a'_{22}$ and subtract the result from (3). Produce another equation by multiplying (2) by $a'_{42}/a'_{22}$ and subtract the result from (4). The result of this is:

$$\begin{bmatrix} a_{11} & a_{12} & a_{13} & a_{14} \\ 0 & a'_{22} & a'_{23} & a'_{24} \\ 0 & 0 & a''_{33} & a''_{34} \\ 0 & 0 & a''_{43} & a''_{44} \end{bmatrix} \begin{Bmatrix} x_1 \\ x_2 \\ x_3 \\ x_4 \end{Bmatrix} = \begin{Bmatrix} c_1 \\ c'_2 \\ c''_3 \\ c''_4 \end{Bmatrix} \begin{matrix} (1) \\ (2) \\ (3) \\ (4) \end{matrix} \tag{10.15}$$

The primes (' and ") indicate that the coefficients have changed. One final run through the procedure will give the required result: an upper triangular matrix. Repeating from above and ignoring all the primes, double primes etc, the result is:

$$\begin{bmatrix} a_{11} & a_{12} & a_{13} & a_{14} \\ 0 & a_{22} & a_{23} & a_{24} \\ 0 & 0 & a_{33} & a_{34} \\ 0 & 0 & 0 & a_{44} \end{bmatrix} \begin{Bmatrix} x_1 \\ x_2 \\ x_3 \\ x_4 \end{Bmatrix} = \begin{Bmatrix} c_1 \\ c_2 \\ c_3 \\ c_4 \end{Bmatrix} \begin{matrix} (1) \\ (2) \\ (3) \\ (4) \end{matrix} \tag{10.16}$$

*Back substitution*

From (4) it is possible to write $x_4 = c_4/a_{44}$. Then the other unknowns can be determined. The general formula can be written:

$$x_i = \frac{1}{a_{ii}}\left[c_i - \sum_{j=i+1}^{n} a_{ij}x_j\right] \quad \text{for } i = n-1, n-2, \ldots 1 \tag{10.17}$$

The procedure is easily converted into computer code, the basis of which is shown in Figure 10.3.

```
! Gaussian elimination code

do k=1,n-1              ! forward elimination
                              ! partial pivoting could go here
 do i=k+1,n
  factor = a(i,k)/a(k,k)
  do j = k+1,n
   a(i,j)=a(i,j)-factor*a(k,j)
  end do
  c(i)=c(i)-factor*c(k)
 end do
end do
x(n)=c(n)/a(n,n)        ! back substitution
do i=n-1,1,-1
 sum=0
 do j=i+1,n
  sum=sum+a(i,j)*x(j)
 end do
 x(i)=(c(i)-sum)/a(i,i)
end do
```

**Figure 10.3** Fortran code segment to carry out Gaussian elimination.

## 10.7 PROBLEMS AND THEIR SOLUTION FOR GAUSS ELIMINATION

There are a number of potential problems with this approach to solving sets of equations.

*Problem 1: division by zero*
The technique relies on calculating a ratio between selected elements at each step and multiplying that ratio by an equation. If the element on the bottom line of the ratio (the pivot element) is zero there is a potential problem that must be dealt with.

*Problem 2: round-off errors*
Since the solution technique relies heavily on previous calculations there is the potential for round-off errors to accumulate and affect the whole solution. This may seem trivial but unfortunately this will always be a problem and, although there are ways to try and reduce its effect, it is always advisable to check the final solution by back-substituting into the original equations to see how well the solution satisfies those equations.

*Problem 3: ill-conditioned systems*
Ill-conditioned systems are where a small change in one of the coefficients results in a large change in the solution. Another way of looking at it is to say that for an

ill-conditioned system a wide range of solutions can approximately satisfy the equations. In these types of situations, because of round off errors, large errors are possible. There are ways of overcoming or at least minimising these problems.

*Technique 1: use of more significant figures*
Many of the problems can be overcome if greater accuracy is possible in the calculations. The use of double precision variables in computer programs can aid in this regard. It does have the disadvantage of doubling the memory requirements.

*Technique 2: pivoting*
As stated earlier, problems can occur where the pivot element is zero or close to zero. One simple way of overcoming this is to search the remaining equations and re-arrange the order of the equations such that the largest coefficient becomes the pivot element. This is called partial pivoting. There are no other changes that have to be made since the order of the equations is of little consequence.

If the columns are also searched for the largest pivot element, then it is called complete pivoting. Complete pivoting has the disadvantage that if the columns are swapped this requires changes in the order of the elements in the X and C matrices. This is often not worth the additional work.

Partial pivoting is demonstrated in Figure 10.4 with code based on pseudocode from Chapra and Canale (1989). If it is to be implemented the partial pivoting could be included in the Gaussian elimination code at the start of the first loop as shown previously.

```
! partial pivoting
pivot = k
 big = abs(a(k,k))
 do ii=k+1,n
  dummy = abs(a(ii,k))
  if (dummy > big)
   big = dummy
   pivot = ii
  end if
 end do
 if (pivot /= k)
  do jj = k,n
   dummy = a(pivot,jj)
   a(pivot,jj) = a(k,jj)
   a(k,jj) = dummy
  end do
  dummy = c(pivot)
  c(pivot) = c(k)
  c(k) = dummy
 end if
```

**Figure 10.4** Fortran code segment to carry out partial pivoting.

## PROBLEMS

**10.1** Write a Fortran program to solve a set of 2 or 3 equations in 2 or 3 unknowns, respectively, using Cramer's rule. The program should be set up so that it can read the input data from a text file and determine the solution. For example, to solve the following set of equations:

$$x_1 + 5x_2 + 2x_3 = 17$$
$$2x_1 + 7x_2 + x_3 = 19$$
$$-x_1 + 3x_2 + 2x_3 = 11$$

set up a file with the following format:

Line 1: General header (text)
Line 2: n (the number of equations (integer))
Line 3: $a_{11}$ $a_{12}$ ... $a_{1n}$ $c_1$
Line 4: $a_{21}$ $a_{22}$ ... $a_{2n}$ $c_2$
.
Line n+2: $a_{n1}$ $a_{n2}$ ... $a_{nn}$ $c_n$

The program should read the data from the file, determine whether there are 2 or 3 unknowns and produce a solution which is written to the screen. At this stage there is no need to consider pivoting.

**10.2** Write a program to solve a set of linear equations using Gauss elimination. The program should read data in the format suggested for Problem 10.1 and produce output on the screen.

**10.3** Investigate the issues with a zero diagonal element by running the program developed in Problems 10.1 and 10.2 on the following problem:

$$x_1 + 5x_2 + 2x_3 = 17$$
$$2x_1 + 7x_2 + x_3 = 19$$
$$-x_1 + 3x_2 = 5$$

There is no need to deal with this problem at present, but how might it be overcome?

**10.4** Develop an Excel spreadsheet that solves a 3 × 3 matrix problem using Cramer's rule. Once the program is working, it would be worth investigating the sorts of errors that can occur in normal use and develop a strategy for dealing with them.

**10.5** Using the Gaussian elimination program written as part of Problem 10.2, test its performance on systems of equations that are not diagonally dominant. For example:

$$12x_1 + 5x_2 + 5x_3 = 37$$
$$9x_1 + 7x_2 + x_3 = 26$$
$$8x_1 + 3x_2 + 2x_3 = 20$$

Implement the partial pivoting and determine if this makes any improvement to the solution. The partial pivoting should also handle the presence of a zero on the diagonal. Try the program with the problem set in Problem 10.3.

**10.6** Use the Gauss elimination program to solve the following set of linear equations (written in matrix form). The matrix is from a truss solution that will be discussed in Chapter 11.

$$\begin{bmatrix} 0.707 & 0 & 0 & -1 & 0 & 0 & 0 & 0 & 0 \\ 0.707 & 0 & 1 & 0 & 0 & 0 & 0 & 0 & 0 \\ 0 & 0 & 0 & 1 & 0.707 & 0 & 0 & -0.707 & 0 \\ 0 & 0 & 0 & 0 & 0.707 & 0 & 1 & 0.707 & 0 \\ 0 & 1 & 0 & 0 & -0.707 & -1 & 0 & 0 & 0 \\ 0 & 0 & -1 & 0 & -0.707 & 0 & 0 & 0 & 0 \\ 0 & 0 & 0 & 0 & 0 & 1 & 0 & 0 & -1 \\ 0 & 0 & 0 & 0 & 0 & 0 & -1 & 0 & 0 \\ 0 & 0 & 0 & 0 & 0 & 0 & 0 & 0.707 & 1 \end{bmatrix} \begin{Bmatrix} F_1 \\ F_2 \\ F_3 \\ F_4 \\ F_5 \\ F_6 \\ F_7 \\ F_8 \\ F_9 \end{Bmatrix} = \begin{Bmatrix} 0 \\ 100 \\ 0 \\ 0 \\ 0 \\ 100 \\ 0 \\ 0 \\ 0 \end{Bmatrix}$$

CHAPTER ELEVEN

# Systems of Equations (Gauss-Seidel Method)

## 11.1 GAUSS-SEIDEL METHOD

The Gauss elimination technique outlined in the previous chapter appears to be a suitable method of solving general sets of equations and there is no hint that, given some well posed equations and care with pivoting, a good solution should not be generally available. However, there is a problem: round-off errors. These are generated with each and every operation and can eventually spoil the solution. Gaussian elimination is thought to be fine for solving up to about 100 equations in 100 unknowns but for larger systems round-off errors usually render the method unsuitable and it is necessary to try different solutions. For such systems approximate or iterative methods can often be used. The advantage is that although the method is an approximate one, it is possible to determine how approximate (or accurate) the solution is required to be and to continue the iterations until that accuracy is attained. The Gauss-Seidel method is one such method. Assume one is trying to solve:

$$[A]\{X\} = \{C\} \qquad (11.1)$$

which, when expanded, can be written:

$$\begin{bmatrix} a_{11} & a_{12} & a_{13} & a_{14} & . & . & a_{1n} \\ a_{21} & a_{22} & a_{23} & a_{24} & . & . & a_{2n} \\ a_{31} & a_{32} & a_{33} & a_{34} & . & . & a_{3n} \\ a_{41} & a_{42} & a_{43} & a_{44} & . & . & a_{4n} \\ . & . & . & . & . & . & . \\ . & . & . & . & . & . & . \\ a_{n1} & a_{n2} & a_{n3} & a_{n4} & . & . & a_{nn} \end{bmatrix} \cdot \begin{Bmatrix} x_1 \\ x_2 \\ x_3 \\ x_4 \\ . \\ . \\ x_n \end{Bmatrix} = \begin{Bmatrix} c_1 \\ c_2 \\ c_3 \\ c_4 \\ . \\ . \\ c_n \end{Bmatrix} \qquad (11.2)$$

If $a_{11}$ is non-zero then the first equation can be re-ordered to isolate $x_1$:

$$x_1 = \frac{c_1 - a_{12}x_2 - a_{13}x_3 - \ldots - a_{1n}x_n}{a_{11}} \tag{11.3}$$

and the other equations can be similarly re-arranged (assuming non-negative diagonal elements in the $A$ matrix):

$$x_2 = \frac{c_2 - a_{21}x_1 - a_{23}x_3 - \ldots - a_{2n}x_n}{a_{22}} \tag{11.4}$$

etc., with the last term being:

$$x_n = \frac{c_n - a_{n1}x_1 - a_{n2}x_2 - \ldots - a_{nn-1}x_{n-1}}{a_{nn}} \tag{11.5}$$

Now, this does not provide a solution as it stands since all the $x_i$ values are unknown. However, if an iterative procedure was commenced by simply guessing $x$ values, this would then lead to new estimates which in turn could be updated. The initial guess need only be something simple like $x_i = 0$ for all $i$.

*Example*
Solve the following set of equations using an approximate method:

$$\begin{aligned} 3x_1 + x_2 + 3x_3 &= 29.20 \\ 1.7x_1 + 9.3x_2 - 2.5x_3 &= 56.54 \\ x_1 + 0.1x_2 + 5.6x_3 &= -7.94 \end{aligned}$$

*Solution*
Start with the guess $x_1 = x_2 = x_3 = 0$ and write the equations as:

$$x_1 = \frac{29.20 - x_2 - 3x_3}{3}, \; x_2 = \frac{56.54 - 1.7x_1 + 2.5x_3}{9.3}, \; x_3 = \frac{-7.94 - x_1 - 0.1x_2}{5.6}$$

The solution proceeds as shown in Table 11.1 with the $x_1$ values being updated first, followed by the $x_2$ and then the $x_3$. Note that in the calculation of the updated values of $x_2$ and $x_3$ there is a choice to be made with regard to the values of the other $x$ values. For example, when calculating a new $x_2$ value either the previous $x_1$ value could be used, or the one that has just been calculated. It is generally preferred to use the most recent values as it can lead to a faster convergence. For the new $x_3$ value each time, the most recent estimates of $x_1$ and $x_2$ are used. This has been done in Table 11.1 so that, for example, the $x_2 = 4.30$ on the second line of the table was calculated using $x_1 = 9.73$ and $x_3 = 0.0$ as these were the two most recent values available for these unknowns at the time. Note that the order in which the iterations are carried out will affect the intermediate values, but that this should not prevent the solution from converging on the correct solution.

**Table 11.1** Solution of the linear equations by the Gauss-Seidel method.

| Iteration | $x_1$ | $x_2$ | $x_3$ |
|---|---|---|---|
| 1 | 0.00 | 0.00 | 0.00 |
| 2 | 9.73 | 4.30 | –3.23 |
| 3 | 11.53 | 3.10 | –3.53 |
| 4 | 12.23 | 2.89 | –3.65 |
| 5 | 12.42 | 2.83 | –3.69 |
| 6 | 12.48 | 2.81 | –3.70 |
| 7 | 12.49 | 2.80 | –3.70 |
| 8 | 12.50 | 2.80 | –3.70 |
| 9 | 12.50 | 2.80 | –3.70 |
| 10 | 12.50 | 2.80 | –3.70 |

The solution is as shown in the last iteration at the end of Table 11.1 and can be checked by substituting into the original equations. It can be seen that the method converges to this solution quite quickly, even from the relatively poor initial guesses.

A Fortran program that will produce the data in Table 11.1 is listed in Figure 11.1. While general, the program has one major flaw: no valid stopping criterion is used, just 10 iterations for each run. Convergence will be discussed later, but at the present it is worth pondering on how convergence should be recognised in this particular case.

```
program gauss_seidel
! program to test gauss seidel method
! written d.walker april 2008
implicit none
real:: a(3,3),x(3),c(3),sum
integer:: i,j,n,its
data a/3.0,1.7,1.0,1.0,9.3,0.1,3.0,-2.5,5.6/
data c/29.2,56.54,-7.94/
n=3
x=0.0            !initial guess is x = 0 for all x
write(*,'(1x,10(f6.3,5x))')(x(i),i=1,n) ! output
do its=1,10      !just do 10 iterations
 do i=1,n        ! do row by row i=1, 2, 3
  sum = c(i)     ! start with x(i)=c(i)
  do j=1,n       ! now take other a(i,j)*x(j) terms
   if(i /= j) sum=sum-a(i,j)*x(j) !use i<>j elements
  end do
  x(i)=sum/a(i,i)
 end do
 write(*,'(1x,10(f6.3,5x))')(x(i),i=1,n) ! output
end do
stop
end program
```

**Figure 11.1** Fortran program to solve Gauss-Seidel method.

## 11.2 CONVERGENCE

The Gauss-Seidel method is similar to the one-point iteration described earlier when dealing with the problem of finding roots. One of the drawbacks of that method was that it was sometimes non-convergent and that it was occasionally slow to converge.

The Gauss-Seidel method also suffers from these drawbacks. If the convergence criteria are studied it is found that for the system to be convergent:

$$|a_{ii}| > \sum_{j \neq i} |a_{ij}| \qquad (11.6)$$

That is, the diagonal elements must be greater than the sum of the absolute values of all the other elements in the row. To illustrate this, consider a set of equations which are similar to the ones shown earlier but where the constant associated with $x_1$ in the first equation has been changed from 3 to 0.3. The new equations are:

$$0.3x_1 + x_2 + 3x_3 = -4.55$$
$$1.7x_1 + 9.3x_2 - 2.5x_3 = 56.54$$
$$x_1 + 0.1x_2 + 5.6x_3 = -7.94$$

It is evident that the equations are now definitely not diagonally dominant. The solution, starting from the same initial guesses, now gives the estimates shown in Table 11.2. The solution does not converge and, in fact, blows up quite quickly.

**Table 11.2** Solution of non-diagonally dominant linear equations using Gauss-Seidel method.

| Iteration | $x_1$ | $x_2$ | $x_3$ |
|---|---|---|---|
| 1 | 0.00 | 0.00 | 0.00 |
| 2 | –15.17 | 8.85 | 1.13 |
| 3 | –56.00 | 16.62 | 8.28 |
| 4 | –153.42 | 36.35 | 25.33 |
| 5 | –389.62 | 84.11 | 66.66 |
| 6 | –962.09 | 199.86 | 166.81 |
| 7 | –2349.52 | 480.40 | 409.56 |
| 8 | –5712.11 | 1160.33 | 997.88 |
| 9 | –13861.75 | 2808.20 | 2423.75 |
| 10 | –33613.29 | 6801.99 | 5879.49 |
| 11 | –81483.38 | 16481.40 | 14254.88 |

Although there are systems where the solution will converge even if not diagonally dominant, it is found that for systems to be solved reliably they should be. Fortunately this is often the case for many engineering problems.

## 11.3 RELAXATION

Relaxation provides a means of speeding up the rate at which the solution converges. After each new estimate of a variable is calculated it is modified by a weighted average of previous estimates.

$$x_i^{new} = \lambda x_i^{new} + (1-\lambda)x_i^{old} \tag{11.7}$$

where $\lambda$ is a weighing factor between 0 and 2. For $\lambda = 1$ there is no weighing and the method is the standard Gauss-Seidel method. For values of the weight between 0 and 1 the modification is called under-relaxation. It is useful for damping out unwanted oscillations as the solution proceeds. For values of the weight between 1 and 2 it is called over-relaxation. This is used where it is assumed that the solution is heading in the right direction but too slowly. Unfortunately the choice of the weight is empirical and must be based on experience.

## PROBLEMS

**11.1** Set up an Excel spreadsheet to apply the Gauss-Seidel method to solving a set of simultaneous equations and determine the solution to the equations:

$$\begin{aligned}3x_1 + x_2 + 3x_3 &= 29.20\\1.7x_1 + 9.3x_2 - 2.5x_3 &= 56.54\\x_1 + 0.1x_2 + 5.6x_3 &= -7.94\end{aligned}$$

Compare the solution to that shown in Table 11.1.

**11.2** Write a Fortran program to solve the same set of equations as in Problem 11.1 using the Gauss-Seidel method. Devise a convergence test that will stop the iterations when the largest change in any of the estimates is no more than 0.01. Be careful: a large negative change should be considered larger than 0.01.

**11.3** Extend the Fortran program to include partial pivoting. As part of the exercise have the program inform the user if any pivoting has occurred. It may be worth noting the number of equations that were re-ordered and including this in the report.

**11.4** Modify the spreadsheet in Problem 11.1 to include relaxation. Investigate the speed of convergence with a range of $\lambda$ values from 0.5 to 1.5.

**11.5** The solution of the forces in a statically determinant truss can be determined using a matrix approach. For example, by taking equilibrium in both the vertical and horizontal directions at sufficient points around the truss, a set of equations can be derived that can be written in matrix form. For example, the truss in Figure 11.2 results in a matrix problem that can be written:

$$\begin{bmatrix} 0.707 & 0 & 0 & -1 & 0 & 0 & 0 & 0 & 0 \\ 0.707 & 0 & 1 & 0 & 0 & 0 & 0 & 0 & 0 \\ 0 & 0 & 0 & 1 & 0.707 & 0 & 0 & -0.707 & 0 \\ 0 & 0 & 0 & 0 & 0.707 & 0 & 1 & 0.707 & 0 \\ 0 & 1 & 0 & 0 & -0.707 & -1 & 0 & 0 & 0 \\ 0 & 0 & -1 & 0 & -0.707 & 0 & 0 & 0 & 0 \\ 0 & 0 & 0 & 0 & 0 & 1 & 0 & 0 & -1 \\ 0 & 0 & 0 & 0 & 0 & 0 & -1 & 0 & 0 \\ 0 & 0 & 0 & 0 & 0 & 0 & 0 & 0.707 & 1 \end{bmatrix} \begin{Bmatrix} F_1 \\ F_2 \\ F_3 \\ F_4 \\ F_5 \\ F_6 \\ F_7 \\ F_8 \\ F_9 \end{Bmatrix} = \begin{Bmatrix} 0 \\ 100 \\ 0 \\ 0 \\ 0 \\ 100 \\ 0 \\ 0 \\ 0 \end{Bmatrix}$$

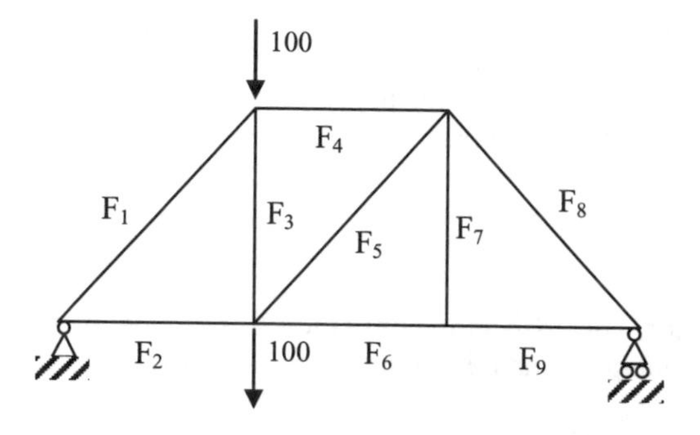


**Figure 11.2** Truss for analysis.

Use an appropriate matrix solution (e.g. Gaussian elimination, Gauss-Seidel) procedure to solve the matrix and hence determine the forces in all members. Note: the matrix, as written, is not diagonally dominant and may cause problems if this aspect is not dealt with.

## CHAPTER TWELVE

# Systems of Equations (LU Decomposition and Thomas Algorithm)

### 12.1 LU DECOMPOSITION

The Gaussian elimination is an intuitively appealing method for solving a matrix. However, there is a more efficient and preferred direct method called LU decomposition. The Gaussian elimination is made up of two stages: forward elimination, and back substitution. Tests have shown that it is the first stage, the forward elimination, that takes most of the computational effort. For 10 equations it represents approximately 80% of the total work and for more than 20 equations it is over 90%. By 100 equation systems it represents virtually 100% of the total effort. Any improvement in the method, therefore, must concentrate on this first step as the substitution steps are relatively cheap. One such method is lower upper (LU) decomposition, which involves three stages:

- decomposition;
- forward substitution;
- back substitution.

Although there are two substitution stages, these are relatively cheap in terms of processing time. The method can be applied to the general problem of solving a set of linear equations that can be written as:

$$[A]\{x\}=\{b\} \tag{12.1}$$

It has been shown that it is possible to determine two matrices, one which is upper triangular (U) and the other lower triangular (L) such that:

$[L][U]=[A]$, that is (for a $4\times 4$ problem):

$$\begin{bmatrix} l_{11} & 0 & 0 & 0 \\ l_{21} & l_{22} & 0 & 0 \\ l_{31} & l_{32} & l_{33} & 0 \\ l_{41} & l_{42} & l_{43} & l_{44} \end{bmatrix}\begin{bmatrix} 1 & u_{12} & u_{13} & u_{14} \\ 0 & 1 & u_{23} & u_{24} \\ 0 & 0 & 1 & u_{34} \\ 0 & 0 & 0 & 1 \end{bmatrix}=\begin{bmatrix} a_{11} & a_{12} & a_{13} & a_{14} \\ a_{21} & a_{22} & a_{23} & a_{24} \\ a_{31} & a_{32} & a_{33} & a_{34} \\ a_{41} & a_{42} & a_{43} & a_{44} \end{bmatrix} \tag{12.2}$$

Therefore $[L][U]\{x\}=\{b\}$ (12.3)

Pre-multiplying both by the inverse of $L$:

$$[L]^{-1}[L][U]\{x\}=[L]^{-1}\{b\} \text{ which can be written } [U]\{x\}=\{z\} \quad (12.4)$$

where $[L]^{-1}\{b\}=\{z\}$ (12.5)

Pre-multiplying by $[L]$ gives:

$$[L]\{z\}=\{b\} \quad (12.6)$$

Therefore it is possible to solve for $z$ in Equation (12.6) by a simple forward substitution and to solve for $\{x\}$ in Equation (12.4) by a simple back substitution.

This may seem like a long way to a solution but the two substitution stages are cheap, and efficient methods of determining $L$ and $U$ have been developed (e.g. Crout decomposition). Based on some tests carried out on randomly generated matrices the results shown in Table 12.1 were obtained. Note that although slower for relatively small problems (less than 100 equations), as the number of equations increases so does the relative advantage of the LU decomposition method.

**Table 12.1** Speed to solve matrix problem using LU decomposition and Gaussian elimination.

| No. of Equations | LU Decomposition (secs) | Gauss Elimination (secs) |
|---|---|---|
| 10 | 0.170 | 0.110 |
| 20 | 0.330 | 0.280 |
| 50 | 1.870 | 1.920 |
| 100 | 11.200 | 12.030 |

Woo and Liu (2004) used LU decomposition in a solution of a finite element model of waves in a harbour. They claimed it was not the most efficient method of solution but was acceptable for systems with up to 5000 nodes.

## 12.2 CROUT DECOMPOSITION

Crout decomposition is an efficient algorithm for determining $L$ and $U$. If the equation is written down the coefficients are easy to determine by working through the matrix multiplication in a convenient manner. To illustrate the general procedure a $4 \times 4$ matrix will be used. This is repeated here for convenience from Equation (12.2) cited earlier.

$$\begin{bmatrix} l_{11} & 0 & 0 & 0 \\ l_{21} & l_{22} & 0 & 0 \\ l_{31} & l_{32} & l_{33} & 0 \\ l_{41} & l_{42} & l_{43} & l_{44} \end{bmatrix} \begin{bmatrix} 1 & u_{12} & u_{13} & u_{14} \\ 0 & 1 & u_{23} & u_{24} \\ 0 & 0 & 1 & u_{34} \\ 0 & 0 & 0 & 1 \end{bmatrix} = \begin{bmatrix} a_{11} & a_{12} & a_{13} & a_{14} \\ a_{21} & a_{22} & a_{23} & a_{24} \\ a_{31} & a_{32} & a_{33} & a_{34} \\ a_{41} & a_{42} & a_{43} & a_{44} \end{bmatrix} \tag{12.7}$$

Multiplying the $L$ and $U$ matrices to determine the first column of the $A$ matrix:

$$l_{11} = a_{11},\ l_{21} = a_{21},\ l_{31} = a_{31},\ l_{41} = a_{41} \text{ or}$$

$$l_{i1} = a_{i1} \text{ for } i = 1,n \tag{12.8}$$

Multiplying the $L$ and $U$ matrices to determine the first row of the $A$ matrix:

$$l_{11} = a_{11},\ l_{11}u_{12} = a_{12},\ l_{11}u_{13} = a_{13},\ l_{11}u_{14} = a_{14} \text{ or}$$

$$u_{1j} = a_{1j} / l_{11} \text{ for j} = 2,n \tag{12.9}$$

All diagonal elements on the U matrix are equal to 1:

$$u_{ii} = 1 \text{ for } i = 1,n \tag{12.10}$$

The procedure then moves back to the L matrix and it is possible to determine that for $j = 2, 3, \ldots, n-1$:

$$l_{ij} = a_{ij} - \sum_{k=1}^{j-1} l_{ik} u_{kj} \qquad \text{for } i = j, j+1, \ldots, n \tag{12.11}$$

$$u_{jk} = \frac{a_{jk} - \sum_{i=1}^{j-1} l_{ji} u_{ik}}{l_{jj}} \text{ for } k = j+1, j+2, \ldots, n \tag{12.12}$$

and

$$l_{nn} = a_{nn} - \sum_{k=1}^{n-1} l_{nk} u_{kn} \tag{12.13}$$

The notation can vary depending on how the algorithm is set out and in this case, the non-standard use of $i$, $j$ and $k$ in Equation (12.13) is due to the use of the $j$ counter for both Equations (12.11) and (12.12). As noted earlier, the solution of the matrix problem [A]{x} = {b} is achieved in two steps:

*Step 1*: Determine {z} in the equation [L]{z} = {b}

*Step 2*: Determine {x} in the equation [U]{x} = {z}

The two substitution steps are achieved using the following algorithms:

*Step 1*

$$z_1 = \frac{b_1}{l_{11}} \quad \text{and} \quad z_i = \frac{b_i - \sum_{j=1}^{i-1} l_{ij} z_j}{l_{ii}} \quad \text{for } i = 2, 3, ..., n \qquad (12.14)$$

*Step 2*

$$x_n = z_n \quad \text{and} \quad x_i = z_i - \sum_{j=i+1}^{n} u_{ij} x_j \qquad \text{for } i = n-1, n-2, ... ,1 \qquad (12.15)$$

*Example*
Solve the following set of equations using LU decomposition:

$x_1 + 5x_2 + 2x_3 = 17$
$2x_1 + 7x_2 + x_3 = 19$
$-x_1 + 3x_2 + 2x_3 = 11$

*Solution*
The equations are written in the form [A]{x}={c}, or:

$$\begin{bmatrix} 1 & 5 & 2 \\ 2 & 7 & 1 \\ -1 & 3 & 2 \end{bmatrix} \begin{Bmatrix} x_1 \\ x_2 \\ x_3 \end{Bmatrix} = \begin{Bmatrix} 17 \\ 19 \\ 11 \end{Bmatrix}$$

The A matrix is then decomposed into [L][U] = [A] giving:

$$\begin{bmatrix} 1 & 0 & 0 \\ 2 & -3 & 0 \\ -1 & 8 & -4 \end{bmatrix} \begin{bmatrix} 1 & 5 & 2 \\ 0 & 1 & 1 \\ 0 & 0 & 1 \end{bmatrix} = \begin{bmatrix} 1 & 5 & 2 \\ 2 & 7 & 1 \\ -1 & 3 & 2 \end{bmatrix}$$

The two substitution steps then give: $z = \begin{Bmatrix} 17 \\ 5 \\ 3 \end{Bmatrix}$ and $x = \begin{Bmatrix} 1 \\ 2 \\ 3 \end{Bmatrix}$

where x is the solution being sought. The solution can be confirmed by substituting in any or all of the original equations.

## 12.3 BANDED MATRICES AND THE THOMAS ALGORITHM

In engineering there are often cases where the solution of a set of $n$ equations produces a highly banded matrix, in many cases with a bandwidth of 3. The solution of the two-dimensional depth-averaged hydraulic flow equations is one particular example that can be formulated in this way. In this case the problem can be written as:

$$\begin{vmatrix} b_1 & c_1 & 0 & 0 & 0 \\ a_2 & b_2 & c_2 & 0 & 0 \\ 0 & a_3 & b_3 & c_3 & 0 \\ . & . & . & . & . \\ 0 & 0 & a_{n-1} & b_{n-1} & c_{n-1} \\ 0 & 0 & 0 & a_n & b_n \end{vmatrix} \begin{Bmatrix} x_1 \\ x_2 \\ x_3 \\ . \\ x_{n-1} \\ x_n \end{Bmatrix} = \begin{Bmatrix} d_1 \\ d_2 \\ d_3 \\ . \\ d_{n-1} \\ d_n \end{Bmatrix} \tag{12.16}$$

That is, in each of the equations there were only three unknowns and at the end points ($i$=1 and $i$=$n$) there were only two due to the application of boundary conditions. In this situation it would be possible to use any of the approaches discussed in the last two chapters to solve the problem but in all cases many of the operations are inefficient as the matrix is already composed mainly of zeros.

The Thomas algorithm makes use of this fact and with a single elimination step followed by back substitution enables a very quick and efficient solution to the problem. The solution technique is so good that in many cases models are sought that lead to a tri-diagonal matrix or a solution technique is modified so that it will be tri-diagonal so as to make use of the algorithm.

The algorithm (in Fortran) is listed in Figure 12.1. Notice the use of double precision (real(8)) variables to help with round-off errors. The subroutine requires as input the four one-dimensional arrays (or vectors) $a$, $b$, $c$ and $d$, of length, $n$, and $n$, the number of equations. It returns the solution in the $d$ array. The original input data in $a$, $b$, $c$ and $d$ are lost in the process.

```
subroutine thomas(a,b,c,d,n)
implicit none
real(8):: a,b,c,d,tk
real:: a(n),b(n),c(n),d(n)
integer:: i,j,n
do j=2,n
 tk=a(j)/b(j-1)
 b(j)=b(j)-tk*c(j-1)
 d(j)=d(j)-tk*d(j-1)
end do
d(n)=d(n)/b(n)
do i=2,n
 j=n+1-i
 d(j)=(d(j)-c(j)*d(j+1))/b(j)
end do
end
```

**Figure 12.1** Thomas algorithm subroutine in Fortran.

## 12.4 EXAMPLE OF A BANDED MATRIX

The one-dimensional heat diffusion equation can be written:

$$\frac{\partial T}{\partial t} = k\frac{\partial^2 T}{\partial x^2} \tag{12.17}$$

where $T$ is the temperature at a point on a metal bar, $t$ is the time and $k$ is the conductivity constant. By the time it has been coded for solution using an implicit finite difference approach (see Chapter 20) the basic equation to be solved at each grid point can be written:

$$-\lambda T_{i-1}^{t+1} + 2(1+\lambda)T_i^{t+1} - \lambda T_{i+1}^{t+1} = \lambda T_{i-1}^{t} + 2(1-\lambda)T_i^{t} + \lambda T_{i+1}^{t} \tag{12.18}$$

where $\lambda = k\ \Delta t/(\Delta x)^2$ and the values at time level $t$ are assumed to be known with the values at time level $t$+1 to be determined at each step. Because both the time and space derivative are centred in time and space the scheme is second-order accurate and the scheme is called the Crank-Nicholson method. If the solution is developed on a grid that is numbered from 0 to $n$, then the set of equations that must be solved at each time step can be written:

$i = 1$: $$2(1+\lambda)T_1^{t+1} - \lambda T_2^{t+1} = \lambda T_0^{t} + 2(1-\lambda)T_1^{t} + \lambda T_2^{t} + \lambda T_0^{t+1} \tag{12.19}$$

$i = 2$: $$-\lambda T_1^{t+1} + 2(1+\lambda)T_2^{t+1} - \lambda T_3^{t+1} = \lambda T_1^{t} + 2(1-\lambda)T_2^{t} + \lambda T_3^{t} \tag{12.20}$$

$i = 3$: $$-\lambda T_2^{t+1} + 2(1+\lambda)T_3^{t+1} - \lambda T_4^{t+1} = \lambda T_2^{t} + 2(1-\lambda)T_3^{t} + \lambda T_4^{t} \tag{12.21}$$

.

.

$i = n$–1: $$-\lambda T_{n-2}^{t+1} + 2(1+\lambda)T_{n-1}^{t+1} = \lambda T_{n-2}^{t} + 2(1-\lambda)T_{n-1}^{t} + \lambda T_n^{t} + \lambda T_n^{t+1} \tag{12.22}$$

When these are put into matrix form, the problem suddenly appears much simpler and the ease of its implementation becomes clear:

$$\begin{bmatrix} 2(1+\lambda) & -\lambda & 0 & 0 & 0 & 0 \\ -\lambda & 2(1+\lambda) & -\lambda & 0 & 0 & 0 \\ 0 & -\lambda & 2(1+\lambda) & -\lambda & 0 & 0 \\ 0 & 0 & . & . & . & 0 \\ . & . & . & . & . & . \\ 0 & 0 & 0 & 0 & -\lambda & 2(1+\lambda) \end{bmatrix} \begin{Bmatrix} T_1^{t+1} \\ T_2^{t+1} \\ T_3^{t+1} \\ \\ \\ T_{n-1}^{t+1} \end{Bmatrix} = \begin{Bmatrix} RHS_1 \\ RHS_2 \\ RHS_3 \\ \\ \\ RHS_{n-1} \end{Bmatrix} \tag{12.23}$$

where the *RHS* expressions are as listed in Equations (12.19) to (12.22). Note that the first and last contain boundary elements ($T_0$ and $T_n$) that must be known *a priori*. Notice also that the matrix is diagonally dominant (the absolute value of the diagonal element is greater than or equal to the absolute sum of the off-diagonal elements) and therefore likely to converge.

It is evident that this system of equations is banded with a bandwidth of three, making it suitable for solution using the Thomas algorithm. As claimed earlier, many such problems in engineering turn out to be in this form, making the algorithm particularly important.

## PROBLEMS

**12.1** Write a Fortran program to solve a matrix problem using LU decomposition. For example:

$$\begin{aligned} x_1 + 5x_2 + 2x_3 &= 17 \\ 2x_1 + 7x_2 \qquad + x_3 &= 19 \\ -x_1 + 3x_2 + 2x_3 &= 11 \end{aligned}$$

The program should read the data from a file and report the results to the screen, including the intermediate matrices, *L* and *U*. It is suggested that the program should expect the file format that has been used in previous problems, namely:

Line 1: General header (text)
Line 2: n (the number of equations (integer))
Line 3: $a_{11}\ a_{12}\ \ldots\ a_{1n}\ c_1$
Line 4: $a_{21}\ a_{22}\ \ldots\ a_{2n}\ c_2$
.
Line n+2: $a_{n1}\ a_{n2}\ \ldots\ a_{nn}\ c_n$

**12.2** Apply the LU decomposition program to the truss problem shown in Chapter 11. Use the program to investigate the importance of diagonal dominance in the solution procedure. As a matter of interest, can the matrix be written in a diagonally dominant form? The matrix is reproduced here for convenience.

$$\begin{bmatrix} 0.707 & 0 & 0 & -1 & 0 & 0 & 0 & 0 & 0 \\ 0.707 & 0 & 1 & 0 & 0 & 0 & 0 & 0 & 0 \\ 0 & 0 & 0 & 1 & 0.707 & 0 & 0 & -0.707 & 0 \\ 0 & 0 & 0 & 0 & 0.707 & 0 & 1 & 0.707 & 0 \\ 0 & 1 & 0 & 0 & -0.707 & -1 & 0 & 0 & 0 \\ 0 & 0 & -1 & 0 & -0.707 & 0 & 0 & 0 & 0 \\ 0 & 0 & 0 & 0 & 0 & 1 & 0 & 0 & -1 \\ 0 & 0 & 0 & 0 & 0 & 0 & -1 & 0 & 0 \\ 0 & 0 & 0 & 0 & 0 & 0 & 0 & 0.707 & 1 \end{bmatrix} \begin{Bmatrix} F_1 \\ F_2 \\ F_3 \\ F_4 \\ F_5 \\ F_6 \\ F_7 \\ F_8 \\ F_9 \end{Bmatrix} = \begin{Bmatrix} 0 \\ 100 \\ 0 \\ 0 \\ 0 \\ 100 \\ 0 \\ 0 \\ 0 \end{Bmatrix}$$

**12.3** By referring to the problem in Chapter 11 where the truss matrix is derived, determine how it is possible to alter the solution to take a variety of externally applied loads.

Notice that it is possible to do this by only changing elements in the right-hand column matrix. Use this to solve the problem shown in Figure 12.2.

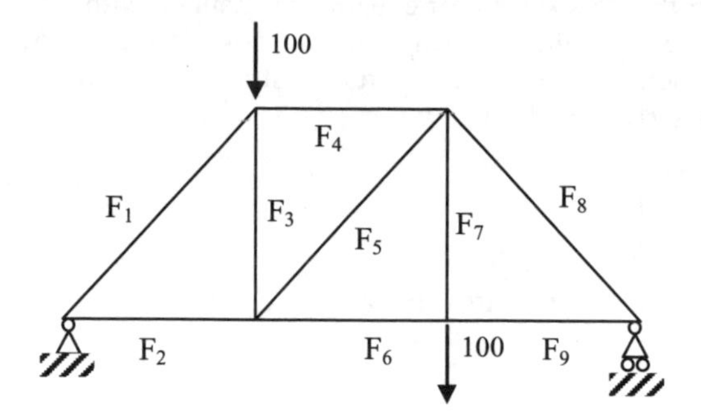


**Figure 12.2** Truss for analysis.

CHAPTER THIRTEEN

# Ordinary Differential Equations (Euler's Method)

## 13.1 INTRODUCTION

In 1671 Isaac Newton developed a solution to the equation:

$$\frac{dx}{dt} = (1+t)x - 3t + t^2 \tag{13.1}$$

by taking an infinite series approach. He found he was able to solve the problem and that the solution depended very much on the particular start conditions (Acheson, 1997). Indeed, the function is very sensitive to the initial value of $x$, heading to either positive or negative infinity as shown in Figure 13.1. The key point to realise with Equation (13.1) is that it defines the gradient of a function at any point in time and/or space rather than the function itself; the problem then is to discover the underlying function.

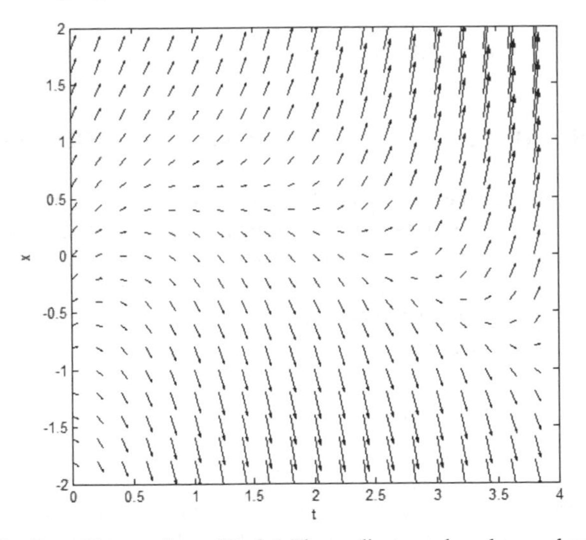


**Figure 13.1** Gradients illustrated on a 2D plot. The gradients are based on evaluating the ordinary differential equation listed as Equation (13.1).

The solution procedure consists of defining the underlying function based on a given start point and the gradients. In Figure 13.1, for example, the problem may be to define the function given a start point of ($t$ = 0.0, $x$ = 1.0). Following the arrows that show the gradient indicates that the function will head upwards (positive $x$) very quickly. A different start point, ($t$ = 0.0, $x$ = –1.0), gives quite a different behaviour and a start point around ($t$ = 0.0, $x$ = 0.5) will be quite different again.

As a matter of interest, the Matlab code to produce Figure 13.1 is listed in Figure 13.2. It uses a non-Matlab plotting routine, `plot_arrow` which was downloaded from the web.

```
% Newton ODE
% plots gradients for dx/dt = (1+t)x + 1 -3t + t^2
% using web-retrieved plot_arrow function
% written david walker, november 2007

clear all;
close all;
scalex=0.05; scaley=0.05; % plotting scales for arrows

t=(0:0.2:4);      % t from 0 to 4, x from -2 to 2
x=(-2:0.2:2);     % plot is t on x axis and x on y axis!!
hold on;          % to stop each arrow being over-written
axis([0 4 -2 2]);    %define plot limits
box on;
xlabel('t'); ylabel('x');    % label axes
rt=size(t,2); rx=size(x,2);   % get sizes of arrays

for i=1:rt
for j=1:rx
 grad=(1+t(i))*x(j)+1-3*t(i)+t(i)*t(i);   % gradient
 plot_arrow(t(i),x(j),t(i)+scalex,x(j)+scaley*grad);
end
end
```

**Figure 13.2** Matlab code to produce Figure 13.1.

## 13.2 ORDINARY DIFFERENTIAL EQUATIONS IN ENGINEERING

Ordinary differential equations (ODEs) come up in a number of fields in engineering, For example, in describing sediment transport in streams or in the sea it is possible to write a differential equation that describes the change in sediment concentration over the water depth. For the simple case, where the conditions are constant in the direction along the bed, the concentration can be assumed to vary only in the vertical, in which case it is possible to write:

$$\frac{dc}{dz} = -\frac{wc}{\varepsilon} \tag{13.2}$$

where $c$ is the concentration, $z$ the vertical axis, $w$ the fall velocity of the particles and $\varepsilon$ the eddy viscosity which is related to the turbulence. The equation comes directly from Fick's law that relates the flux of a material to the product of a diffusion coefficient and the concentration gradient. It is worth noting that in reality there are often bedforms (ripples, dunes) that lead to variations in the horizontal direction as well (see, for example, Figure 13.3) but this does not detract from the value of being able to handle the simple situation, even if it is taken as a precursor to a more accurate model.

**Figure 13.3** Sediment suspensions over a rippled bed. Photo: P. Nielsen.

A typical concentration profile, for the case of uniform conditions in the horizontal, based on solution of Equation (13.2), is shown in Figure 13.4.

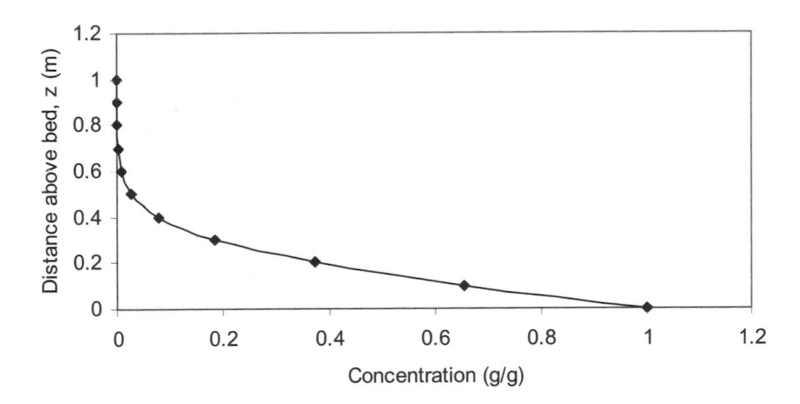


**Figure 13.4** Sediment concentration profile based on solution of an ordinary differential equation.

If the fall velocity, $w$, and eddy viscosity, $\varepsilon$, are constant, or a simple function of $z$, it is possible to integrate this expression (using standard calculus) and solve for $c$, but there are many situations where these conditions may not hold and solution by analytical means is simply not possible. However, knowing a starting point (the concentration at the bed) and the slope everywhere (given by Equation (13.2)), it should be possible to determine the form of the equation. That is essentially the topic of this and the next chapter.

A second example that will be used to illustrate the solution methods is called the logistic growth law. It is generally expressed as a differential equation, and can be written:

$$\frac{dE}{dt} = \alpha E(E_f - E) \qquad (13.3)$$

where $E$ is the population or general dimension that is being modelled, $\alpha$ is a growth coefficient, $t$ is time, and $E_f$ is the final stable value. The solution can be determined analytically by integration:

$$E = \frac{E_f E_0}{E_0 + (E_f - E_0)\exp(-\alpha E_f t)} \qquad (13.4)$$

It is therefore possible to plot the changing values of $E$ against time (see Figure 13.5), but what happens if the differential equation is given in such a form that cannot be integrated?

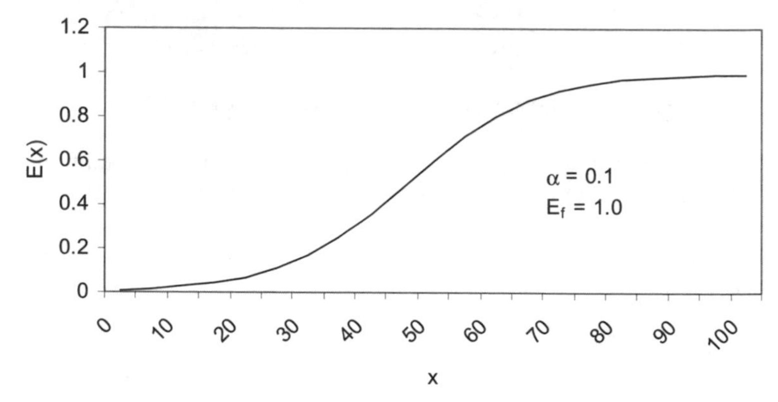


**Figure 13.5** The logistic growth law function.

The logistic growth law equation is first-order since the highest order of the derivatives is one: (*dE/dt*). A second-order differential equation would contain a second derivative – for example, the equation describing the position x of a mass-spring system can be written:

$$m\frac{d^2x}{dt^2} = c\frac{dx}{dt} + kx = 0 \qquad (13.5)$$

This can be reduced to a first-order system by defining a new variable that is itself a derivative. In the case of the mass-spring system it is possible to define:

$$y = \frac{dx}{dt} \qquad (13.6)$$

and re-writing the main equation in terms of $y$ gives:

$$m\frac{dy}{dt}+cy+kx=0 \tag{13.7}$$

These two equations therefore form a first-order set of equations. Thus a second-order ODE has been reduced to a first-order ODE. This technique also works for even higher-order equations. Therefore, since many ODE problems can be reduced to a first-order problem the emphasis of this chapter will be on first-order equations.

## 13.3 PRE-COMPUTER METHODS OF SOLVING ODES

Before numerical methods were available ODEs were usually solved analytically. Unfortunately there were equations with no exact integral and therefore no exact solution. In this case it was sometimes possible to obtain an approximate solution by linearising the equation to be solved. A linear ODE can be written:

$$a_n(x)y^{(n)}+...+a_1(x)y'+a_0(x)y=f(x) \tag{13.8}$$

where $y^{(n)}$ is the $n^{\text{th}}$ derivative of $y$ with respect to $x$ and the $a$ and $f$ values are specified functions of $x$. The equation is linear because there are no products or non-linear functions of the dependent variable y and its derivatives. Linear ODEs are important because they can be solved analytically. For example, the equation of a pendulum can be written:

$$\frac{d^2\theta}{dt^2}+\frac{g}{l}\sin\theta=0 \tag{13.9}$$

This equation is non-linear because of the term sin(θ). Now if it was θ instead of sin(θ) then the equation would be linear. A solution is formed realising that for small angles θ approximately equals sin(θ). (Example: 1°=0.0175 radians, sin(1°)=0.0175; 5°=0.0873 radians, sin(5°)=0.0872; 10°=0.1745 radians, sin(10°)=0.1736) The equation can be approximated as:

$$\frac{d^2\theta}{dt^2}+\frac{g}{l}\theta=0 \tag{13.10}$$

The equation, therefore, has been linearised and a solution can be found analytically. Therefore in the pre-computer days it was possible to solve some ODEs but general solutions were not possible. Numerical methods allow a much more general approach to the problem.

## 13.4 ONE-STEP METHODS

As an introduction to many of the methods that will be covered in this and the next chapter, consider an equation:

$$\frac{dy}{dx} = f(x, y) \tag{13.11}$$

In solving it the general process is to start at a given point on the function and attempt to follow the slope to a new point. The general formula that is followed is:

$$y_{i+1} = y_i + mh \tag{13.12}$$

where $m$ is the slope of the line, $h$ the step size and the subscripts refer to the value of the function. In non-mathematical terms, the new value is equal to the old value plus the slope of the function multiplied by the step size. All the one-step methods follow that general formula, the difference being how the slope is estimated.

The general procedure, therefore, is to start at a known or defined starting point and then take a step along the function using the estimate of the slope and the chosen step. This is then followed by further steps until the required section of the function is determined. The process is illustrated in Figure 13.6. It is evident that the key to this procedure is the estimate of the slope of the function and the size of the step taken. There is the usual trade-off between a large step size which will determine the solution in a relatively small number of steps and the loss of accuracy that a large step will likely bring.

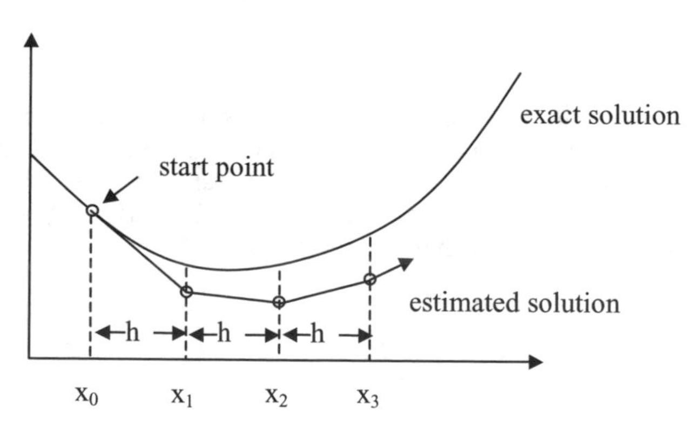


**Figure 13.6** The basic procedure for one-step methods of solving ODEs.

## 13.5 EULER'S METHOD

The Euler method uses the first derivative as defined by the formula at the point as the estimate of the slope. Therefore:

$$y_{i+1} = y_i + f(x_i, y_i)h \tag{13.13}$$

where *f(x,y)* is the given differential equation. In fact, the formula can be written slightly differently as:

$$y_{i+1} = y_i + y'h \tag{13.14}$$

and in this form starts to look like part of a Taylor series expansion. The full Taylor series (see Appendix A) is:

$$y_{i+1} = y_i + y'h + \frac{y''}{2!}h^2 + ... \tag{13.15}$$

where the pattern continues with the higher order derivatives.

Therefore, the Euler formula truncates the series with an order of error that is proportional to $h^2$, written $O(h^2)$. This error is called the local error as it applies at each and every step as the solution develops. The global error, which applies to the final solution, is $O(h)$ since, with the error accumulating on each iteration, the number of operations is inversely proportional to the step size, $h$. It is also evident that if the function is linear the method will give exact results since in this case the second and higher derivatives are zero and the Euler equation will match exactly the Taylor expansion.

Using Euler's method on the logistic growth law function defined earlier with $E_0$ = 0.01, and using a time-step of 5 seconds gives the results shown in Figure 13.7.

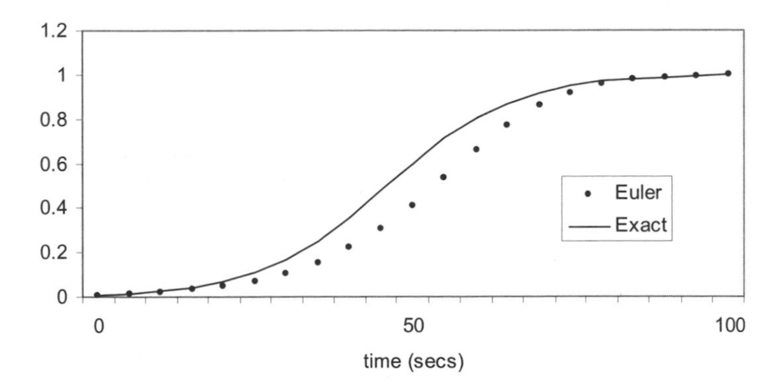


**Figure 13.7** Solution of the logistic growth law by Euler's method and a time step of 5 s.

In this case the maximum error occurs near $t$=35 seconds. By reducing the time step it is possible to improve the accuracy, but this improvement in accuracy comes at a cost in terms of calculation effort and time. The results for a range of time steps are shown in Table 13.1 where the error at $t$=35 seconds is used as a common point to compare the different solutions.

It is worth noting in the results that as the time steps reduce by a factor of 2 or 5 the error also reduces (approximately) at that same rate. For example, halving the timestep from 1.0 seconds to 0.5 seconds reduces the error from 10.24% to 5.31%. This is as expected for the scheme as it is first order accurate.

**Table 13.1** Effect of varying the time step with Euler's method.

| Time step (secs) | Estimated solution at $t = 35$s | Error % |
|---|---|---|
| 5.0 | 0.1539 | 38.58 |
| 1.0 | 0.2250 | 10.24 |
| 0.5 | 0.2373 | 5.31 |
| 0.1 | 0.2479 | 1.09 |
| Exact | 0.2507 | |

## 13.6 CHAOTIC BEHAVIOUR IN SOLUTIONS

Acheson (1997) gives an example where Euler's method is shown to produce results that appear chaotic. The equation to be solved is:

$$\frac{dx}{dt} = t - x^2 \tag{13.16}$$

with starting values $x_0 = 0.0$ or thereabouts. Interestingly enough, the solution seems to proceed smoothly initially and only becomes chaotic for steps of 0.05 at approximately $t = 420$ secs. A plot of the solution is shown in Figure 13.8.

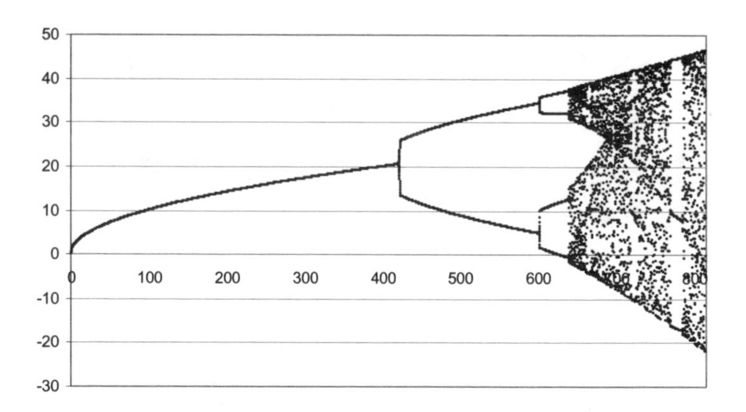


**Figure 13.8** Solution for a particular function using Euler's method and a step = 0.05.

## 13.7 MODIFIED EULER METHOD

The Euler method, while conceptually simple, has a serious problem: it assumes that the gradient of the function over the total step length is equal to the gradient at the start of the step. Given that the gradient may be varying continuously over the step this is a possible source of error. A way of improving the Euler method is to use it to estimate the slope at the mid-point of the interval and then assume that this slope is applicable over the whole interval. In this case, the mid-point value of the function is estimated using the standard Euler approach:

$$y_{i+1/2} = y_i + f(x_i, y_i)\frac{h}{2} \qquad (13.17)$$

and the slope at the mid-point is calculated at this point and used in place of the slope at the start of the interval. The estimate of the new point, therefore, is:

$$y_{i+1} = y_i + f(x_{i+1/2}, y_{i+1/2})h \qquad (13.18)$$

The utilisation of the slope from the mid-point is better than from the end-point and the method has an error of $O(h^2)$. The results for a range of time steps are shown in Table 13.2.

**Table 13.2** Effect of varying time step with the modified Euler's method.

| Time step (secs) | Estimated solution at t = 35s | Modified Euler Error % | Standard Euler Error % |
|---|---|---|---|
| 5.0 | 0.2368 | 5.53 | 35.58 |
| 1.0 | 0.2250 | 0.28 | 10.24 |
| 0.5 | 0.2505 | 0.073 | 5.31 |
| 0.1 | 0.2506 | 0.003 | 1.09 |
| Exact | 0.2507 | | |

## PROBLEMS

**13.1** Sketch by hand the gradients as defined by the equation Newton was working on in 1671:

$$\frac{dx}{dt} = (1+t)x + 1 - 3t + t^2$$

for a range of values between $x$=0 and $x$=1 and $t$=0 and $t$=4. This should verify Figure 13.1.

**13.2** Reproduce the results shown in Table 13.1 for the logistic growth law formula by applying Euler's method. For this question it may be convenient to set up an Excel spreadsheet where the time step is a named variable that can be changed as required. Use $E_0 = 0.01$ as the start value at $t = 0$.

**13.3** Solve the sediment concentration equation:

$$\frac{dc}{dz} = -\frac{wc}{\varepsilon}$$

in a channel of depth 1.0 metre (that is, the solution should start at $z = 0$ m and proceed in steps to the water surface where $z = 1.0$ m) for the situations where:

(a) $c_0 = 1.0$ at $z = 0.0$ m, $\varepsilon = 1.0$ m$^2$/s, $w = 0.15$ m/s (particle size $\cong$ 1 mm);
(b) $c_0 = 1.0$ at $z = 0.0$ m, $\varepsilon = 1.0$ m$^2$/s, $w = 0.01$ m/s (particle size $\cong$ 0.1 mm).

Comment on the predicted concentration profile and explain the behaviour if possible.

**13.4** Develop solutions to the ODE:

$$\frac{dx}{dt} = t - x^2$$

with a starting value $x_0 = 0.0$ at $t$=0. Explore the solution behaviour and see if it is possible to reproduce the chaotic behaviour for a time step of 0.05 s.

CHAPTER FOURTEEN

# Ordinary Differential Equations (Heun and Runge-Kutta Methods)

## 14.1 INTRODUCTION

In the previous chapter it was shown that a number of phenomena that are relevant to engineering can be described in terms of ordinary differential equation (ODEs). The methods that were described to solve such problems had the advantage that they were intuitively obvious and gave a clear idea of the solution methodology. However, the low-order of accuracy associated with the measures makes them unlikely candidates in the solution of real problems. The methods in this chapter will build on the earlier methods but will be shown to be far superior. The basic operation is still the same, but it will be seen that much more attention is given to estimating the slope to be used in the step from one point to the next.

## 14.2 HEUN'S METHOD

Although it would be possible to improve Euler's method by including higher derivative terms in the equation this can not always be done as it requires the function itself to be differentiable. This is often not possible and methods that do not require higher-order terms, but nevertheless result in higher accuracies, have been developed.

One obvious drawback with Euler's method is that the slope that is calculated at the start of the interval is assumed to hold over the entire step. One way to improve this is to calculate the slope at the start and the end of the interval and use the average. The process is accomplished in two steps. In the first step an initial estimate of the new value, $y'_{i+1}$, is made:

$$y'_{i+1} = y_i + f(x_i, y_i)h \tag{14.1}$$

It is evident that this estimate is in fact the point that the simple Euler method would predict as the next point on the function. Heun's method uses this to estimate the slope of the function at this end-point and then combines that with the slope at the start point to give an improved estimate of the slope over the total step. The final estimate for the next point can be written:

$$y_{i+1} = y_i + \frac{f(x_i, y_i) + f(x_{i+1}, y'_{i+1})}{2} h \tag{14.2}$$

The method is called a predictor-corrector. When the method is applied to the same growth function described in the previous chapter the results are as listed in Table 14.1. As a reminder, the logistic growth law equation can be written:

$$\frac{dE}{dt} = \alpha E(E_f - E) \tag{14.3}$$

where $E$ is the population or general dimension that is being modelled, $\alpha$ is a growth coefficient, $t$ is time, and $E_f$ is the final stable value. Notice that reducing the step by an order of two (from 1.0 to 0.5) reduces the error by a factor of four, indicating that the global error is $O(h^2)$.

**Table 14.1** Effect of varying the time step with Heun's method.

| Time step (secs) | Estimated solution at t = 35s | Error % |
|---|---|---|
| 5.0 | 0.2353 | 6.13 |
| 1.0 | 0.2498 | 0.32 |
| 0.5 | 0.2504 | 0.08 |
| 0.1 | 0.2506 | 0.003 |
| Exact | 0.2506 | |

When programming this, and other higher-order methods, in Excel there is the potential for the numerous function evaluations at different points in time and space to make coding so complicated that it becomes almost impossible to follow. The coding in Excel, therefore, is made much easier and more efficient by the inclusion of an Excel Visual Basic module to define the function. In the function listed in Figure 14.1, the values of $\alpha$, $E_f$, and $E$ are passed in the function argument list. The use of a function makes coding much simpler and is easily worth the extra effort involved. It is interesting to note that for a worksheet with hundreds of lines of formulae the inclusion of the module slows the calculations considerably; a penalty worth enduring.

```
Function my_f(alpha, ef, e)

' function to evaluate logistic growth law
' written d walker, february 2008
' alpha = growth rate, Ef = final value

my_f = alpha * e * (ef - e)
End Function
```

**Figure 14.1** Excel Visual Basic module to plot exact solution to logistic growth law function.

## 14.3 RUNGE-KUTTA METHODS

Runge-Kutta methods enable high-orders of accuracy without the need to calculate higher derivatives. The most general form of the equation is fourth-order accurate. It can be written:

$$y_{i+1} = y_i + \frac{h}{6}(k_1 + 2k_2 + 2k_3 + k_4) \tag{14.4}$$

where

$$\begin{aligned} k_1 &= f(x_i, y_i) \\ k_2 &= f(x_i + 0.5h, y_i + 0.5hk_1) \\ k_3 &= f(x_i + 0.5h, y_i + 0.5hk_2) \\ k_4 &= f(x_i + h, y_i + hk_3) \end{aligned} \tag{14.5}$$

The method was applied to the same growth function described earlier and the results are listed in Table 14.2 and plotted for a time step of 5 seconds in Figure 14.2. The rapid reduction in the percentage error as the time step is reduced is evident and, for this particular equation, the accuracy of the Runge-Kutta method is such that it really requires more than four significant figures in the output.

**Table 14.2** Effect of varying the time step with a fourth-order Runge-Kutta method.

| Time step (secs) | Estimated solution at t = 35s | Error % |
|---|---|---|
| 5.0 | 0.2505 | 0.066 |
| 1.0 | 0.2507 | 0.0001 |
| 0.5 | 0.2507 | 0.000 009 |
| 0.1 | 0.2507 | 0.000 000 01 |
| Exact | 0.2507 | |

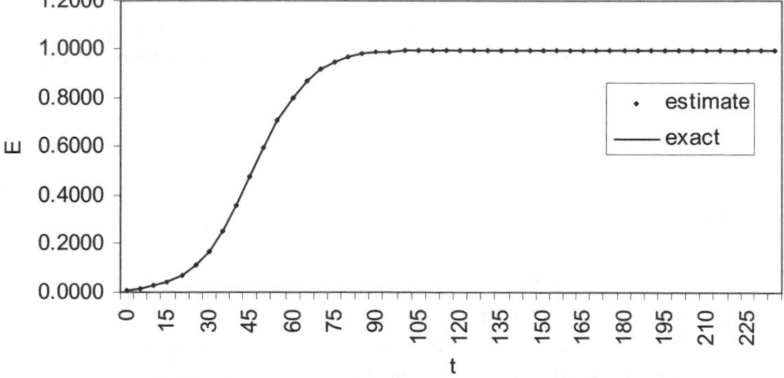


**Figure 14.2** The logistic growth law function.

## 14.4 COMPUTATIONAL EFFORT

The methods have been presented to date without any regard to the processing effort required to obtain the solution. Although the higher-order methods have the advantage in terms of accuracy they also require more function calculations and it may be that the extra work involved is not worth it. By defining a term:

$$Effort = n_f \frac{b-a}{h} \tag{14.6}$$

where $n_f$ is the number of function calls made per iteration, $a$ and $b$ are the limits of the calculation and $h$ is the step size, it is possible to determine the most efficient methods in terms of computer time. In Table 14.3 the effort required to achieve errors of less than 1% (approximately) is shown for the four methods outlined in the chapters so far. The Runge-Kutta is clearly superior, at least for the particular equation being used as an example.

**Table 14.3** Effort required to achieve approximately 1% error in the estimate at t=35 seconds.

| Method | Time step | Effort |
|---|---|---|
| Euler | 0.1 | 350 |
| modified | 2 | 35 |
| Heun's | 2 | 35 |
| Runge-Kutta | 10 | 14 |

## 14.5 ADAPTIVE STEP METHODS

In the methods demonstrated so far there are a couple of characteristics that make them less than ideal. Firstly, the methods have all used a step size that, once selected, has remained constant. There will be times when the shape or slope of the function means there is an advantage in being able to change the step size. For example, in the logistic growth law a smaller step as the gradient changes rapidly would lead to a better solution, whereas quite large steps could be taken once the function had reached its final plateau. This should lead to a more efficient solution by optimising the number of iterations required.

There is a more serious flaw in the methods covered so far: without an exact answer to compare the numerical predictions with, it is impossible to know how accurate a particular method is. After all, these methods will be used for precisely the reason that an exact analytical solution is not known, so how can accuracy be determined?

Methods that deal with both of these issues are called adaptive step methods. In these, the step size is constantly being assessed and varied if necessary to achieve a given level of accuracy. The Runge-Kutta-Fehlberg method is a variation on the standard Runge-Kutta scheme. The method uses two different Runge-Kutta schemes of different orders of accuracy; in this case a fourth-order scheme and a fifth-order scheme. The difference between the two estimates gives an estimate of the error at each step and the time step can be adjusted accordingly.

One possible disadvantage with this method might be the large number of function evaluations necessary at each step to estimate the fourth and fifth-order values. The Fehlberg variation gets around this by using the fourth-order estimates to help evaluate the fifth-order estimates. The method can be written:

*Fourth-order scheme*

$$y_{i+1} = y_i + h\left(\frac{25}{216}k_1 + \frac{1408}{2566}k_3 + \frac{2197}{4104}k_4 - \frac{1}{5}k_5\right) \tag{14.7}$$

*Fifth-order scheme*

$$y_{i+1} = y_i + h\left(\frac{16}{135}k_1 + \frac{6656}{12825}k_3 + \frac{28561}{56430}k_4 - \frac{9}{50}k_5 + \frac{2}{55}k6\right) \tag{14.8}$$

where

$$k_1 = 2f(x_i, y_i) \tag{14.9}$$

$$k_2 = f(x_i + \frac{1}{4}h, y_i + \frac{1}{4}hk_1) \tag{14.10}$$

$$k_3 = f(x_i + \frac{3}{8}h, y_i + \frac{3}{32}hk_1 + \frac{9}{32}hk_2) \tag{14.11}$$

$$k_4 = f(x_i + \frac{12}{13}h, y_i + \frac{1932}{2197}hk_1 - \frac{7200}{2197}hk_2 + \frac{7296}{2197}hk_3) \tag{14.12}$$

$$k_5 = f(x_i + h, y_i + \frac{439}{216}hk_1 - 8hk_2 + \frac{3680}{513}hk_3 - \frac{845}{4104}hk_4) \tag{14.13}$$

$$k_6 = f(x_i + \frac{1}{2}h, y_i - \frac{8}{27}hk_1 + 2hk_2 - \frac{3544}{2565}hk_3 + \frac{1859}{4104}hk_4 - \frac{11}{40}hk_5) \tag{14.14}$$

The error estimate is obtained by subtracting the two:

$$E_a = h\left(\frac{1}{360}k_1 - \frac{128}{4275}k_3 - \frac{2197}{75240}k_4 + \frac{1}{50}k_5 + \frac{2}{55}k_6\right) \tag{14.15}$$

Therefore, to solve the ODE, use the fourth-order scheme and after each step estimate the error. Then a method of adjusting the time step must be determined based on the strategy that if the error is smaller than required, the step can be increased, and if the error is larger than required, the step must be reduced. Care must be taken that the scheme does not go into an infinite loop where the step is repeatedly increased and decreased without actually getting anywhere!

Methods such as this – and there are others – seem at first glance like a lot of work and therefore probably not worth the extra effort in programming. However, there are a number of situations where schemes such as this lead to a very quick and accurate solution that would have taken much longer using a conventional Runge-Kutta scheme and a small step for the whole solution.

## 14.6 INTEGRATION USING ODE TECHNIQUES

It should be noted that the methods outlined in this chapter are also applicable to integration. This is because the solution of:

$$I = \int_a^b f(x)dx \tag{14.16}$$

and

$$\frac{dy}{dx} = f(x) \tag{14.17}$$

are equivalent where the differential equation is solved for $y(b)$ given the initial condition $y(a) = 0$. Therefore it is possible to consider fourth-order Runge-Kutta methods along with Simpson's rule and the trapezoidal rule as possible methods for integrating functions.

## PROBLEMS

**14.1** Reproduce the results shown in Table 14.1 for the logistic growth law formula by applying Heun's method. In this case $E_0 = 0.01$. For this question it may be convenient to set up an Excel spreadsheet where the time step is a named variable that can be changed as required. Define the function using a Visual Basic module.

**14.2** Use the Runge-Kutta algorithm to solve the logistic growth law formula. It is at this stage that the need for Visual Basic modules will become apparent. As a first attempt it may be worth attempting a solution without using a function. Next, try defining a function and using it in the evaluation of the $k_1$, $k_2$, $k_3$ and $k_4$ values. Reproduce the results shown in Table 14.2.

**14.3** Solve the function defined by the equation Newton was working on in 1671:

$$\frac{dx}{dt} = (1+t)x + 1 - 3t + t^2$$

for starting values of:

(a) $t = 0, x = -1$
(b) $t = 0, x = 1$
(c) $t = 2, x = 0.$

CHAPTER FIFTEEN

# Finite Difference Modelling (Introduction)

## 15.1 INTRODUCTION

There are a multitude of engineering problems that can be expressed in terms of partial differential equations (PDEs) – that is, differential equations where partial rather than full derivatives are employed. Some examples include:

- flow through porous media;
- flow under dams;
- heat conduction on plates and wires;
- stresses in underground excavations;
- waves on water (Figure 15.1), vibrating strings;
- ocean wave energy growth and dissipation;
- sediment transport in rivers and on the coast;
- global weather forecasting;
- dispersion, smoke stacks.

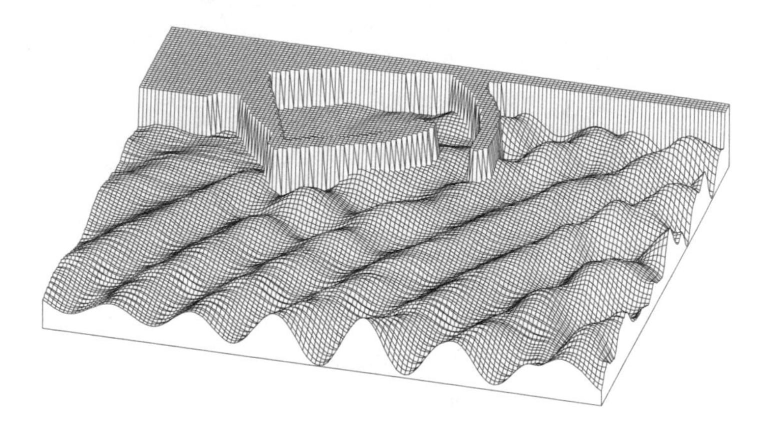

**Figure 15.1** Finite difference model of waves approaching a marina.

In some cases there are analytical solutions to these equations. That is, it is possible to integrate the equations and determine an exact solution. In most cases this is not possible, and it is necessary to employ an approximate method. One of the two most common methods is finite difference methods.

*Example: transport of pollutants*

In a long river, where a pollutant is being released in such a way that it is spread evenly over the cross section, it is possible to write a partial differential equation that describes the concentration at any point downstream. The governing equation is the transport equation and can be written:

$$\frac{\partial c}{\partial t} + u\frac{\partial c}{\partial x} = K\frac{\partial^2 c}{\partial x^2} \tag{15.1}$$

where $u$ is the convective velocity (constant), $c$ is the concentration of the material, $x$ is the axis running down the river, and $K$ is the diffusion/dispersion coefficient. If the release is constant in time it is possible to develop a solution for the concentration downstream:

$$\frac{c}{c_0} = \frac{1}{2}e^{\frac{ux}{K}}erfc\left[\frac{x+ut}{2\sqrt{Kt}}\right] + \frac{1}{2}erfc\left[\frac{x-ut}{2\sqrt{Kt}}\right] \tag{15.2}$$

where $c_0$ is the constant upstream concentration, $c$ is the concentration at point $x$ and time $t$, $x$ is the distance downstream (or upstream for $-x$), $t$ is time, $u$ is the stream velocity, $K$ the dispersion coefficient, and $erfc(x)$ is the complementary error function. A typical solution is shown in Figure 15.2, where a snapshot of the concentration along a length of river is shown at various times following the start of the release of a material into the river. Note that as time goes by the concentration of the material in the river extends further and further from the release point.

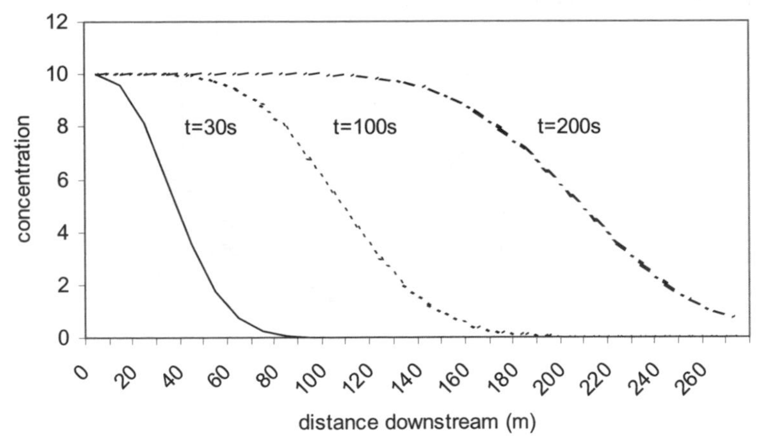


**Figure 15.2** Concentration of material in a river at various times following release at a point.

A second way to display the results is to plot the concentration at a point downstream of the release point as time goes by. This is done in Figure 15.3 where a point 100 metres from the release point is shown to have no concentration of material at all until approximately 40 seconds after release when the material first arrives, and concentration is at the maximum by approximately 160 seconds after release.

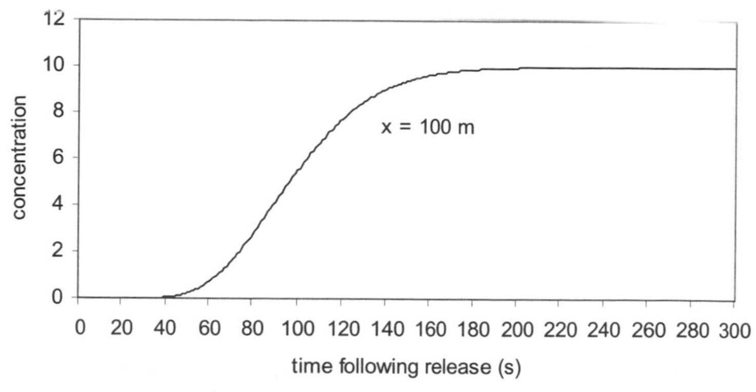


**Figure 15.3** Concentration of material in a river at a point 100 metres downstream of the release point.

If the upstream concentration varies or is intermittent, then there is no general analytical solution and one of the more feasible ways to determine concentrations is to develop a finite difference approximation to the governing equations and to use this to determine a solution. In this case the analytical solution is useful for providing a check to the accuracy of the approximate solution.

## 15.2 FIRST-ORDER FINITE DIFFERENCE APPROXIMATIONS

In Appendix A, details of the Taylor series are given and it can be seen that if a function of a single variable, $f(.)$, is continuous and sufficiently smooth, by which it can be assumed it has continuous derivatives, its Taylor series about some point, $x$, provides an infinite series which converges to the value of the function at points sufficiently close to $x$. The series is a polynomial in the distance, $h$, between $x$ and the nearby point, $x+h$, and the coefficients are given by the value of the function and its derivatives at the point $x$. This can be expressed as:

$$f(x+h) = f(x) + h.f'(x) + \frac{h^2}{2!}f''(x) + \frac{h^3}{3!}f'''(x) + \ldots \qquad (15.3)$$

where $f(x)$ is the basic function, $f'(x)$ the first derivative, $f''(x)$ the second derivative, etc., and $h$ is an increment which is generally taken to be small and positive. It is also true that if $h$ is small the later terms in the expression tend to be very small and eventually insignificant on the basis that if $h$ is small then $h^2$ is even smaller and $h^3$ smaller still. Therefore it would not be unreasonable to omit some of the later terms, referred to as the higher-order terms, and consider the expression as an approximation rather than an equality:

$$f(x+h) \approx f(x) + h.f'(x) + \frac{h^2}{2!} f''(x) \tag{15.4}$$

It is possible, therefore, to derive an approximation for the derivative by re-arranging the equation. This gives:

$$f'(x) \approx \frac{f(x+h) - f(x)}{h} - \frac{h}{2!} f''(x) \tag{15.5}$$

Finally, ignoring the second derivative term leads to:

$$f'(x) = \frac{f(x+h) - f(x)}{h} + O(h) \tag{15.6}$$

where the $O(h)$ is an error term of order $h$. That is, the largest omitted term is multiplied by $h$. Subsequent terms would be multiplied by $h^2$, $h^3$ etc., and would be smaller given that $h$ is small. The error is also referred to as a first-order error since it contains $h$ to the first power. What this means is that a derivative at a point can be estimated (remember there are errors caused by the terms that have been ignored) using the value of the function at that point ($x$) and a point some small distance from the point ($x+h$). For example, to estimate the first derivative of $f(x) = x^4$ at $x$=2, using the fact that $f(2)$ = 16, and $f(2.1)$ = 19.448:

$$f'(x) \approx \frac{19.448 - 16}{0.1} = 34.481$$

The exact value, based on elementary calculus, is 32.0 so the error is 7.75%. If, instead of having $h$ = 0.1, it was possible to use $h$ = 0.05, then using $f(2)$=16.0 and $f(2.05)$=17.661 gives a value of 33.220. By halving the $h$ value, the error has been halved also, a feature of first-order accurate schemes.

## 15.3 HIGHER-ORDER FINITE DIFFERENCE APPROXIMATIONS

First-order schemes, while convenient to derive, are generally not used due to the fact that their errors are relatively large. By some simple manipulations it is possible to derive higher-order (and therefore more accurate) estimates. To do this, the procedure is to write Taylor series expressions for $f(x+h)$ and $f(x-h)$ and then subtract one from the other. For example, the first can be written:

$$f(x+h) = f(x) + hf'(x) + \frac{h^2}{2!} f''(x) + \frac{h^3}{3!} f'''(x) + \ldots \tag{15.7}$$

and the second, which is similar but for the change in sign with $h$:

$$f(x-h) = f(x) - hf'(x) + \frac{h^2}{2!}f''(x) - \frac{h^3}{3!}f'''(x) + ... \tag{15.8}$$

Subtracting the second from the first gives:

$$f(x+h) - f(x-h) = 2hf'(x) + 2\frac{h^3}{3!}f'''(x) + ... \tag{15.9}$$

Now, re-arranging to get the term for the first derivative on the left-hand side of the equation, and then truncating the error terms, gives:

$$f'(x) = \frac{f(x+h) - f(x-h)}{2h} + O(h^2) \tag{15.10}$$

The result is a new estimate for the derivative. This is called a central difference because the estimate at a point $x$ is made using information from either side, that is at $x+h$ and $x-h$. This is shown in Figure 15.4. The scheme is second-order accurate because the most significant term that is omitted has $h^2$ in it.

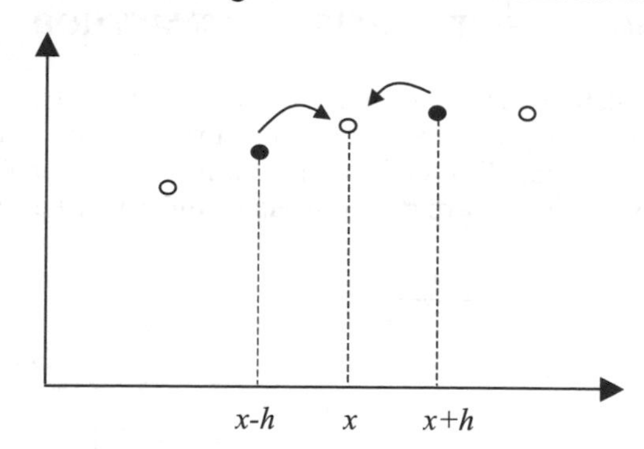


**Figure 15.4** The data used to determine the central difference approximation for the first derivative.

Reworking the example used previously shows the effect of the second-order scheme in improving the accuracy of the estimate. For $h = 0.1$ the estimate can be written:

$$f'(x) \approx \frac{19.448 - 13.032}{2(0.1)} = 32.08$$

which has an error of 0.25% – a significant improvement. Reducing the step to $h = 0.05$ leads to an error of 0.06%. Hence, for a second-order scheme halving the step leads to a four-times improvement in the accuracy. Not only are second-order schemes more accurate, reducing the space step leads to significantly greater accuracy compared to the first-order approximation.

## 15.4 SECOND DERIVATIVE APPROXIMATIONS

Writing down the forward and backward estimates of the Taylor expansion and adding them leads to an estimate for the second derivative to second-order accuracy. This is because the addition of the two terms removes the two first-order terms associated with the first derivative. The final approximation is:

$$f''(x) = \frac{f(x+h) - 2f(x) + f(x-h)}{h^2} \qquad (15.11)$$

It should be noted that the verification of the estimates for the various derivatives has involved a rather artificial device: the use of known functions such as $f(x)=x^4$. This may lead to some confusion about the application of the method. Generally, engineers will be dealing with numerical data rather than functions. The power of the finite difference approximation is that it deals with discrete data rather than functions. The use of the functions has just been to show the accuracy of the method. This will be explored further in the following section.

## 15.5 APPLICATION OF FINITE DIFFERENCE METHOD

The solution of equations using finite differences involves two basic operations which are shown in Figure 15.5: discretisation of the solution area, and approximation of the partial differential equation. In this case Laplace's equation is used for the demonstration. A solution procedure for it will be set out in Chapter 16.

$$\nabla^2 \phi = \frac{\partial^2 \phi}{\partial x^2} + \frac{\partial^2 \phi}{\partial y^2} = 0$$

(a) Solution domain for partial differential equation showing the equation and a continuous domain.

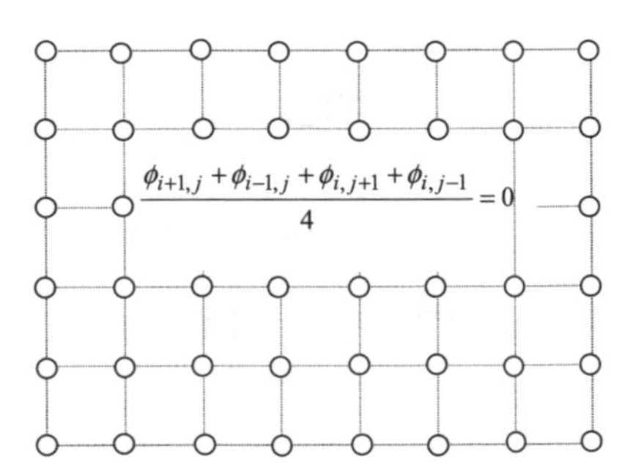


(b) Solution domain for FDM showing schematic of grid, and the finite difference approximation to the partial differential equation.

**Figure 15.5** Representation of difference between the solution of the full PDE and its FD approximation.

In Figure 15.5 (a) the solution is developed over a continuous area according to a second-order (in this case) differential equation. In Figure 15.5 (b) the area has been split into discrete points and the equation approximated to make use of those points. Since the solution area is split into a number of discrete points the values in the finite difference solution will be a series of points rather than a continuous function.

The solution area might be a continuous river, a section of the coast, or a flat metal plate. In each case, if a finite difference solution is to be attempted the area is divided into a number of points, usually equally spaced.

## 15.6 SELECTING THE SPACE STEP

The question naturally arises as to how many points a solution space should be split into. It should be noted that this has little to do with choosing a nice number such as 100 or choosing a nice spacing between the points such as 1.0 metre. The subdivision is based on the expected solution and the accuracy and resolution that is required. Consider, for example, the situation where groundwater fluctuations near the coast are being modelled from a tidally driven boundary. Nielsen (1990) derived an analytical solution to a simplified set of equations. A particular solution in Figure 15.6 shows the change in pressure head with increasing distance from the coast.

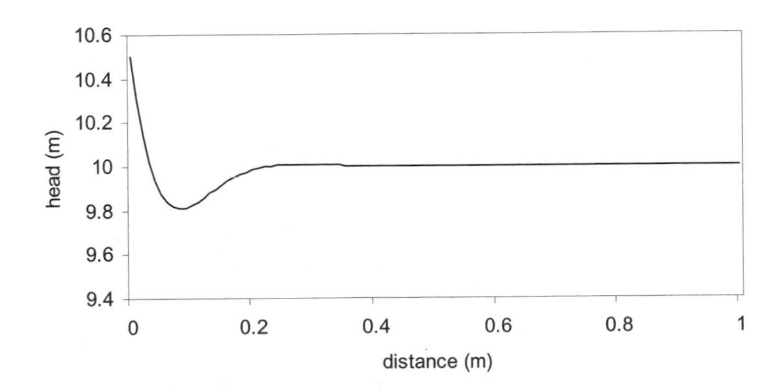


**Figure 15.6** Particular solution to a groundwater elevation problem.

The important point to consider is that any finite difference solution will be represented just by the points that are selected and these will depend critically on the spacing that is chosen. Also the derivatives are calculated based on these discrete points. The situation in Figure 15.7 illustrates what happens if the distance spacing is too large: important features of the system are lost as the spacing simply does not allow the fine changes in the values to be represented in sufficient detail. This would lead to a poor solution.

## 15.7 SELECTING THE TIME STEP

In many ways the programmer has some flexibility in selecting the space step, but its value does have some limitations brought on by the need to model the system accurately. The only other variable in the solution is the time step that is taken in each iterative step towards a solution.

It is not surprising that there are limits to how large this value can be, and those limits are generally to do with the stability of the solution. It is not proposed to go into great detail at this stage on the stability analysis; it should be sufficient to know that there are likely to be limits to $\Delta t$, and that these will be based on stability of the solution and on accuracy.

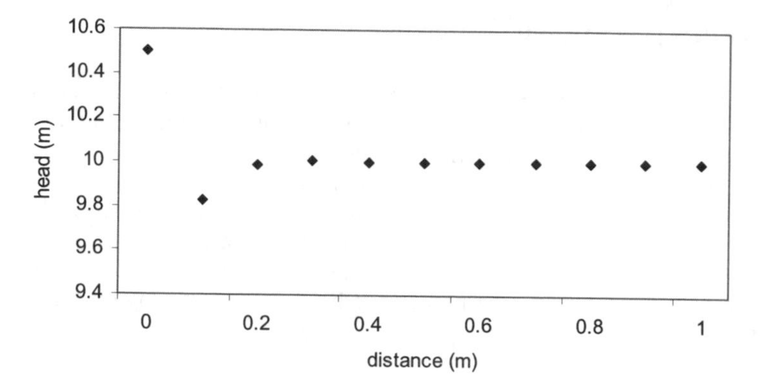


**Figure 15.7** A finite difference representation where the grid spacing is too large resulting in a lack of sufficient detail.

## PROBLEMS

**15.1** Use Equation (15.6) (the one-sided difference approximation) and Equation (15.10) (central difference approximation) forms of the first derivative to determine the accuracy of the estimates for the case where the derivative of the function $f(x)=e^x$ is required at $x=1$. Use space steps of 0.1 and 0.05.

**15.2** For the following data estimate the rate of acceleration ($dv/dt$) of the vehicle at $t = 4$ seconds:

| Time (s) | 0.0 | 2.0 | 4.0 | 6.0 |
|---|---|---|---|---|
| Velocity (m/s) | 0.0 | 10.4 | 25.0 | 41.6 |

Questions 3, 4 and 5 form a set, and work on a quantity $c=c(x,t)$ that is a function of time ($t$) and space ($x$). It could be thought of as a concentration which might vary with time or with location in a river or a pond. To do this consider that data values have been recorded at a number of different locations over a period of time.

**15.3** Write a first-order (one-sided) finite difference approximation for the following partial derivative: $\dfrac{\partial c}{\partial t}$

**15.4** Write a second-order (central difference) finite different approximation for the following partial derivative: $\frac{\partial c}{\partial x}$

**15.5** Write a second-order (central difference) finite difference approximation for the following second-order partial derivative: $\frac{\partial^2 c}{\partial x^2}$

**15.6** Write the finite difference approximation that would apply to an array of elements $f_i$, $i = 1,n$, for the second derivative, as shown in Equation (15.11):

$$f''(x) = \frac{f(x+h) - 2f(x) + f(x-h)}{h^2}$$

**15.7** Consider the waves approaching the marina as shown in Figure 15.1: suggest a rationale for selecting the space step in a finite difference solution to this problem.

**15.8** Laplace's equation can be written: $\frac{\partial^2 T}{\partial x^2} + \frac{\partial^2 T}{\partial y^2} = 0$ where $T$ is a quantity that is varying over a two-dimensional $(x,y)$ solution space. Derive a finite difference approximation for the equation assuming a grid of points has been set out where a general point is given as $T(i,j)$ which has immediate neighbours $T(i+1,j)$, $T(i-1,j)$, $T(i,j+1)$ and $T(i,j-1)$.

## CHAPTER SIXTEEN

# Finite Difference Modelling (Laplace's Hot Plate)

### 16.1 INTRODUCTION

The solution of partial differential equations can be split into a number of quite different approaches depending on the types of equations involved. The general formula for second-order linear PDEs can be written:

$$A\frac{\partial^2 u}{\partial x^2} + B\frac{\partial^2 u}{\partial x \partial y} + C\frac{\partial^2 u}{\partial y^2} + D = 0 \qquad (16.1)$$

where $x$ and $y$ can be spatial coordinates, or one in space and one in time. Furthermore, $u$ will generally be a function of $x$ and $y$, and $A$, $B$, $C$ and $D$ are constants. The type of problem and therefore the type of solution can be determined based on the quantity $B^2 - 4AC$. The possibilities are listed in Table 16.1.

**Table 16.1** Types of PDE and how they are categorised.

| $B^2 - 4AC$ | Category | Examples |
|---|---|---|
| $< 0$ | Elliptic | Laplace equation (steady state in 2D), flow under dams, heated plates, loaded plates. |
| $= 0$ | Parabolic | Heat conduction (time variable in 1D), heated wires, flow of pollutants in rivers, smoke diffusion. |
| $> 0$ | Hyperbolic | Wave equation (time variable in 1D), vibrating wires, water waves. |

Elliptic equations are usually time-independent (steady-state) problems in two spatial dimensions where known boundary conditions lead to a solution in the internal region of interest. Examples include flow under dams and heat conduction on a plate. They are also called boundary value problems.

Parabolic equations are often time-dependent with an additional space dimension. Heat conduction along a wire and the transport equation are examples of parabolic equations, which are also called initial value problems. Parabolic and hyperbolic equations have issues with the stability of the solutions and great care must be taken since the finite difference scheme that is chosen may not lead to a solution.

Hyperbolic equations are similar in some ways to parabolic equations but since the unknown is characterised with a second derivative with respect to time the solutions oscillate; an example is the wave equation.

The solutions of finite difference equations can also be split, logically, into those that approach a steady state and those that do not. Steady-state problems have partial differential equations (PDEs) that do not contain any time derivatives. In this case there is a solution that does not vary in time. The solution procedure may be iterative, in which case the solution is approached from an initial starting value. For example, the concentration of dye in a river where the dye is being released at a constant rate will approach a steady state and the equation and solution method will reflect this.

Transient problems (those that involve changes in time) have time as one of the derivatives in the PDE. For example, the concentration of dye in a river where the release rate is varying in time will also vary in time and the solution technique will have to take account of this. For this type of problem there may be multiple (intermediate) answers and each may be of interest in that they will be the actual solution at a particular time.

## 16.2 ELLIPTIC FINITE DIFFERENCE TECHNIQUES

The solution of elliptic FD equations will be illustrated with the Laplace equation. The equation can be written:

$$\nabla^2 T = 0 \qquad \text{or} \qquad \frac{\partial^2 T}{\partial x^2} + \frac{\partial^2 T}{\partial y^2} = 0 \tag{16.2}$$

Using central differences on the separate terms in the equation gives:

$$\frac{\partial^2 T}{\partial x^2} = \frac{T_{i+1,j} - 2T_{i,j} + T_{i-1,j}}{\Delta x^2} + O(\Delta x^2) \tag{16.3}$$

$$\frac{\partial^2 T}{\partial y^2} = \frac{T_{i,j+1} - 2T_{i,j} + T_{i,j-1}}{\Delta y^2} + O(\Delta y^2) \tag{16.4}$$

Substituting these into the original equation, and ignoring the error terms, gives:

$$\frac{T_{i+1,j} - 2T_{i,j} + T_{i-1,j}}{\Delta x^2} + \frac{T_{i,j+1} - 2T_{i,j} + T_{i,j-1}}{\Delta y^2} \tag{16.5}$$

Collecting terms (and assuming $\Delta x = \Delta y$ ) leads to:

$$T_{i+1,j} + T_{i-1,j} + T_{i,j+1} + T_{i,j-1} - 4T_{i,j} = 0 \tag{16.6}$$

The layout of the points over a two-dimensional grid is illustrated in Figure 16.1.

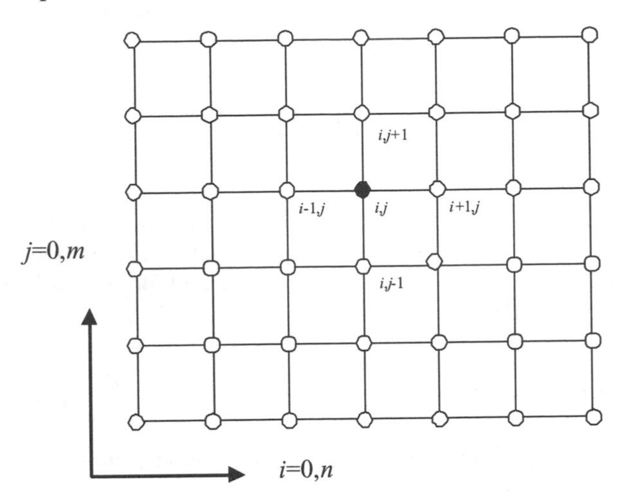


**Figure 16.1** Two-dimensional grid for solution of Laplace equation.

Referring to the grid in Figure 16.1, it is possible to consider points as either internal or external (on the boundary). For the internal points it will always be possible to apply the equation, since each point has the four neighbours, namely $(i+1, j)$, $(i-1, j)$, $(i, j-1)$ and $(i ,j+1)$, necessary to apply the equation at any particular point $(i,j)$. On the boundary this is not the case since there will be one or two missing points. For this reason boundary values must be known *a priori*. This explains why this is called a boundary value problem; the solution is driven by what is happening at the boundary. In this simple example the boundary conditions present no particular problem but in general the specification of boundary conditions can be the most difficult part of a problem.

If the equation is written for each internal point there will be $(m-1)(n-1)$ equations in $(m-1)(n-1)$ unknowns. A simple 12 x 12 grid (10 internal divisions in each direction) would lead to 100 equations in 100 unknowns. A more detailed 100 x 100 grid leads to almost 10,000 equations in 10,000 unknowns.

**Weather Models**

In 1991 global weather models had grown to the extent such that the solution involved 45,000 grid points and 31 standard elevations giving a model with 5,000,000 variables. The solution involved 5,000,000 equations in 5,000,000 unknowns. – E.N. Lorenz (1993)

Elimination methods for solving matrices (e.g. Gaussian elimination, discussed in Chapter 10) are suitable for up to about 100 equations. For larger systems the problems of round-off errors accumulating make such methods unsuitable. It is better to use approximate methods such as the Gauss-Seidel (also called Liebmann's method when used for PDEs; this is discussed in Chapter 11).

If the Gauss-Seidel method is applied to this problem it is necessary to isolate one of the unknowns on the LHS and have the others on the RHS. The equation is re-written:

$$T_{i,j} = \frac{T_{i+1,j} + T_{i-1,j} + T_{i,j+1} + T_{i,j-1}}{4} \tag{16.7}$$

and then solved iteratively. As with the regular Gauss-Seidel method the procedure continues until the system converges. It is instructive at this point to see exactly what the equation is saying: at the final solution each point will be the average of its four immediate neighbours.

## 16.3 WORKED EXAMPLE

A plate is being heated on four sides at different temperatures as shown in Figure 16.2.

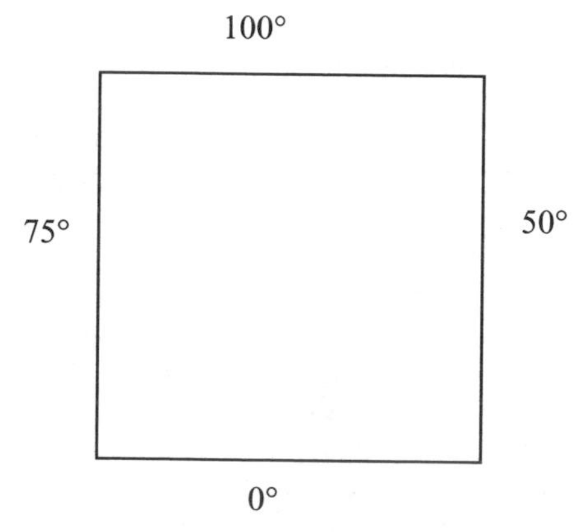


**Figure 16.2** Heated plate example.

To commence the solution the plate is divided into a number of cells. In this case the decision is taken to divide the area into four strips in each direction (giving three internal points in each direction). The solution grid is shown in Figure 16.3 together with the starting temperature guesses. Note that the initial guesses are required for the unknown interior points. For want of a better guess, 0° is used for all points. Although not a particularly good guess in many respects, the decision is based partly on wanting to illustrate that the method will converge, even from a poor start point. The starting values for all points are shown in Table 16.2. Note that corner points are not required and are ignored in the solution.

**Table 16.2** Starting values for the hot plate example.

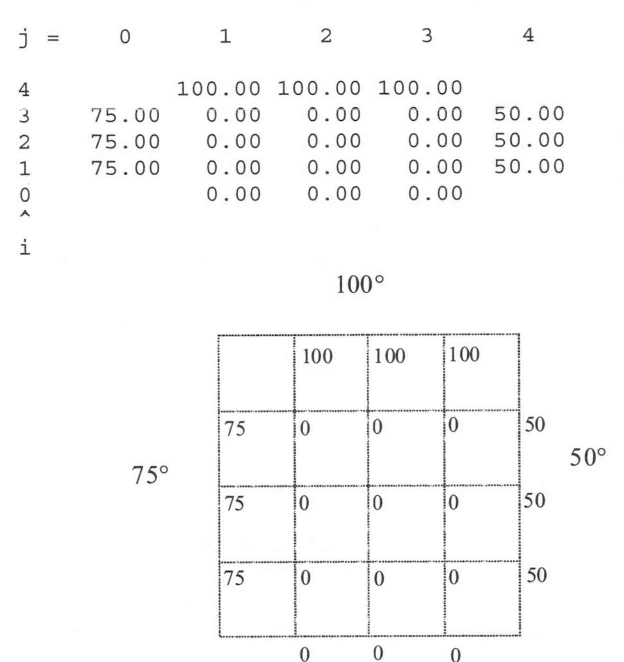

| i \ j = | 0 | 1 | 2 | 3 | 4 |
|---|---|---|---|---|---|
| 4 | | 100.00 | 100.00 | 100.00 | |
| 3 | 75.00 | 0.00 | 0.00 | 0.00 | 50.00 |
| 2 | 75.00 | 0.00 | 0.00 | 0.00 | 50.00 |
| 1 | 75.00 | 0.00 | 0.00 | 0.00 | 50.00 |
| 0 | | 0.00 | 0.00 | 0.00 | |



**Figure 16.3** Heated plate divided into a finite number of points.

The equation is then applied point by point over the grid and the values gradually change until the results have converged to a final solution as shown in Table 16.3. The exact order of proceeding through the grid is unimportant. It will, of course, affect the intermediate values that are calculated, but the final solution is independent of the solution procedure. As a matter of interest, the final results came after 29 iterations.

**Table 16.3** Final values for the hot plate example.

| i \ j = | 0 | 1 | 2 | 3 | 4 |
|---|---|---|---|---|---|
| 4 | | 100.00 | 100.00 | 100.00 | |
| 3 | 75.00 | 78.52 | 76.03 | 69.59 | 50.00 |
| 2 | 75.00 | 63.09 | 56.14 | 52.37 | 50.00 |
| 1 | 75.00 | 42.80 | 33.18 | 33.87 | 50.00 |
| 0 | | 0.00 | 0.00 | 0.00 | |

It should be noted that the solution here is steady-state or time-independent. It is not possible to look at the intermediate steps on the way to the solution and try and see how the heat is flowing through the plate. The intermediate steps have no meaning by themselves. They are simply incomplete steps on the way to a solution.

## 16.4 CODING LIEBMANN'S METHOD IN FORTRAN

The basic equation that must be solved can be written:

$$T_{i,j} = \frac{T_{i+1,j} + T_{i-1,j} + T_{i,j+1} + T_{i,j-1}}{4} \qquad (16.8)$$

If the elements in the $x$ direction start at 1 and finish at $n$, and the elements in the $y$ direction start at 1 and finish at $m$, then the core of a solution can be written in Fortran as shown in Figure 16.4.

```
do i = 2,n-1
 do j = 2,m-1
  t(i,j)=(t(i+1,j)+t(i-1,j)+t(i,j+1)+t(i,j-1))/4.0
 end do
end do
```

**Figure 16.4** Fortran code segment to implement Liebmann's method.

The reason that the loops start at 2 and end one element from the end ($n$–1 and $m$–1) is that the end points are boundaries and must be known *a priori*. The problem with the code as written is that it updates the points but does not allow any estimate of whether the values are changing or whether the solution has converged. This will generally be necessary. For this the old value must be stored temporarily and then compared to the new estimate. For each sweep through the grid only the largest change (or error) needs to be considered. A better section of code, therefore, is shown in Figure 16.5.

```
maxchange = 0.0          ! initialise the maximum change to zero
do i = 2,n-1
 do j = 2,m-1
  old = t(i,j)           ! store the previous value

  t(i,j)=(t(i+1,j)+t(i-1,j)+t(i,j+1)+t(i,j-1))/4.0

  change = abs(t(i,j)-old)
  if(change > maxchange) then     ! store value if it is larger
   maxchange = change
  endif
 end do
end do
```

**Figure 16.5** Fortran code segment to implement Liebmann's method checking for convergence.

The code sweeps through the grid once and at the end has the maximum change that occurred at any of the internal points that were updated. The whole section of code could be run multiple times, but this would not allow the error to be tracked. The optimum approach would be to enclose the section of code in a do-while loop as shown Figure 16.6.

Note that it is necessary to set a parameter to an arbitrarily high value so that the do-while loop starts for the first time.

```
maxchange = 10.0
do while(maxchange >= 0.01)

  !(insert update code segment)

end do
```

**Figure 16.6** Fortran code segment to implement convergence into Liebmann's method.

## PROBLEMS

**16.1** Determine what type of equation the transport equation is: elliptic, parabolic or hyperbolic. Does it fit the general description given in the chapter?

**16.2** Solve the heated plate problem using hand calculations to implement Liebmann's Method. Assume the plate is 1.0 metre square and divide it into a small but convenient number of grids. The boundary conditions are as follows:

top edge: 100° C
left: 75° C
right: 50° C
bottom: 0° C

Do sufficient iterations to see that the solution is developing as expected.

**16.3** Now set up a solution for the heated plate from Problem 16.2 on an Excel spreadsheet. To begin with, try implementing the solution on the same grid as that shown in this chapter. An issue is likely to arise in the implementation of this solution that is worth dealing with here. Since each grid point is determined by the solution at adjacent grid points and those points in turn are based on the original point there arises what Excel calls a "circular reference". Normally this might indicate that an error had been made in the cell formulae but in this case it is exactly what is required. To overcome the error message (and more importantly to allow the solution to proceed) go to Menu–Tools–Calculation and turn on Iteration by ticking the box.

**16.4** Extend the solution of Problem 16.3 to a larger grid – for example, $10 \times 10$. An issue that arises is the work involved in resetting the worksheet to solve the larger problem. This is perhaps satisfactory for relatively small grids but consider the effort and difficulty involved in moving to a $1000 \times 1000$ grid.

**16.5** Write a Fortran program to solve the same heated plate problem as described in Problem 16.2. Reproduce the results shown in the chapter (realising that there may be some variation depending on the exact stopping criteria used) and then extend the problem by solving for a $10 \times 10$ grid. In this case the changes should be relatively easy to implement, particularly if the program has been written making use of variables to describe the size of the grid and to implement the solution.

The only issue will be reporting the output on the screen. Attempt to write the output to a file and then import it into Excel to plot the final results.

**16.6** How important is the initial guess likely to be in determining the final solution? Should there be a strategy for determining a better one than simply setting the whole grid to the same arbitrary value?

**16.7** What is the order of accuracy of the Liebmann's method?

**16.8** Would the solution method be suitable if one of the boundaries was varying in time, say, being heated by the sun and so having a sinusoidal pattern with a period of 24 hours? Why?

CHAPTER SEVENTEEN

# Finite Difference Modelling (Solution of Pure Convection Equation)

## 17.1 INTRODUCTION

The one-dimensional transport equation can be written:

$$\frac{\partial c}{\partial t} + u\frac{\partial c}{\partial x} = K\frac{\partial^2 c}{\partial x^2} \tag{17.1}$$

where $u$ is the convective velocity which is considered constant at each step in the solution, $c$ is the concentration of the material, $x$ is the axis (perhaps running down the river), and $K$ is the diffusion/dispersion coefficient. In some cases relevant to engineering practice a material will be in an environment where there is no diffusive transport – for example, a sampling drogue buoy in a long river. In this case the equation reduces to:

$$\frac{\partial c}{\partial t} + u\frac{\partial c}{\partial x} = 0 \tag{17.2}$$

where $c$ is the concentration of the material or the temperature of the substance, $x$ is the axis, and $u$ is the velocity of the fluid in which the material is being transported. To solve this equation using the finite difference method two steps are undertaken:

1. the solution space is digitised into a number of discrete points rather than a continuum; and
2. the equation is approximated using standard approximations derived on the basis of a Taylor expansion of the function.

The way this is carried out is set out in the next two sections before a full description of the final solution.

## 17.2 SELECTING THE SPACE STEP SIZE

The digitisation of the solution space for this simple example is achieved by dividing the solution space into a number of equally spaced points. For example, if dye is diffusing along a pipe that is 100 metres long it may be convenient to divide the pipe into 101 equally spaced points (a point on each end means that for 100 elements there are 101 points). This is shown in Figure 17.1, although the discussion in Chapter 15 on grid sizes is particularly relevant.

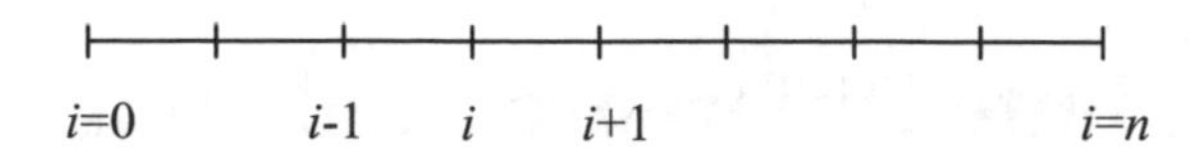


**Figure 17.1** Discrete solution space for finite difference model. The points are numbered from $i=0$ to $i=n$ and at any point along the line there are the general points $i-1$, $i$ and $i+1$ that form neighbours.

Note that one general approach is to number the nodes sequentially from 0 (or 1) to $n$ where $n$ depends on the total number used in the solution, and that at any point it is possible to label three neighbouring points as $i-1$, $i$ and $i+1$. This will become important in the development of the solution.

## 17.3 DERIVING THE FINITE DIFFERENCE APPROXIMATION

Using the results determined from the Taylor expansion of the function at a point, estimates for the two derivatives in Equation (17.2) can be written as forward difference approximations:

$$\frac{\partial c}{\partial t} = \frac{c_i^{t+1} - c_i^t}{\Delta t} + O(\Delta t) \tag{17.3}$$

and

$$\frac{\partial c}{\partial x} = \frac{c_i^t - c_{i-1}^t}{\Delta x} + O(\Delta x) \tag{17.4}$$

Adding the two parts of the approximation and re-arranging the equation gives:

$$c_i^{t+1} = c_i^t + \frac{u\Delta t}{\Delta x}(c_{i-1}^t - c_i^t) \tag{17.5}$$

where it is evident that the concentration at a point in time is equal to the concentration one time step earlier plus a proportion of the change in concentration over a spatial step. Therefore, if the information is available at the old time level the new time level can be determined in terms of the old information and known values of $u$, $\Delta t$ and $\Delta x$. This new time level then becomes the old time level and the process is repeated as necessary. The scheme is illustrated in Figure 17.2.

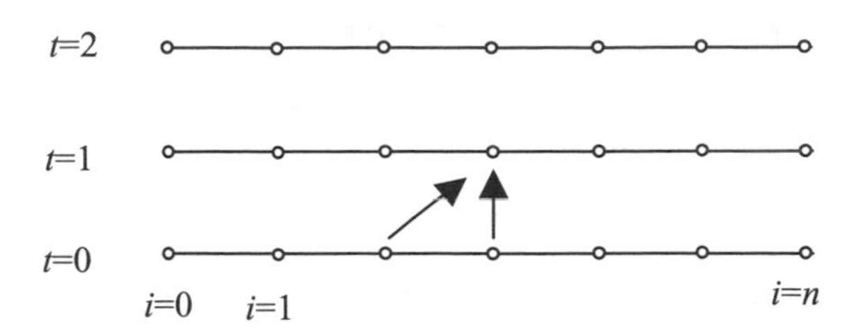


**Figure 17.2** Solution procedure for heat equation. The arrows indicate the points that contribute to a new typical point at the new time level.

## 17.4 WORKED EXAMPLE IN EXCEL

Take as an example the situation where the convection equation is being solved where the velocity of the flow is $u$=1.0 m/s, a time step has been selected, $\Delta t$=1.0 s and a space step has also been specified, $\Delta x$=1.0 m. It is required to determine the behaviour over the first 11 seconds. The initial concentration is represented as a triangular spike with a peak of 1.0 at $x$=3.0 m.

In Excel a grid of points is set out to represent time (going down Column A) and space (going along Row 10). This is shown in Figure 17.3 where three names (*dt*, *dx*, *u*) have been defined (Insert – Name – Create) and the $x$ and $t$ values are defined in terms of these variables. For example, each $x$ distance is defined as the previous distance plus *dx*. This makes the sheet efficient and allows for easy changes in time and distance steps.

Microsoft Excel - Convection Model.xls

File Edit View Insert Format Tools Data Window Help Acrobat

NORMDIST = =B10+dx

| | A | B | C | D | E | F | G | H | I | J | K |
|---|---|---|---|---|---|---|---|---|---|---|---|
| 1 | Convection Model | | | | | | | | | | |
| 2 | Walker - March 2005 | | | | | | | | | | |
| 3 | | | | | | | | | | | |
| 4 | dt | 1 | s | | | | | | | | |
| 5 | dx | 1 | m | | | | | | | | |
| 6 | u | 1 | m/s | | | | | | | | |
| 7 | | | | | | | | | | | |
| 8 | | | | | | | | | | | |
| 9 | | | | | | | | | | | |
| 10 | t | 0 | =B10+dx | 2 | 3 | 4 | 5 | 6 | 7 | 8 | 9 |
| 11 | 0 | | | | | | | | | | |
| 12 | 1 | | | | | | | | | | |
| 13 | 2 | | | | | | | | | | |
| 14 | 3 | | | | | | | | | | |
| 15 | 4 | | | | | | | | | | |
| 16 | 5 | | | | | | | | | | |
| 17 | 6 | | | | | | | | | | |

$$e_i^{t+1} = e_i^t + \frac{u\Delta t}{\Delta x}(e_{i-1}^t - e_i^t)$$

**Figure 17.3** First phase of solution in Excel. Note use of *dx* in setting up distances in the *x* direction (along Row 10).

In the next phase, the boundary and initial conditions are input. The state at $t = 0$ is a single spike at $x = 3.0$ with values of 0.5 at $x = 2.0$ and $x = 4.0$. Assuming zero concentration upstream means that for all time the concentration at $x = 0.0$ m is 0.0. This is shown in Figure 17.4.

Microsoft Excel - Convection Model.xls

File Edit View Insert Format Tools Data Window Help Acrobat

J6 =

| | A | B | C | D | E | F | G | H | I | J | K |
|---|---|---|---|---|---|---|---|---|---|---|---|
| 1 | Convection Model | | | | | | | | | | |
| 2 | Walker - March 2005 | | | | | | | | | | |
| 3 | | | | | | | | | | | |
| 4 | dt | 1 | s | | | | | | | | |
| 5 | dx | 1 | m | | | | | | | | |
| 6 | u | 1 | m/s | | | | | | | | |
| 7 | | | | | | | | | | | |
| 8 | | | | | | | | | | | |
| 9 | | | | | | | | | | | |
| 10 | t | 0 | 1 | 2 | 3 | 4 | 5 | 6 | 7 | 8 | 9 |
| 11 | 0 | 0 | 0 | 0.5 | 1 | 0.5 | 0 | 0 | 0 | 0 | 0 |
| 12 | 1 | 0 | | | | | | | | | |
| 13 | 2 | 0 | | | | | | | | | |
| 14 | 3 | 0 | | | | | | | | | |
| 15 | 4 | 0 | | | | | | | | | |
| 16 | 5 | 0 | | | | | | | | | |
| 17 | 6 | 0 | | | | | | | | | |

$$e_i^{t+1} = e_i^t + \frac{u\Delta t}{\Delta x}(e_{i-1}^t - e_i^t)$$

**Figure 17.4** Second phase of solution in Excel. The values at $t = 0$ have been input as have the values for $x = 0.0$. These are done by hand, although most have been copied automatically by dragging.

In the third phase the first finite difference approximation is entered in the top left-hand cell of the solution space. With the way the grid has been numbered this is at $t = 1$ s and $x = 1.0$ m. The formula uses the values at the last time step ($t = 0$ s) and the two neighbours ($x = 0$, $x = 1$) as specified in the finite difference formula. This is shown in Figure 17.5.

Microsoft Excel - Convection Model.xls

File Edit View Insert Format Tools Data Window Help Acrobat

NORMDIST = =C11+u*dt/dx*(B11-C11)

| | A | B | C | D | E | F | G | H | I | J | K |
|---|---|---|---|---|---|---|---|---|---|---|---|
| 1 | Convection Model | | | | | | | | | | |
| 2 | Walker - March 2005 | | | | | | | | | | |
| 3 | | | | | | | | | | | |
| 4 | dt | 1 | s | | | | | | | | |
| 5 | dx | 1 | m | | | | | | | | |
| 6 | u | 1 | m/s | | | | | | | | |
| 7 | | | | | | | | | | | |
| 8 | | | | | | | | | | | |
| 9 | | | | | | | | | | | |
| 10 | t | 0 | 1 | 2 | 3 | 4 | 5 | 6 | 7 | 8 | 9 |
| 11 | 0 | 0 | 0 | 0.5 | 1 | 0.5 | 0 | 0 | 0 | 0 | 0 |
| 12 | 1 | 0 | =C11+u*dt/dx*(B11-C11) | | | | | | | | |
| 13 | 2 | 0 | | | | | | | | | |
| 14 | 3 | 0 | | | | | | | | | |
| 15 | 4 | 0 | | | | | | | | | |
| 16 | 5 | 0 | | | | | | | | | |
| 17 | 6 | 0 | | | | | | | | | |

$$e_i^{t+1} = e_i^t + \frac{u\Delta t}{\Delta x}(e_{i-1}^t - e_i^t)$$

**Figure 17.5** Third phase of solution in Excel. The first formula is added. Note that the use of the defined names allows an equation that can be read much more easily than one that referred to cells B4, B5 and B6.

Once entered, the formula is copied to the right and then filled down. This is an excellent example of the power of relative addressing used by Excel which is exactly aligned with the relative addressing used in the finite difference method. The result is shown in Figure 17.6.

Microsoft Excel - Convection Model.xls

| | A | B | C | D | E | F | G | H | I | J | K | |
|---|---|---|---|---|---|---|---|---|---|---|---|---|
| 1 | Convection Model | | | | | | | | | | | |
| 2 | Walker - March 2005 | | | | | | | | | | | |
| 3 | | | | | | | | | | | | |
| 4 | dt | 1 | s | | | | | | | | | |
| 5 | dx | 1 | m | | | | | | | | | |
| 6 | u | 1 | m/s | | | | | | | | | |
| 7 | | | | | | | | | | | | |
| 8 | | | | | | | | | | | | |
| 9 | | | | | | | | | | | | |
| 10 | t | 0 | 1 | 2 | 3 | 4 | 5 | 6 | 7 | 8 | 9 | |
| 11 | 0 | 0.00 | 0.00 | 0.50 | 1.00 | 0.50 | 0.00 | 0.00 | 0.00 | 0.00 | 0.00 | 0 |
| 12 | 1 | 0.00 | 0.00 | 0.00 | 0.50 | 1.00 | 0.50 | 0.00 | 0.00 | 0.00 | 0.00 | 0 |
| 13 | 2 | 0.00 | 0.00 | 0.00 | 0.00 | 0.50 | 1.00 | 0.50 | 0.00 | 0.00 | 0.00 | 0 |
| 14 | 3 | 0.00 | 0.00 | 0.00 | 0.00 | 0.00 | 0.50 | 1.00 | 0.50 | 0.00 | 0.00 | 0 |
| 15 | 4 | 0.00 | 0.00 | 0.00 | 0.00 | 0.00 | 0.00 | 0.50 | 1.00 | 0.50 | 0.00 | 0 |
| 16 | 5 | 0.00 | 0.00 | 0.00 | 0.00 | 0.00 | 0.00 | 0.00 | 0.50 | 1.00 | 0.50 | 0 |
| 17 | 6 | 0.00 | 0.00 | 0.00 | 0.00 | 0.00 | 0.00 | 0.00 | 0.00 | 0.50 | 1.00 | 0 |

$$e_i^{t+1} = e_i^t + \frac{u\Delta t}{\Delta x}(e_{i-1}^t - e_i^t)$$

**Figure 17.6** Filled-in solution to convection equation. Note that as time goes by (moving down the rows) the spike of concentration moves gradually in the downstream direction.

The results are plotted in Figure 17.7. Notice that the spike of pollution is progressing at 1.0 m/s as expected along in the x direction. Note also that the shape of the spike is unchanged. It is worth noting that this is not realistic behaviour for a pollutant, but it serves as a useful test of the numerical model.

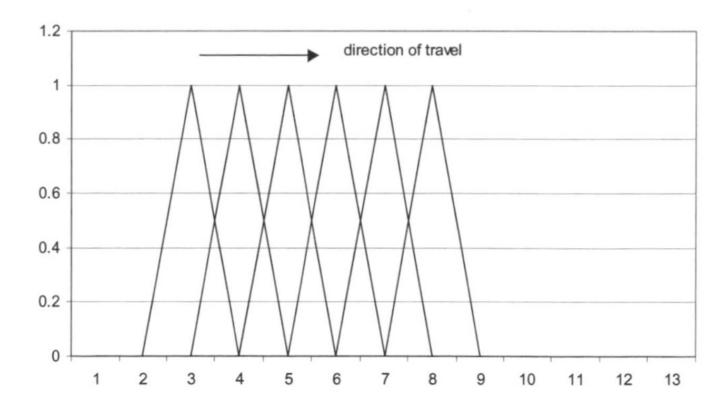


**Figure 17.7** Pure convection with $u$=1.0 m/s. The peak moves along the river unchanged.

## 17.5 STABILITY, ACCURACY AND OTHER ISSUES

The solution shown in Figure 17.7 demonstrates that the method can work very well, but there are issues that must be investigated. These are highlighted most easily by simply changing the velocity of the flow in the problem. An equally valid change would be to reduce the time step and leave the velocity the same. If the velocity is reduced to 0.75 m/s the solution changes. The results are shown in Figure 17.8 and plotted in Figure 17.9.

Microsoft Excel - Convection Model.xls

File Edit View Insert Format Tools Data Window Help Acrobat

B7 =

| | A | B | C | D | E | F | G | H | I | J | K | |
|---|---|---|---|---|---|---|---|---|---|---|---|---|
| 1 | Convection Model | | | | | | | | | | | |
| 2 | Walker - March 2005 | | | | | | | | | | | |
| 3 | | | | | | | | | | | | |
| 4 | dt | 1 | s | | $e_i^{t+1} = e_i^t + \frac{u\Delta t}{\Delta x}(e_{i-1}^t - e_i^t)$ | | | | | | | |
| 5 | dx | 1 | m | | | | | | | | | |
| 6 | u | 0.75 | m/s | | | | | | | | | |
| 7 | | | | | | | | | | | | |
| 8 | | | | | | | | | | | | |
| 9 | | | | | | | | | | | | |
| 10 | t | 0 | 1 | 2 | 3 | 4 | 5 | 6 | 7 | 8 | 9 | |
| 11 | 0 | 0.00 | 0.00 | 0.50 | 1.00 | 0.50 | 0.00 | 0.00 | 0.00 | 0.00 | 0.00 | 0 |
| 12 | 1 | 0.00 | 0.00 | 0.13 | 0.63 | 0.88 | 0.38 | 0.00 | 0.00 | 0.00 | 0.00 | 0 |
| 13 | 2 | 0.00 | 0.00 | 0.03 | 0.25 | 0.69 | 0.75 | 0.28 | 0.00 | 0.00 | 0.00 | 0 |
| 14 | 3 | 0.00 | 0.00 | 0.01 | 0.09 | 0.36 | 0.70 | 0.63 | 0.21 | 0.00 | 0.00 | 0 |
| 15 | 4 | 0.00 | 0.00 | 0.00 | 0.03 | 0.15 | 0.45 | 0.69 | 0.53 | 0.16 | 0.00 | 0 |
| 16 | 5 | 0.00 | 0.00 | 0.00 | 0.01 | 0.06 | 0.23 | 0.51 | 0.65 | 0.44 | 0.12 | 0 |
| 17 | 6 | 0.00 | 0.00 | 0.00 | 0.00 | 0.02 | 0.10 | 0.30 | 0.54 | 0.59 | 0.36 | 0 |

**Figure 17.8** Output from finite difference model where $u = 0.75$ m/s.

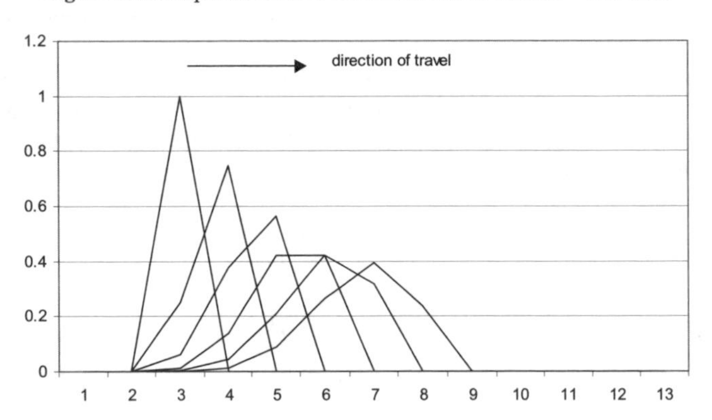


**Figure 17.9** Results of the convection model with $u$=1.0 m/s.

The pollutant is moving and spreading as might be expected but in this case the diffusion (spreading) is due to the low order of accuracy of the finite difference approximation used. If one looks at the peak of the spike it has not travelled as far (as would be expected due to the reduced velocity) but more noticeably the peak of the spike has been reduced and the whole spike has flattened. Although diffusion is a real enough phenomenon, in this case it is purely numerical. To overcome this problem it is advisable to work with a higher-order scheme, and to keep the time step as large as possible.

There is, however, a limit to how large the time step can be. If a new velocity is tried – for example, 1.5 m/s – there is quite a different behaviour. The results as developed on the Excel worksheet are shown in Figure 17.10.

Microsoft Excel - Convection Model.xls

| | A | B | C | D | E | F | G | H | I | J | K | |
|---|---|---|---|---|---|---|---|---|---|---|---|---|
| 1 | Convection Model | | | | | | | | | | | |
| 2 | Walker - March 2005 | | | | | | | | | | | |
| 3 | | | | | | | | | | | | |
| 4 | dt | 1 | s | | | | | | | | | |
| 5 | dx | 1 | m | | | | | | | | | |
| 6 | u | 1.5 | m/s | | | | | | | | | |
| 7 | | | | | | | | | | | | |
| 8 | | | | | | | | | | | | |
| 9 | | | | | | | | | | | | |
| 10 | t | 0 | 1 | 2 | 3 | 4 | 5 | 6 | 7 | 8 | 9 | |
| 11 | 0 | 0.00 | 0.00 | 0.50 | 1.00 | 0.50 | 0.00 | 0.00 | 0.00 | 0.00 | 0.00 | 0 |
| 12 | 1 | 0.00 | 0.00 | -0.25 | 0.25 | 1.25 | 0.75 | 0.00 | 0.00 | 0.00 | 0.00 | 0 |
| 13 | 2 | 0.00 | 0.00 | 0.13 | -0.50 | -0.25 | 1.50 | 1.13 | 0.00 | 0.00 | 0.00 | 0 |
| 14 | 3 | 0.00 | 0.00 | -0.06 | 0.44 | -0.63 | -1.13 | 1.69 | 1.69 | 0.00 | 0.00 | 0 |
| 15 | 4 | 0.00 | 0.00 | 0.03 | -0.31 | 0.97 | -0.38 | -2.53 | 1.69 | 2.53 | 0.00 | 0 |
| 16 | 5 | 0.00 | 0.00 | -0.02 | 0.20 | -0.95 | 1.64 | 0.70 | -4.64 | 1.27 | 3.80 | 0 |
| 17 | 6 | 0.00 | 0.00 | 0.01 | -0.13 | 0.78 | -2.25 | 2.11 | 3.38 | -7.59 | 0.00 | 5 |

$$e_i^{t+1} = e_i^t + \frac{u\Delta t}{\Delta x}(e_{i-1}^t - e_i^t)$$

**Figure 17.10** Output from convection equation where $u = 1.5$ m/s. Note that the solution becomes unstable very quickly!

The results are also plotted in Figure 17.11. This highlights another potential problem: that of an unstable solution. For an explanation related to the physical reality, the instability can be thought of as arising from the fact that at the given velocity it is possible for the information in one grid to travel more than one grid spacing in a given time step. If this happens instability can arise. There is a quantity defined as $u\Delta t/\Delta x$ which is called the Courant number, *C*. Stability analysis can be carried out on the finite difference scheme which will place an upper limit on the Courant number. It is generally necessary that $C \leq 1$, and in many cases $C \leq 0.5$, for a stable solution.

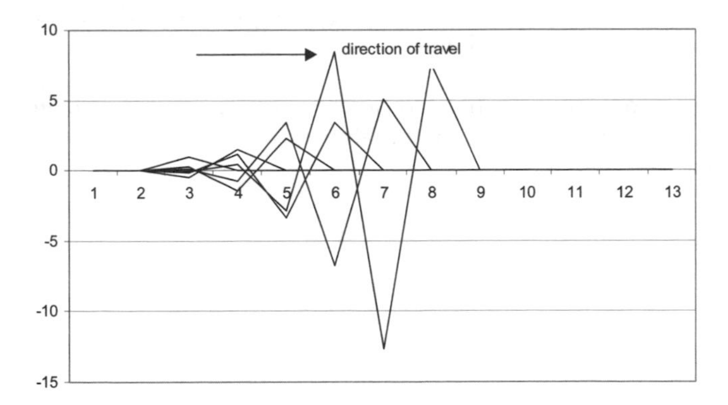


**Figure 17.11** Solutions that look like this highlight one of the major problems with finite difference methods. They are not all stable.

The programmer must be aware, therefore, that in selecting the space step and time step there will be issues of solution stability. Given that river velocities are generally beyond the control of the modeller, and that the space step is usually based on other considerations, the most usual stability assessment is in terms of the time step that is selected. It may seem apparent from the workings above that if the Courant number could be made to equal 1.0 then the scheme would work perfectly. Unfortunately this is not always possible since the velocity $u$ is not generally the same throughout the scheme and the trouble of generating a grid for a unique velocity is also generally not feasible. Generally the Courant number is kept as large as possible since this leads to a quicker solution and reduces numerical diffusion.

## PROBLEMS

**17.1** Set up the solution to the one-dimensional convection equation as set out in this topic in Excel. Attempt to reproduce the results and, in doing so, verify that the scheme is stable for Courant numbers less than or equal to 1.0.

**17.2** Modify the solution derived for Problem 17.1 to investigate the effect of abrupt changes in the starting concentration. For example, in the example given in the chapter a triangular shape was used and this has a very distinct peak. See if starting with a more rounded shape leads to better solutions with less numerical diffusion, especially for situations where the Courant number is less than 1.0.

**17.3** Modify the solution derived for Problem 17.1 to investigate the effect of reducing the time step when the Courant number is already less than 1.0. It should be possible to show that there are few benefits and that the numerical diffusion increases, leading to a poorer solution. To do this calculate and compare the solutions after (say) 10 seconds using time steps of 1.0, 0.5, 0.2, 0.1, 0.01.

**17.4** Why is it necessary to know the upstream and downstream boundary conditions for this solution technique?

**17.5** Of the problems with finite difference method highlighted in this chapter, numerical diffusion and solution instability, which is the more likely to have serious consequences?

CHAPTER EIGHTEEN

# Finite Difference Modelling (Solution of Pure Diffusion Equation)

## 18.1 INTRODUCTION

The one-dimensional transport equation can be written:

$$\frac{\partial c}{\partial t}+u\frac{\partial c}{\partial x}=K\frac{\partial^2 c}{\partial x^2} \qquad (18.1)$$

where $u$ is the convective velocity, $C$ is the concentration of the material, $x$ is the axis (perhaps running down the river), and $K$ is the diffusion/dispersion coefficient. In some cases relevant to engineering practice a material will be in an environment where there is no convective transport – for example, pollutant released into a long and narrow pipe with zero flow, or where heat may be diffusing along a metal wire driven purely by temperature differences. In this case the equation reduces to:

$$\frac{\partial c}{\partial t}=K\frac{\partial^2 c}{\partial x^2} \qquad (18.2)$$

where $C$ is the concentration of the material or the temperature of the substance, $x$ is the axis, and $K$ is the diffusion/dispersion coefficient. To solve this equation using the finite difference method two steps are undertaken:

1. the solution space is digitised into a number of discrete points rather than a continuum; and
2. the equation is approximated using standard approximations derived on the basis of a Taylor expansion of the function.

The way this is carried out is set out in the next two sections before a full description of the final solution.

## 18.2 SELECTING THE SPACE STEP SIZE

The discretisation of the solution space is achieved simply by dividing the area (location rather than a strict area since it may be a length of wire or a volume of fluid) into a number of equally spaced points. For example, if dye was diffusing along a pipe that was 100 metres long it may be convenient to divide the pipe into 101 equally spaced points (a point on each end means that for 100 elements there are 101 points). This is shown in Figure 18.1, although the discussion in Chapter 15 on grid sizes is particularly relevant.

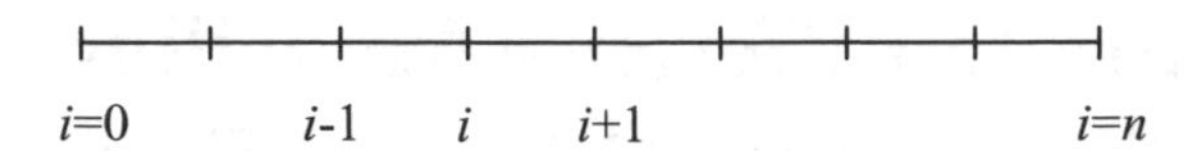


**Figure 18.1** Discrete solution space for finite difference model. The points are numbered from $i$=0 to $i$=$n$ and at any point along the line there are the general points $i$–1, $i$ and $i$+1 that are neighbours.

Note that one general approach is to number the nodes sequentially from 0 to $n$ where the $n$ depends on the total number used in the solution, and that at any point it is possible to label three neighbouring points as $i$–1, $i$ and $i$+1. This will become important in the development of the solution. The question of how many points the solution space should be divided into is an important one and will be dealt with in a later section. For the moment, assume that it is possible to make this choice.

## 18.3 DERIVING THE FINITE DIFFERENCE APPROXIMATION

If the pure diffusion equation is used to solve temperature along a metal bar it may be convenient to write the equation in terms of a temperature, $T$. The heat equation can be written:

$$\frac{\partial T}{\partial t} = k\frac{\partial^2 T}{\partial x^2} \tag{18.3}$$

where $T$ is the temperature at a point on a metal bar, $t$ is the time and $k$ is the conductivity constant. Using the results determined from the Taylor expansion of the function at a point, estimates for the two derivatives can be written as:

$$\frac{\partial T}{\partial t} = \frac{T_i^{t+1} - T_i^t}{\Delta t} + O(\Delta t) \tag{18.4}$$

and

$$\frac{\partial^2 T}{\partial x^2} = \frac{T_{i+1}^t - 2T_i^t + T_{i-1}^t}{(\Delta x)^2} + O(\Delta x^2) \tag{18.5}$$

Note that the time derivative is a forward difference approximation of first-order accuracy and that the space derivative is a second-order accurate second derivative that is centred. Together these form what is referred to as a FTCS (forward time, centre space) scheme and should be overall approximately first-order accurate since this is the worst of the various orders of errors in any of the terms in the equation. Notice that in writing the finite difference approximation the subscript ($i$) refers to the space step and the superscript ($t$) refers to time. This is a common notation and should be followed. Substituting these two approximations into the heat equation gives:

$$\frac{T_i^{t+1} - T_i^t}{\Delta t} = \frac{T_{i+1}^t - 2T_i^t + T_{i-1}^t}{\Delta x^2} \qquad (18.6)$$

and re-arranging to get the unknowns (those at the $t$+1 time level) on the left, and knowns (those at the $t$ time level) on the right gives:

$$T_i^{t+1} = T_i^t + \lambda(T_{i+1}^t - 2T_i^t + T_{i-1}^t) \qquad (18.7)$$

where $\lambda = k\,\Delta t/(\Delta x)^2$. The equation gives an explicit formula on how to calculate the temperature at a particular point, $i$, at the new time level, $t$+1, if the values at neighbouring points ($i$–1, $i$, $i$+1) are known at the old time level, $t$. In physical terms it is evident that the temperature at a point and at a particular time is equal to the temperature at that point at the previous time step plus a part of the temperature increase between that point and the neighbouring points.

This type of problem is often referred to as an initial value problem. In other words the solution is governed by what the initial conditions are. Change these and the solution will change too. The general solution technique is to specify everything at some initial time level, which for the purposes of the problem will be referred to as the $t$=0 time level, and then to step forward in time to the $t$=1 time level by applying the equation point by point for all the $i$ points until the entire length of the solution space has been done. Then this time level effectively becomes the old time level and the procedure is repeated until all are calculated at the $t$=2 level and so on. In this notation any quantity at the $t$ time level will be known and everything at the $t$+1 time level will be unknown. The way the solution proceeds is illustrated in Figure 18.2. Since everything on the RHS is known and the only unknown is on the LHS the equation can be solved explicitly for all values at time $t$+1.

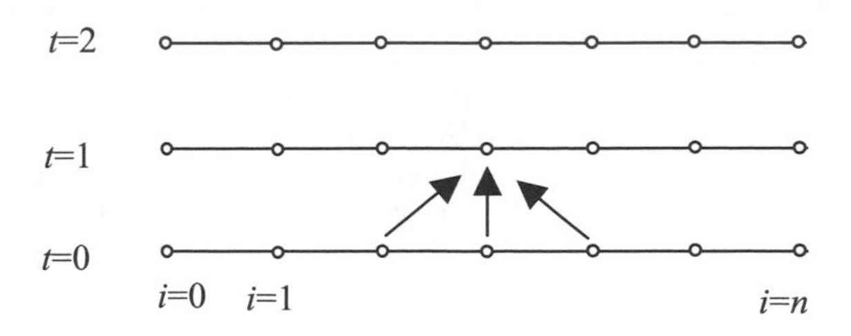


**Figure 18.2** Solution procedure for heat equation. The arrows indicate the points that contribute to a new typical point at the new time level.

## 18.4 BOUNDARY CONDITIONS

The explicit scheme to update the points to the new time level set out in Equation (18.7) appears to provide an easy solution to the problem. But what happens near the solution boundaries? The situation is illustrated in Figure 18.3 where the situation at $i$=1 is shown to be quite workable, but a problem appears when $i$=$n$. Note at this point that there is no neighbour at $i$=$n$+1 and there is a problem applying Equation (18.7). Points at $i$=0 are similarly affected and together form a common problem: how to handle solution boundaries.

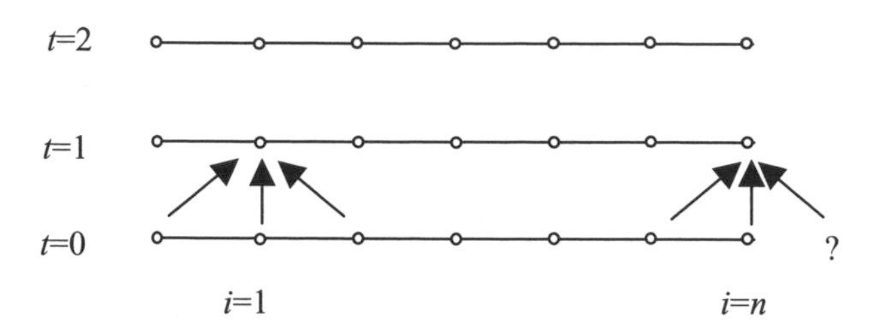


**Figure 18.3** Solution procedure for heat equation where a boundary causes problems. Note that at $i$=$n$ there is no neighbour to use at $i$=$n$+1.

In this case, and many like it, it is necessary to be able to specify the boundary values throughout time without recourse to calculating them using Equation (18.7) as part of the solution. There are basically three types of boundary conditions that can be used in this type of solution:

1. a Dirichlet boundary, which involves a fixed value;
2. a Neumann boundary, which makes use of a specified derivative;
3. a Robin boundary, which is a combination of the other two.

*Dirichlet boundary*

In some situations the solution at the boundary will be known at all times and can simply be specified. For example, in the case of groundwater models (that solve Laplace's equation for hydraulic head throughout a solution domain) where a boundary exists at a lake or reservoir the head at that boundary will be the known water level in the water body and can simply be set as this fixed value. This presents the simplest case in terms of its implementation.

For a heated bar, it may be that the temperature of one end where the heat is applied will be known for all time and, again, this may be used in the solution. It is also possible that the temperature at both ends will be known, which gives the easiest situation to deal with. In this case the boundary could be specified as:

$$\left.\begin{aligned} T_0^t &= C_0 \\ T_n^t &= C_n \end{aligned}\right\} \text{for all t} \qquad (18.8)$$

where $C_0$ and $C_n$ are the temperatures at the left and right hand end of the bar.

**Environmental Modelling Applications**

Perlin and Kit (2002) give an example of a Dirichlet boundary in the solution of wave-current interactions in coastal situations. The authors were keen to include the effects of turbulence, and a component of this was the effect of wind-driven shear on the water surface. In the solution of the turbulent kinetic energy *k*, which involves a PDE, the authors argued that it was possible to specify *k* values directly for the surface of the flow in terms of the wind velocity, thus implementing a Dirichlet boundary.

Raubenheimer et al. (1999) modelled groundwater fluctuations near a tidal beach and, knowing that the oscillations died out about 100 metres from the coast, set an inland boundary some distance from this and used a constant water-table level as the boundary condition (a Dirichlet boundary condition).

*Neumann boundary*

In some situations it is not possible to set an actual value at a boundary, but instead a gradient or derivative is known. For example, in the case of groundwater models a known inflow ($Q$) can be represented, using Darcy's law, as $Q = f(dh/dx)$, that is as a function of the head gradient at the boundary.

In other situations it may be possible to model a floating boundary by enforcing a zero gradient boundary. This might be used in the case of the downstream boundary of a river where pollutant concentration is being modelled. In this case a zero gradient downstream boundary allows the concentration to be driven up by the flows and while not strictly rigorous allows a reasonable solution to develop.

Farlow (1993) points out that Neumann boundaries imply no net gain or loss from the solution area. For example, in a groundwater flow model, a Neumann boundary applied to model the flow of water across boundaries will need to be such that there is no net gain or loss of water in the area being modelled.

For the heated bar problem it might be assumed that there was a zero gradient at the cool end of the bar. In this case the temperature at the boundary point would be equal to its immediate neighbour inside the bar. This condition could be written as:

$$T_n^t = T_{n-1}^t \quad \text{for all } t \tag{18.9}$$

*Robin boundary*

Farlow (1993) gives details of boundaries of the mixed type where there is both a gradient condition and some fixed value that also applies. This type of boundary is potentially useful in a number of situations but is less common and will not be covered here. It is worth knowing, however, that it exists and can be implemented should the situation call for it.

## 18.5 CODING THE HEAT EQUATION IN FORTRAN

The basic equation to be programmed can be written (as before) as:

$$T_i^{t+1} = T_i^t + \lambda(T_{i+1}^t - 2T_i^t + T_{i-1}^t) \quad (18.10)$$

where $\lambda = k\ \Delta t/(\Delta x)^2$. The numbering of the elements can be based on the preferences of the programmer but a standard scheme would have the *x* axis going from *i*=0 to *i*=*n*. In this case the two extremes are boundary values and must be supplied.

Although the solution steps forward in time it is usual to save only two basic time steps: the new value at time *t*+1 and the old value at time *t*. Once the whole grid has been updated to time *t*+1, this becomes the old time level and the solution proceeds. The basic steps are shown in Figure 18.4. The missing parts of the code are the assignment of boundary conditions and the code necessary to keep track of the time.

```
! update points

do i = 1,n-1
 t(i,2)=t(i,1)+lambda*(t(i+1,1)-2*t(i,1)+t(i-1,1))
end do
! move to new time level

 t(:,1) = t(:,2) !carry out using matrix capability
```

**Figure 18.4** Fortran code to implement a solution to the heat equation.

## 18.6 HEATED BAR SIMULATION

A bar, initially at 0° for its whole length, is heated at one end to 100° and to 40° at the other. The aim is to determine how the temperature of the bar changes in time. Assume that the bar is 10 cm long and that it is split into 1 cm sections for the finite difference representation. The solution starts with the initial temperatures being (100, 0, 0, 0, 0,..., 0, 40) so, for example, if a two-dimensional array temp is defined where the first dimension is the time level and the second is the space the start conditions can be written as shown in Figure 18.5.

```
! set all values to 0
temp(0,:)  = 0
temp(0,0)  = 100  ! set boundary conditions
temp(0,n)  = 40
```

**Figure 18.5** Fortran code to move the solution to the next time level.

A program can be written to calculate all the elements *temp*(1,1) to *temp*(1,*n*–1) since *temp*(*t*,0) and *temp*(*t*,*n*) are both boundary values and stay constant at 100° and 40° respectively. Once that is complete it is possible to calculate *temp*(2,1) to *temp*(2,*n*–1) and so on until the desired time level has been reached.

The results of the simulation, calculated in Excel, are shown in Figure 18.6.

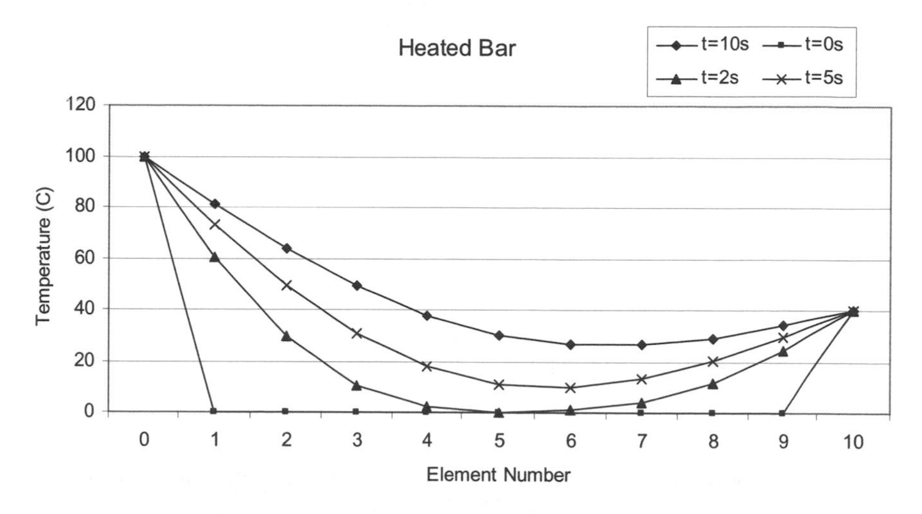


**Figure 18.6** Heated bar results plotted over time. It is not shown in the plot but after some time the solution settles on a final steady state with no further changes.

*Example*
Solve the transport equation for the situation where the concentration of dye in a pipe that is full of water is maintained at a steady value. The water in the pipe is at rest, and the concentration of dye is maintained at 1.0 g/g at the injection point. Assume the pipe is 10 m long and that the dye is injected at the mid-point. Use a diffusion coefficient of 1 $m^2$/s.

*Solution*
The situation can be modelled using the one-dimensional diffusion (or heat) equation where the finite difference approximation can be written as:

$$T_i^{t+1} = T_i^t + \lambda(T_{i+1}^t - 2T_i^t + T_{i-1}^t) \qquad (18.11)$$

where $\lambda = k\,\Delta t/(\Delta x)^2$. In this case the $k$ value is the diffusion coefficient = 1.0 $m^2$/s. For convenience, and assuming it will be small enough, select $\Delta x$ = 0.1 m. A time step, $\Delta t$ = 1.0 s, will be tried.

A spreadsheet is set up with key parameters defined as variables in the top left-hand corner of the sheet using `Insert - Name - Create`. Time runs down Column A and the distance along the pipe increases over a row. The sheet is shown in Figure 18.7.

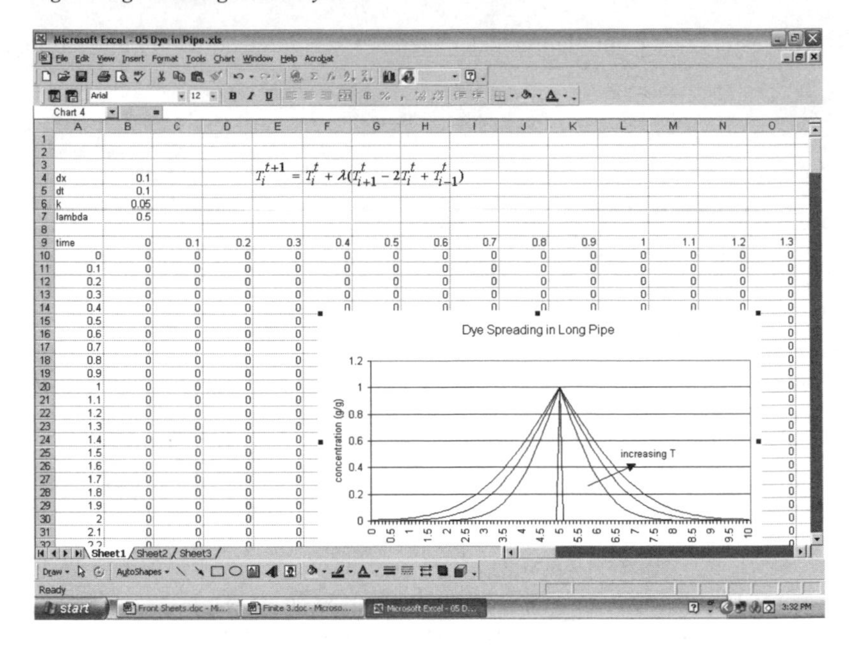


**Figure 18.7** Excel spreadsheet solution for dye in pipe.

The final solution, having selected various times at which to show the solution, is plotted in Figure 18.8.

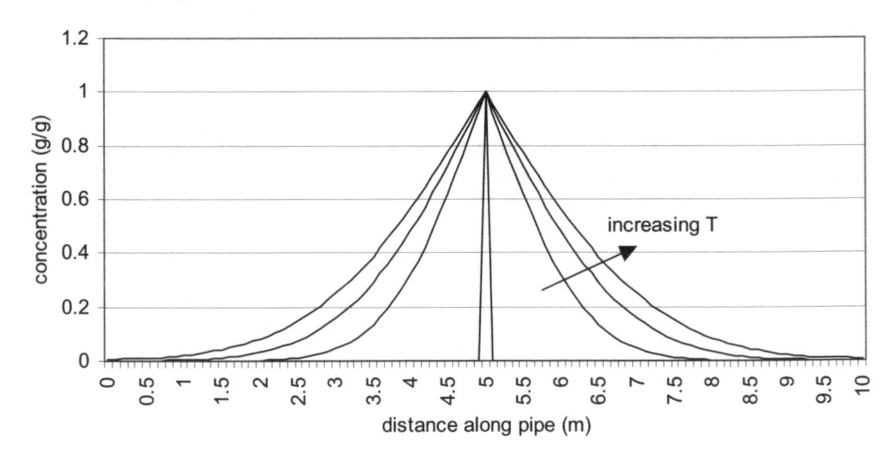


**Figure 18.8** Excel plot of solution for dye in pipe.

## PROBLEMS

**18.1** Use Excel to reproduce the temperature plot shown in Figure 18.6. For the solution try using $\Delta t = 0.5$ s, $k = 0.8$ cm$^2$/s, and $\Delta x = 1.0$ cm. The bar is 10 cm long.

**18.2** Develop a solution using Fortran where an initially cold bar (T=20°C along its length) is heated by a flame which increases the temperature in the middle to 100°C instantaneously and maintains that temperature. See Figure 18.9 for the physical layout. The length of the bar is 100 cm, and the heat conductivity parameter $k$ = 0.835 $cm^2/s$. Use a Neumann boundary for either end of the bar, so that there will be a (convenient) zero gradient in temperature at the end points. (As a way to get started it may be sensible to assume that the temperature remains at 20°C at the ends until the rest of the solution is behaving properly.)

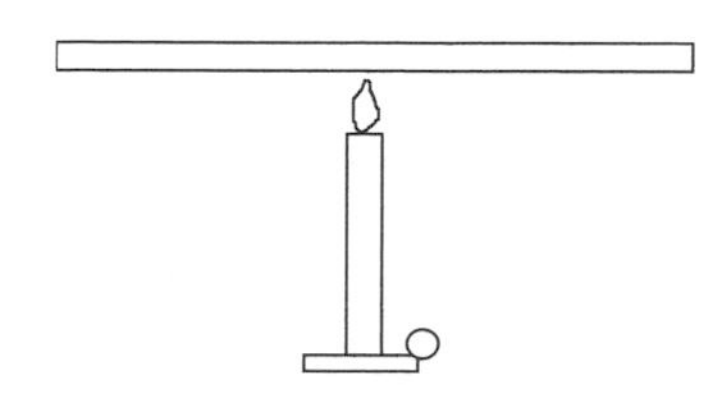

**Figure 18. 9** Metal bar with heat applied to centre.

**18.3** In an earlier chapter the solution of a heated plate was carried out by setting up a matrix of temperatures and iterating to a solution. Why is that not possible for these heat diffusion problems?

**18.4** Higher-order finite difference approximations generally use information from a larger number of neighbouring points. What are the advantages and disadvantages of this approach?

**18.5** One of the most difficult aspects of modelling heat transfer is determining the diffusion coefficient for the material in question. How could this be estimated from experiments?

**18.6** In a paper on sediment transport over ripples (Zedler and Street, 2006) the authors describe a part of their finite difference model: "Central differences were employed to represent the settling term $\partial(wC)/\partial x$ ... throughout most of the domain, whereas forward differences were employed in the ... cells near the bed."

Why would they use central differences in some places and forward differences near the bed?

## CHAPTER NINETEEN

# Finite Difference Modelling (Solution of Full Transport Equation)

### 19.1 INTRODUCTION

The one-dimensional transport equation can be written:

$$\frac{\partial c}{\partial t}+u\frac{\partial c}{\partial x}=K\frac{\partial^2 c}{\partial x^2} \tag{19.1}$$

where $u$ is the convective velocity (a constant), $c$ is the concentration of the material, $x$ is the axis (perhaps running down the river), and $K$ is the diffusion/dispersion coefficient. To solve this equation using the finite difference method two steps are undertaken:

1. the solution space is discretised into a number of discrete points rather than a continuum; and
2. the equation is approximated using standard approximations derived on the basis of a Taylor expansion of the function.

The way this is carried out is set out in the next two sections before a full description of the final solution.

### 19.2 DERIVING THE FINITE DIFFERENCE APPROXIMATION

Using the results determined from the Taylor expansion of the function at a point, estimates for the three derivatives can be written as:

$$\frac{\partial c}{\partial t}=\frac{c_i^{t+1}-c_i^t}{\Delta t}+O(\Delta t) \tag{19.2}$$

$$u\frac{\partial c}{\partial x}=u\frac{c_{i+1}^t-c_{i-1}^t}{2\Delta x}+O(\Delta x^2) \tag{19.3}$$

and

$$K\frac{\partial^2 c}{\partial x^2}=K\frac{c_{i+1}^t-2c_i^t+c_{i-1}^t}{\Delta x^2}+O(\Delta x^2) \tag{19.4}$$

Note that the time derivative is a forward difference approximation of first-order accuracy and the two space derivatives are both second-order accurate due to the fact that the estimate is centred at the point, $i$, where the derivatives are being estimated. Together this forms an FTCS (forward time, centre space) scheme and should be overall approximately first-order accurate since this is the lowest of the various orders of errors in any of the terms in the equation. Substituting these approximations into the transport equation and re-arranging gives:

$$c_i^{t+1}=(s+0.5C)c_{i-1}^t+(1-2s)c_i^t+(s-0.5C)c_{i+1}^t \tag{19.5}$$

where subscripts refer to the space coordinate, superscripts to time, and $s$ and $C$ (the Courant number) are defined as:

$$s=\frac{K\Delta t}{(\Delta x)^2} \tag{19.6}$$

$$C=\left(\frac{u\Delta t}{\Delta x}\right) \tag{19.7}$$

The equation gives an explicit formula to calculate the concentrations at a particular point, $i$, at the new time level, $t$+1, if the values at neighbouring points ($i$–1, $i$, $i$+1) are known at the old time level, $t$. The way the solution proceeds is illustrated in Figure 19.1.

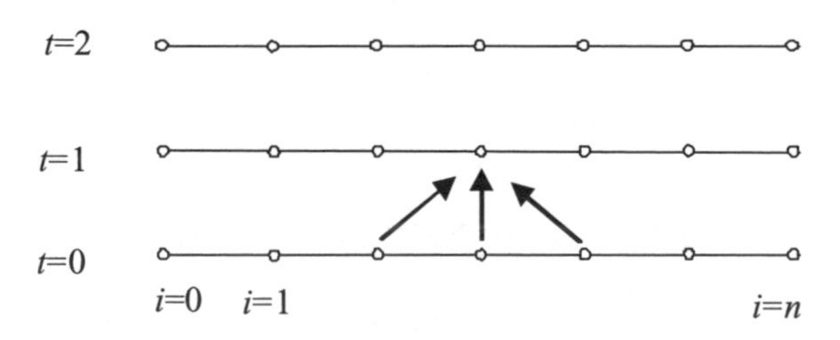


**Figure 19.1** Solution procedure for the transport equation. The arrows indicate the points that contribute to a new typical point at the new time level.

## 19.3 VERIFICATION OF FINITE DIFFERENCE APPROXIMATION

In the case of a constant upstream concentration (representing a pipe pouring into a narrow river as shown in Figure 19.2, for example) there is an exact analytical solution. It can be written (Kullenberg, 1971):

$$\frac{c}{c_0} = \frac{1}{2} e^{\frac{ux}{K}} erfc\left[\frac{x+ut}{2\sqrt{Kt}}\right] + \frac{1}{2} erfc\left[\frac{x-ut}{2\sqrt{Kt}}\right] \tag{19.8}$$

where $c$ is the concentration at point $x$ and time $t$, $x$ is the distance downstream (or upstream for $-x$), $t$ is the time, $U$ is the stream velocity, $K$ is the dispersion coefficient and $erfc(x)$ is the complementary error function. [Note: At the time of writing the Excel function ERFC() only worked with positive arguments. Be warned!]

**Figure 19.2** Discharge into a river, where upstream concentration is constant.

To illustrate the comparison a simulation was undertaken using the following parameters: $c_0 = 1.0$; $u = 1.0$ m/s; $K = 2.0$ m$^2$/s; $\Delta t = 0.1$ s; and $\Delta x = 1.0$ m. The solution for the two methods of calculation is shown in Figure 19.3. On it the general behaviour of the pollutant is shown by plotting the finite difference approximation for the situation after 2, 5 and 10 seconds. The comparison with the analytical solution is shown only for 10 seconds. The results indicate that in this case the finite difference model is able to reproduce the analytical solution to a high degree of accuracy.

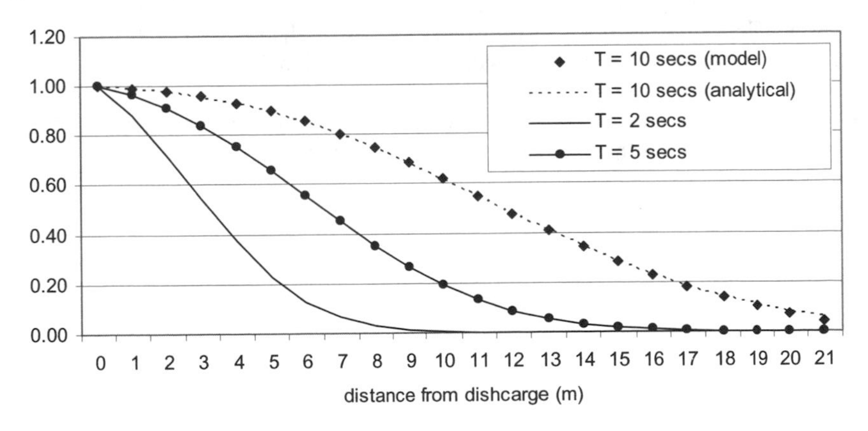


**Figure 19.3** Comparison of analytical and finite difference approximation for concentration at certain times following the start of a release.

## 19.4 CASE STUDY: THERMAL RIVER POLLUTION

McIntyre et al. (2004) describe a study where heat pollution in a Chinese river was modelled using a one-dimensional version of the transport equation. The equation can be written:

$$\frac{dT}{dt} = -v\frac{dT}{dx} + D\frac{d^2T}{dx^2} + \Phi \tag{19.9}$$

where $T$ is the water temperature, $D$ the diffusion coefficient, $v$ the river velocity, $t$ the time, $x$ the distance along the model and $\Phi$ a source and sink term that took account of solar input and losses to the environment through the bed of the river and to the atmosphere. The authors applied the model in a stochastic analysis where many thousands of runs were undertaken as part of a Monte Carlo simulation to develop statistical quantities related to temperature along the length of the river. For this reason the model had to be fast and versatile in terms of its solution under a range of velocities and diffusion coefficients. This was achieved by having an adaptive time-step routine in the model so that the maximum possible time step was used in the solution, as long as the accuracy did not suffer.

## PROBLEMS

**19.1** Derive the finite difference approximation and show that the explicit formula given in Equation (19.5) is correct.

**19.2** Develop an Excel spreadsheet to verify the results shown in Figure 19.3. (Take care: the fact that the error function has errors in it causes problems unless this is taken into account.)

**19.3** It would have been possible to use a first-order forward approximation for the first space derivative in the transport equation rather than the second-order centred one. What are the reasons that this would have been done?

**19.4** In the explicit schemes, there is only ever one unknown at the new time level. What could be done if there were more than one unknown?

CHAPTER TWENTY

# Finite Difference Modelling (Alternate Schemes)

## 20.1 INTRODUCTION

So far in this book the schemes that have been developed to solve partial differential equations (PDEs) such as the transport, heat or convection equation have been derived in such a way that there is a single unknown in each equation. These explicit schemes are easy to code and implement, but there are other ways of tackling the problem. Firstly, there are other ways of constructing the individual approximations that lead to different explicit schemes. Secondly, there are schemes that can be developed where each equation contains more than one unknown. These schemes, which are called implicit, require a matrix solution for their implementation. This adds a level of complication for the programmer but there are significant advantages with this approach.

## 20.2 ADVANCED EXPLICIT METHODS

If one considers the time-stepping procedure in the solution of the transport equation it is apparent that the single step from the time level $t$ to the next level, $t+1$ is similar to the simple Euler solution method in the solution of ordinary differential equations. One of the problems with the Euler approach is that it uses only information from the current time level to predict the state at the next time level. More advanced methods work in a predictor-corrector fashion where a prediction about the next time level is made, but then knowledge from that initial prediction is used to calculate an improved prediction. This idea is also applicable in explicit finite difference applications. Bhallamudi and Chaudhry (1991) give an example of application of the MacCormack method of explicit finite differences where the movement from one time level to the next is carried out in two stages, one a predictor and the second a corrector. The application is beyond the scope of this text but modellers should be aware of the extensions that are possible.

There are other ways of forming the approximations that lead to different schemes. These are outlined briefly in Section 20.5. Before that is done, it is worth looking at implicit schemes.

## 20.3 IMPLICIT SCHEMES

The finite difference schemes shown so far have been referred to as explicit since at each point the next value in time at the point is calculated in terms of known values surrounding the point. It is also possible to derive schemes where there may be several unknowns at each point. This leads to a matrix solution and the schemes are called implicit. Although more difficult to program, implicit schemes have significant advantages particularly with respect to stability. In many situations the scheme can be derived in terms of three unknowns per point, which leads to a tri-diagonal matrix. This can be solved using the Thomas algorithm. As an example, the heat conduction equation can be written:

$$\frac{\partial T}{\partial t} = k\frac{\partial^2 T}{\partial x^2} \tag{20.1}$$

Taking derivatives about the time $t$+1/2 the time derivative can be written as:

$$\frac{\partial T}{\partial t} = \frac{T_i^{t+1} - T_i^t}{\Delta t} + O(\Delta t^2) \tag{20.2}$$

which is effectively centred at the mid-time ($t$+1/2) step leading to a second-order approximation. The second derivative in space can be written as:

$$\frac{\partial^2 T}{\partial x^2} = \frac{1}{2}\left(\frac{T_{i+1}^t - 2T_i^t + T_{i-1}^t}{\Delta x^2} + \frac{T_{i+1}^{t+1} - 2T_i^{t+1} + T_{i-1}^{t+1}}{\Delta x^2}\right) + O(\Delta x^2) \tag{20.3}$$

which is also second-order accurate. Combining these, and sorting terms gives:

$$-\lambda T_{i-1}^{t+1} + 2(1+\lambda)T_i^{t+1} - \lambda T_{i+1}^{t+1} = \lambda T_{i-1}^t + 2(1-\lambda)T_i^t + \lambda T_{i+1}^t \tag{20.4}$$

where $\lambda = k\ \Delta t/(\Delta x)^2$. Because both the time and space derivative are centred in time and space the scheme is second-order accurate and the scheme is called the Crank-Nicholson method.

## 20.4 SOLVING IMPLICIT METHODS

If the Crank-Nicholson method is to be solved on a grid where the elements are numbered 1 to $n$ with 1 and $n$ being boundary values the equations are of the form:

$$\begin{aligned}
i=2 \quad & -\lambda T_1^{t+1} + 2(1+\lambda)T_2^{t+1} - \lambda T_3^{t+1} = \lambda T_1^t + 2(1-\lambda)T_2^t + \lambda T_3^t \\
i=3 \quad & -\lambda T_2^{t+1} + 2(1+\lambda)T_3^{t+1} - \lambda T_4^{t+1} = \lambda T_2^t + 2(1-\lambda)T_3^t + \lambda T_4^t \\
i=4 \quad & -\lambda T_3^{t+1} + 2(1+\lambda)T_4^{t+1} - \lambda T_5^{t+1} = \lambda T_3^t + 2(1-\lambda)T_4^t + \lambda T_5^t
\end{aligned} \tag{20.5}$$

When written in matrix form (and assuming that $n = 6$) the equations would be:

$$\begin{bmatrix} 2(1+\lambda) & -\lambda & 0 & 0 \\ -\lambda & 2(1+\lambda) & -\lambda & 0 \\ 0 & -\lambda & 2(1+\lambda) & -\lambda \\ 0 & 0 & -\lambda & 2(1+\lambda) \end{bmatrix} \begin{Bmatrix} T_2 \\ T_3 \\ T_4 \\ T_5 \end{Bmatrix} = \begin{Bmatrix} R_2 + \lambda T_1 \\ R_3 \\ R_4 \\ R_5 + \lambda T_6 \end{Bmatrix} \tag{20.6}$$

where the $R_2$, $R_3$ etc. are based on the known values at the old time level. For example:

$$R_2 = \lambda T_1^t + 2(1-\lambda)T_2^t + \lambda T_3^t \tag{20.7}$$

The boundary conditions appear on the right-hand side (RHS) in the first and last row of the matrix. The solution of this matrix problem can be tackled by any of the standard matrix solution techniques although it should be noted that the matrix is sparse, and in fact tri-diagonal (3 elements across, centred on the diagonal) and diagonally dominant. This means it can be solved using the Thomas algorithm. The Thomas algorithm assumes the matrix is in the form:

$$\begin{bmatrix} b_1 & c_1 & 0 & 0 \\ a_2 & b_2 & c_2 & 0 \\ 0 & a_3 & b_3 & c_3 \\ 0 & 0 & a_4 & b_4 \end{bmatrix} \begin{Bmatrix} x_1 \\ x_2 \\ x_3 \\ x_4 \end{Bmatrix} = \begin{Bmatrix} d_1 \\ d_2 \\ d_3 \\ d_4 \end{Bmatrix} \tag{20.8}$$

Therefore, in Fortran the general statements to set up and solve the matrix would be as shown in Figure 20.1.

```
do i=2,n-1
 a(i-1)= -lambda
 b(i-1)= 2*(1+lambda)
 c(i-1)  = -lambda
 d(i-1)=lambda*t(i-1,1)+2*(1-lambda)*t(i,1)+lambda*t(i+1,1)
end do
                    !  add boundary conditions
d(1)  = d(1)+lambda*t(1,2) ! left b.c.
d(n-2)=d(n-2)+lambda*t(n,2) !right b.c.
                    !  solve using Thomas algorithm
call thomas(a,b,c,d,n-2)
                    ! get answers from d array
do i=2,n-1
 t(i,2)  = d(i-1)
end do
```

**Figure 20.1** Fortran code segment to solve the Crank-Nicholson implicit method.

In this case care must be taken with the indices of the *a, b, c* and *d* arrays, realising that the unknown temperatures have *i* values that go from 2 to *n*–1 but the elements of the arrays ($a,b,c,d$) must go from 1 to $n$–2. Two time levels are in play: $t(i,1)$ is the old time level and $t(i,2)$ the new time level. After all the new time levels are calculated the new level becomes the old level and the process continues.

It is evident that the implicit scheme is more complicated to program, but there are significant advantages mainly to do with accuracy and efficiency. Nowadays it would be most common for modellers to use implicit schemes for these two very important reasons. This is discussed further in the next section.

## 20.5 ACCURACY AND THE FINITE DIFFERENCE

It must be remembered that the solutions derived using the finite difference method are approximate. Just how approximate can be observed occasionally if there are analytical solutions with which to compare them, but in many cases the very reason for using finite differences is because there are no analytical solutions. In these cases it is important to have an understanding of the likely accuracy and efficiency of the scheme in use. In general finite difference schemes should be tested for:

- convergence (solution approaches solution of PDE as $\Delta x$ and $\Delta t$ approach 0);
- consistency (FD equation approaches PDE as $\Delta x$ and $\Delta t$ approach 0); and
- stability (errors do not grow).

There are a number of standard schemes that have been derived to solve the problem and these will be listed and discussed. The reason for this is to show that it is not possible (or sensible) to generate schemes simply using a knowledge of the Taylor series and assume they will work properly. In a lot of cases there will be significant problems and more importantly the problems may not be immediately obvious.

*Scheme 1: FTCS*

The pure diffusion equation, such as that listed as Equation (20.1), can be approximated by:

$$T_i^{t+1} = sT_{i-1}^t + (1-2s)T_i^t + sT_{i+1}^t \tag{20.9}$$

where $s = \dfrac{\alpha \Delta t}{(\Delta x)^2}$ and $\alpha$ is the diffusion coefficient which is the same as $k$ in Equation (20.1). This standard scheme is stable for $s \leq 0.5$.

*Scheme 2: Richardson CTCS*

This will be discussed in more detail in the next section, but the lesson of the Richardson scheme is an important one. In the standard FTCS scheme it can be shown that the order of accuracy is only first-order because of the first-order time derivative. It makes sense that if this could be second-order the overall scheme should be better. But it is not!

$$\frac{T_i^{t+1} - T_i^{t-1}}{2\Delta t} = \alpha \frac{T_{i+1}^t - 2T_i^t + T_{i-1}^t}{\Delta x^2} \tag{20.10}$$

Two approximations, both of which are second-order accurate, lead to an overall scheme that is unconditionally unstable for all time steps.

*Scheme 3: Dufort Frankel*
This improved scheme is similar to that of Richardson, but the following replacement is made:

$$T_i^t = \frac{1}{2}\left(T_i^{t+1} + T_i^{t-1}\right) \tag{20.11}$$

This then leads to:

$$\frac{T_i^{t+1} - T_i^{t-1}}{2\Delta t} = \alpha \frac{T_{i+1}^t - \left(T_i^{t+1} + T_i^{t-1}\right) + T_{i-1}^t}{\Delta x^2} \tag{20.12}$$

or

$$T_i^{t+1} = \frac{2s}{1+2s}\left(T_{i-1}^t + T_{i+1}^t\right) + \frac{1-2s}{1+2s} T_i^{t-1} \tag{20.13}$$

This is stable for all $s$, but requires that $\Delta t << \Delta x$ for consistency.

*Scheme 4: classical implicit*
The first of the implicit schemes gives three unknowns in each line of the matrix. It is derived using the following approximations:

$$\frac{T_i^{t+1} - T_i^t}{\Delta t} = \alpha \frac{T_{i+1}^{t+1} - 2T_i^{t+1} + T_{i-1}^{t+1}}{\Delta x^2} \tag{20.14}$$

so that

$$-sT_{i-1}^{t+1} + (1+2s)T_i^{t+1} - sT_{i+1}^{t+1} = T_i^t \tag{20.15}$$

The scheme is unconditionally stable, but requires a matrix solution.

*Scheme 5: Crank Nicholson (implicit)*
The Crank Nicholson method is unconditionally stable, but again, requires a matrix solution. This classic scheme can be written:

$$\frac{T_i^{t+1} - T_i^t}{\Delta t} = \frac{\alpha}{2}\left(\frac{T_{i+1}^{t+1} - 2T_i^{t+1} + T_{i-1}^{t+1}}{\Delta x^2} + \frac{T_{i+1}^t - 2T_i^t + T_{i-1}^t}{\Delta x^2}\right) \tag{20.17}$$

## 20.6 LEWIS FRY RICHARDSON

It has been seen in the last section that a numerical approximation developed by Richardson to solve the heat equation is unconditionally unstable. At the time of its development this was not realised since according to Roache (1982) "the instability did not manifest itself in Richardson's sample calculations only because of the small number of time-step calculations performed".

Richardson was involved in a number of ventures during his time. As a practising Quaker, Lewis Richardson was a non-combatant during World War I and drove an ambulance on the front line. In between those duties he spent many hours attempting to predict the weather over the coming six hours for the whole of Europe using a hand-driven solution to the governing equations. Unfortunately his solutions quickly diverged from what was likely to happen, and in fact Lorenz (1993) has noted, "Richardson correctly attributed his failure to inadequacies in the initial wind measurements, although subsequent analysis has shown that his procedure would have produced serious although less drastic errors even with perfect initial data."

Despite the failure, the basic idea of developing a solution to the governing equations using data at a regular grid of points was a significant piece of work. Nor was he content to develop just the method of solution – he also foresaw a weather centre where 64,000 people (referred to as "computers", working by hand) would be able to "produce a weather forecast more rapidly than the weather itself could advance" (Lorenz, 1993).

## PROBLEMS

**20.1** Modify an earlier solution of the heat equation to use the Richardson scheme. Verify whether it is in fact unconditionally unstable.

**20.2** Modify an earlier solution of the heat equation to use the Dufort Frankel method. Verify its stability and accuracy performance and compare these to the suggested values.

**20.3** Write a program to solve the heat equation using the Crank Nicholson implicit scheme. Verify its performance – in particular, its ability to use much larger time steps than the explicit scheme and still remain stable.

CHAPTER TWENTY ONE

# Probability and Statistics (Descriptive Statistics)

## 21.1 INTRODUCTION

Statistics are everywhere: in the news, weather summaries, health reports, commercial marketing and even on the sides of cereal boxes. A clear understanding of statistics is vital for engineers and this chapter aims to set out some basic statistical concepts. This is done, not in general terms, but by outlining the analysis of a particular dataset. This dataset, given in Table 21.1, shows the cricket attendances of the Boxing Day test matches at the Melbourne Cricket Ground (MCG) and the observed weather on those same days. After all, what topics for the discussion of statistics could be more common than the weather, or more popular than sport? It is worth remembering that just as types of data can be different, so too can be the techniques used to investigate them and for this reason the following discussion serves only as an example.

## 21.2 DATA EXPLORATION

The first step in analysing any data is to get an overview of it, which includes an appreciation of any additional information not summarised by the data itself (usually relating to the method of data collection) and an understanding of possible inconsistencies. The data in Table 21.1 show the attendances of five-day cricket matches at the MCG against various countries (England, West Indies, South Africa, and others) that start on Boxing Day (26 December) each year.

To encourage a comparison of like with like, this dataset has already been distilled from a larger pool of attendance records at the MCG covering rock concerts, football matches and all other types of cricket (e.g. one-day matches). Some of the reasons for selecting the data presented in Table 21.1 are:

(a) concerts may have higher capacity by allowing crowd onto the playing surface;
(b) attendances at other cricket matches will depend on the day which they are held on (weekday vs weekend) whereas the Boxing Day match occurs during a general holiday period; and
(c) match attendances can depend on the duration of the event (test matches can last up to five days), thus only the Boxing Day attendance and the average attendances are shown to ensure a more consistent comparison.

**Table 21.1** Cricket attendance, rainfall and temperature records for Boxing Day (BD) Test Matches against Australia at the Melbourne Cricket Ground.

| Year | Opponent | BD Attend. | Match Ave. Attend. | BD Rain (mm) | Match Ave. Rain (mm/day) | BD Max. Temp. (°C) | Match Ave. Max. Temp. (°C) |
|---|---|---|---|---|---|---|---|
| 1950 | England | 35697 | 47799 | 0 | 0.125 | 33 | 27.28 |
| 1952 | S. Africa | 12675 | 24063 | 2 | 2.02 | 22.7 | 23.92 |
| 1968 | W. Indies | 18766 | 28344 | 5.8 | 1.575 | 16.7 | 23.18 |
| 1974 | England | 77167 | 50150 | 0.8 | 3.12 | 26.6 | 24.8 |
| 1975 | W. Indies | 85661 | 55689 | 0 | 0.2 | 31.3 | 28.12 |
| 1980 | NZ | 28671 | 16549 | 0 | 2.84 | 29 | 28.02 |
| 1981 | W. Indies | 39982 | 33700 | 0 | 0.2 | 22.6 | 23.28 |
| 1982 | England | 63900 | 42976 | 0 | 0.4 | 28 | 26.64 |
| 1983 | Pakistan | 40240 | 22322 | 0 | 0 | 19.1 | 26.92 |
| 1984 | W. Indies | 25555 | 19434 | 3.2 | 1.04 | 18.9 | 24.94 |
| 1985 | India | 18146 | 13152 | 0.6 | 0.2 | 20.4 | 24.38 |
| 1986 | England | 58203 | 35939 | 0 | 0 | 21.6 | 24.2 |
| 1987 | NZ | 51087 | 25437 | 0.8 | 3.72 | 29.2 | 25.88 |
| 1988 | W. Indies | 24246 | 21682 | 48.6 | 9.88 | 17.9 | 20.1 |
| 1990 | England | 49763 | 25906 | 0 | 0 | 20.9 | 26.98 |
| 1991 | India | 42494 | 22342 | 0 | 0.05 | 35.6 | 26.52 |
| 1992 | W. Indies | 28397 | 16664 | 5 | 4.16 | 21.6 | 22.72 |
| 1993 | S. Africa | 15604 | 9713 | 4.4 | 25.68 | 14.9 | 17.02 |
| 1994 | England | 31367 | 28898 | 0 | 0 | 28.4 | 27.54 |
| 1995 | Sri Lanka | 55239 | 22678 | 0 | 0 | 19.6 | 19.94 |
| 1996 | W. Indies | 72821 | 43890 | 0 | 0 | 19.6 | 21.36 |
| 1997 | S. Africa | 73812 | 32036 | 0 | 0 | 21.9 | 23.48 |
| 1998 | England | 61580 | 39758 | 1.8 | 5.6 | 23.5 | 21.48 |
| 1999 | India | 49082 | 26911 | 5.4 | 14.12 | 24.1 | 20.56 |
| 2000 | W. Indies | 73233 | 33325 | 0 | 0.4 | 24.9 | 20.5 |
| 2001 | S. Africa | 61796 | 38256 | 1.2 | 4.8 | 16 | 23.28 |
| 2002 | England | 64189 | 35532 | 0 | 0.2 | 21.3 | 29.2 |
| 2003 | India | 62613 | 35932 | 0 | 0 | 22.1 | 28.56 |
| 2004 | Pakistan | 61552 | 32270 | 0 | 5.15 | 22.5 | 21.56 |
| 2005 | S. Africa | 71910 | 38467 | 0.6 | 0.2 | 26.7 | 31.94 |
| 2006 | England | 89155 | 81450 | 10.4 | 2.16 | 17.8 | 22.32 |
| 2007 | India | 68465 | 41540 | 0 | 0 | 25.6 | 30.5 |

Despite the careful selection, there are still possible inconsistencies which should be noted, although they are not further investigated here.

(a) The MCG stands have undergone significant changes over time resulting in a decreased maximum crowd capacity since the 1970s (approx. 120,000 down to

100,000). The standing room has been decreased for safety reasons, but as a compensating factor, the size of the grandstands has been increased.

(b) Metropolitan temperature records can be subject to heat island effects where increased urbanisation over decades can cause spurious trends in the observed temperatures (Camilloni and Barros, 1997).

(c) Daily rainfall measurements can be subject to numerous biases when not carefully recorded (Viney and Bates, 2004).

One of the best methods used to become familiar with a dataset is to visualise the data using simple plots. Computer packages such as Matlab and Excel have graphics capabilities that are best suited for this task, whereas it can be time-consuming to create graphics from scratch using a language such as Fortran or C.

An *x-y* plot can be used in a various ways; for example, Figure 21.1 is a time-series plot of attendances. There are gaps in this series (e.g. 1950s). However, these are not missing values; it is only since 1980, with the exception of 1989, that it has been an annual event. The plot shows a slight increase over this period, but there is also considerable scatter in the data.

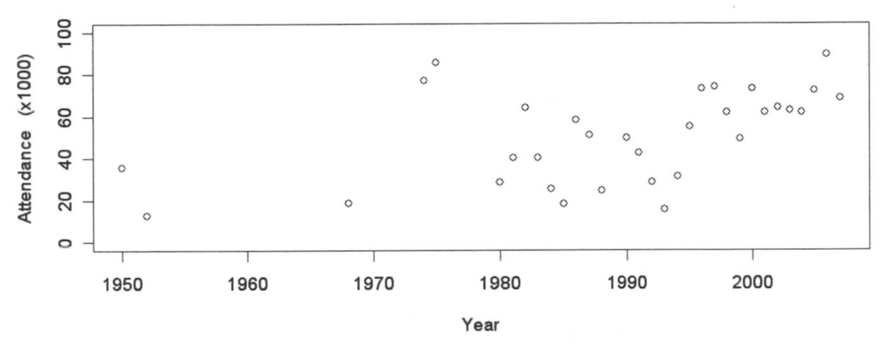


**Figure 21.1** Time series of Boxing Day attendances at the MCG.

Figure 21.2 shows scatter plots of Boxing Day attendance against the maximum daily temperature and the daily rainfall total respectively. In Figure 21.2(a) there does not appear to be a strong relationship between the temperature and the attendance as there is considerable scatter at all temperatures. In contrast, in Figure 21.2(b) there appears to be a decrease in attendance with increasing rainfall, which would have been forecast, with a single anomalous observation when the rainfall was 10.4 mm. This was the test match against England in 2006 when Australia was competing to regain the Ashes. Notice that rainfall occurs over a large range of values and a log-scale has been used, with zero rainfall plotted as a special case (as $\log(0)=-\infty$, zero cannot be shown on a log-scale).

Given the outlying value in the rainfall scatter plot, it is worthwhile comparing the crowd attendance for different opponents. Figure 21.3 is a box plot of the attendance with respect to four of the opposition teams.

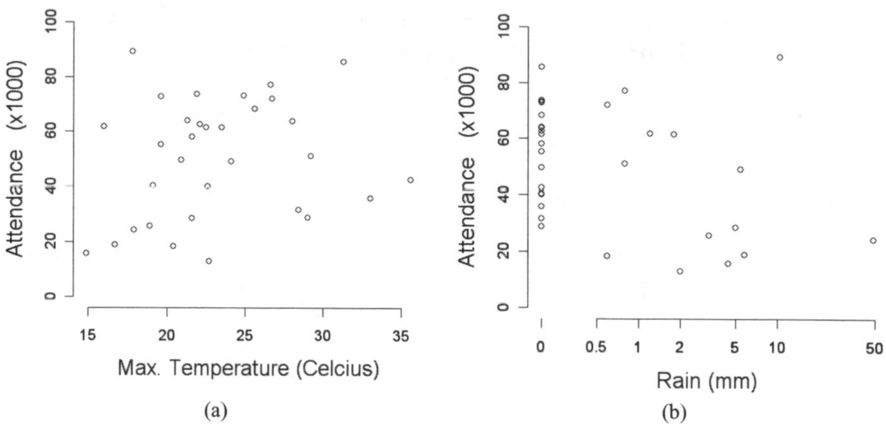


**Figure 21.2** Scatter plots of Boxing Day attendance with respect weather variables (a) daily maximum temperature, and (b) daily rainfall total.

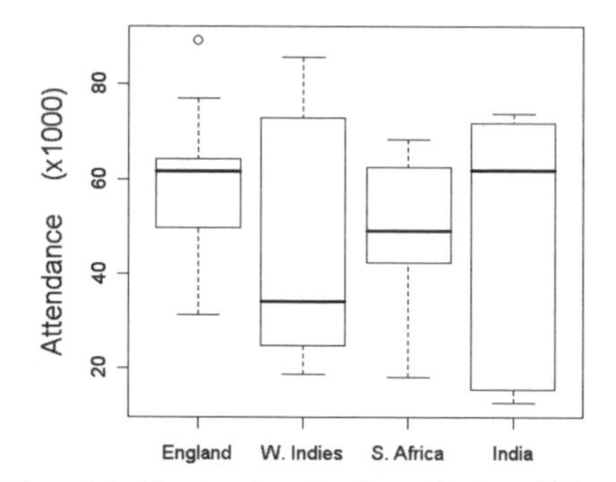


**Figure 21.3** Box plot of boxing day attendance for four different opponents.

Consider the box plot for the West Indies in detail. The first step is to sort the attendance data in ascending order (18766, 24246, 25555, 28397, 39982, 72821, 73233, 85661). The median is the middle value, but as the number of data is even the middle is between the 4th and 5th data values. Therefore the average of 28397 and 39982 is taken, which equals 34189.5. In general if there are $n$ data, the median is the $(n+1)/2$th datum when they are put in ascending order. The median is the value below which 50% of the data lie. The median is shown as the thicker horizontal line within the box. The upper and lower ends of the box are the upper and lower quartile respectively. The lower quartile is the value below which 25% lie. It is calculated as the $(n+1)/4$th value. For the West Indies data this is the 2.25th position, which is calculated as 24246 + 0.25*(25555–24246). The upper quartile is the value below which 75% of the data lie and is calculated as the $3(n+1)/4$th value. It follows that 50% of the data lies within the ends of the box. It is conventional to show any data that lie more than 1.5 times the inter-quartile range (i.e. the upper minus the lower quartile) from the ends of the box as distinct points. There are no such data for the West Indies, but there are for the England attendances: the 2006 match previously mentioned. The dashed lines are referred to as whiskers and extend to the largest and smallest data not shown individually.

Figure 21.4 is a bar chart of the attendance data on Boxing Day for all of the opposing teams. A bar chart is a count of the number of data points that land within a certain bin width. The bin widths are chosen to cover the entire range of the data and must have an equal increment. Here the bin width is 10,000 attendees and the starting point is taken as 10,000 as it is below and close to the lowest observation. The crowd size is approximately evenly spread between 10,000 and 90,000, but with the most common crowd size occurring between 60,000 and 70,000 people (7 occurrences) and with only a few crowds above 80,000. A bar chart is commonly referred to as histogram (e.g. Matlab has the command `hist`), but in this book the term histogram is reserved for a bar chart that is scaled to have an area of one.

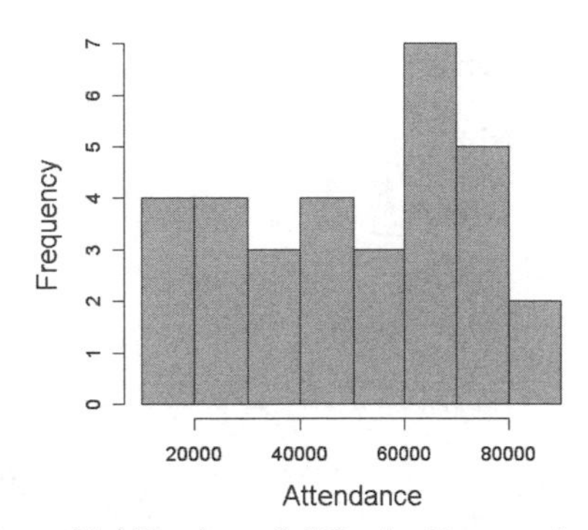


**Figure 21.4** Bar chart of all Boxing Day attendances.

## 21.3 SUMMARY STATISTICS

Having viewed the data, the next step is to use one or more statistics to encapsulate the overall information. The up-side of a statistic is that it is easier to look at one number instead of 100 or 1000, but the down-side is that it is only a single number and it will be necessary to calculate various statistics to provide a measure of spread and asymmetry as well as the mean value.

Consider a set of $n$ numbers, $\{x_1,x_2,\ldots,x_n\}$ that are to be analysed in a Fortran program. The minimum and maximum of the data set are two very basic statistics and a method to calculate them is shown in Figure 21.5.

```
min = x(1)
do i = 2,n
  if(x(i)<min) then
    min = x(i)
  end if
end do
```

```
max = x(1)
do i = 2,n
  if(x(i)>max) then
    max = x(i)
  end if
end do
```

**Figure 21.5** Algorithm to find minimum or maximum of a data set.

The algorithm for the minimum starts assuming the first value as the smallest. The loop passes over each element of the array and checks whether the $i^{th}$ number is less than the current minimum. If it is, then the minimum is updated and the search continues until the last data point. The algorithm for the maximum is similar.

The mean ($m$), standard deviation ($s$) and skewness ($k$) are three simple statistics that are often estimated from a dataset. The formulae are shown along with some Fortran code used to calculate them in Figure 21.6. In these examples the standard deviation requires the mean to be calculated beforehand, similarly the skewness requires the standard deviation and the mean.

```
sum = 0
do i = 1,n
  sum = sum + x(i)
end do
mean = sum/real(n)
```

$$m = \frac{1}{n}\sum_{i=1}^{n} x_i \tag{21.1}$$

```
sum2 = 0.0
do i = 1,n
  sum2= sum2 + (x(i)-mean)**2.0
end do
stdv = sqrt(sum2/real(n-1))
```

$$s = \sqrt{\frac{1}{n-1}\sum_{i=1}^{n}\left(x_i - m\right)^2} \tag{21.2}$$

```
sum3 = 0.0
do i = 1,n
  sum3= sum3 + (x(i)-mean)**3.0
end do
skew = sum3/stdv**3.0/real(n)
```

$$k = \frac{\frac{1}{n}\sum_{i=1}^{n}\left(x_i - m\right)^3}{s^3} \tag{21.3}$$

**Figure 21.6** Equations for calculating mean, standard deviation and skewness.

There are several explanations for the use of $n$–1 in Equation (21.2). For the moment, it is only pointed out that $n$–1 gives an undefined value with a sample size $n = 1$ because both the numerator and the denominator equal zero. This is appropriate if the standard deviation is used to indicate the variability in the underlying population the sample is drawn from.

The skewness is a measure of asymmetry. It is easy to show that $\Sigma(x_i - m) = 0$ for $i=1..n$, but if the deviations from the mean are cubed any data in the tails of the distribution will make relatively larger contributions. If they are in the right-hand tail the contribution is positive, if in the left-hand tail, it is negative. The skewness is then obtained as the average cubed deviation made non-dimensional by dividing by the standard deviation cubed. A skewness with $|k| > 1$ is quite noticeable.

The boxplot in Figure 21.3 incorporated statistics for the lower quartile, median and upper quartile and it would be useful to compute these also. To do this it is necessary to sort the data. In packages such as Excel and Matlab there is typically a sort function that can be used. A basic but slow algorithm for sorting data in Fortran is given in Figure 21.7.

```
do  ! keep on bubbling
  swaps = 0
  do i = 2,n  ! do one pass of the array
    if(x(i)<x(i-1))then ! compare two values
      temp=x(i); x(i)=x(i-1); x(i-1)=temp !swap if needed
      swaps = swaps + 1 ! count the swaps
    end if
  end do
  if(swaps==0) exit ! there are no more swaps
end do
```

**Figure 21.7** Bubble sort algorithm.

This algorithm is called bubble sort because the small numbers "bubble to the top" by successive swaps between adjacent values in the array. It would be worth testing this code by following it by hand with a small dataset. Having sorted the data, the $p$% point is estimated as the $(n+1)(p/100)^{th}$ point, using interpolation if needed.

The attendance, temperature and rain for Boxing Day are summarised in Table 21.2. The mean crowd is about 50,000, but there is a large standard deviation of about 22,000. The distribution of attendees is approximately symmetric since the skewness is near zero. Assuming Boxing Day is a typical summer's day in Melbourne, one might expect a temperature of 23°C. Half of the time the rainfall is below 0.3 mm but the distribution is highly skewed and rainfall has been observed up to 25.7 mm on this day.

**Table 21.2** Summary statistics for Boxing Day at the MCG.

| Statistic | Attendance | Temp. (°C) | Rain (mm) |
|---|---|---|---|
| No Data | 32 | 32 | 32 |
| Minimum | 12675 | 14.9 | 0 |
| 0.25 Quartile | 30693 | 19.6 | 0 |
| Median | 53163 | 22.3 | 0.30 |
| 0.75 Quartile | 65258 | 26.6 | 3.3 |
| Maximum | 89155 | 35.6 | 25.7 |
| Mean | 50408 | 23.3 | 2.75 |
| Std Dev. | 21699 | 5.0 | 5.3 |
| Skewness | –0.15 | 0.61 | 3.22 |

These statistics can also be used to compare subgroups of the data. As an example, the mean Boxing Day attendances and the mean attendance for the entire test match are summarised in Table 21.3 for the various cricketing teams.

**Table 21.3** Summary statistics for Boxing Day at the MCG.

| Opponent | No. Matches | Mean Boxing Day Attendance | Mean Match Attendance | Ratio |
|---|---|---|---|---|
| England | 9 | 59002 | 43157 | 1.37 |
| West Indies | 8 | 46083 | 31591 | 1.46 |
| South Africa | 5 | 47159 | 28507 | 1.65 |
| India | 5 | 48160 | 27975 | 1.72 |
| New Zealand | 2 | 39879 | 20993 | 1.90 |
| Pakistan | 2 | 50896 | 27296 | 1.86 |
| Sri Lanka | 1 | 55239 | 22678 | 2.44 |

Table 21.3 shows that the attendance on Boxing Day is higher than the average attendance irrespective of the opposition team. Notably, while England, Sri Lanka and Pakistan all had high crowds on Boxing Day, England was able to maintain a higher attendance record than the other two teams. A similar comparison shows that the West Indies have a better retention of sports-goers than the South African team. While these comparisons are valid, one should take care that the number of data points used to compute the statistics is very low. In short the more data the less influence an individual observation has on any given statistic.

## PROBLEMS

**21.1** For the Boxing Day rainfall data draw a box plot. Explain the difference between the mean and the median. (Hint: The mean is the centre of gravity if the data are considered as point masses.)

**21.2** Plot the rain versus the temperature and also log(rain + 0.1) against temperature. Which plot is preferable? Is there evidence of a relationship between rainfall and temperature?

**21.3** Compute the mean and standard deviation of attendance on dry and rainy days.

**21.4** Plot a bar chart of the observed rainfall. Is the skewness noticeable?

**21.5** The following formula is sometimes given for hand calculation of the standard deviation:

$$s = \sqrt{\frac{1}{n-1}\left(\sum_{i=1}^{n} x_i^2 - \frac{1}{n}\left(\sum_{i=1}^{n} x_i\right)^2\right)}$$

(a) Calculate the standard deviation of {1,2,3,4,5} using this formula and the formula given in Equation (21.2) .
(b) What is the advantage of this formula?
(c) In Excel calculate the standard deviation of the following 5 estimates of the speed of light (m/s) {299 792 458.351, 299 792 458.021, 299 792 458.138, 299 792 458.251, 299 792 458.283} using the above formula and the formula in Equation (21.2).
(d) Compare the result from (c) with the in-built Excel `stdev()` function.
(e) What is the disadvantage of the above formula? Explain the discrepancies.
(f) Why should this formula not be programmed?
(g) Suggest an adaptation that might be suitable for a program.

**21.6** Write a bubble sort algorithm to sort the attendance data and compute the median value.

CHAPTER TWENTY TWO

# Probability and Statistics (Population and Sample)

## 22.1 INTRODUCTION

One aspect of probability and statistics concentrates on accurately summarising and describing patterns in an individual data sample. Another more common aspect is to use a data sample to make general inferences about the underlying population it has been taken from. For example, a group of university students might be surveyed about the amount and type of beer they consume on a weekly basis. The aim of this study might be to infer results about the student body as a whole; for example, whether some types of students drink more than others or whether university students consider brand X beer to be trendy. These inferences might be useful to a brewery that wants to improve flagging sales of brand X beer.

If inferences are to be made about a population from a sample, it is desirable for the sample to be representative. In a simple random sampling scheme every member of the population has an equally likely chance of being picked. Stratified sampling is a more complex random sampling strategy that also has this property. For example, a survey of students may require a quota to be met for each discipline (medicine, science, arts, etc.) where the quotas are in keeping with the ratio of students in each discipline in the student population. However, it is not essential that every member of the population has the same chance of being in a sample. This is provided that their probability of being in the sample is known and the result is weighted in an appropriate way.

In some cases the population is clearly physically defined, but in others it may be imaginary and typically infinite. For example, the record of a company's beer sales over the past 50 years can be considered as a random sample of all possible sales (past, present, future). The size of the sample is important, since the larger the sample, the more precise will be the estimates of the population parameters. Simply increasing the size does not remedy every problem though, as it does not avoid the issue of bias. Biases can occur if the sample is in some way not suitably representative of the population. Issues of bias can be subtle, therefore data samples and strategies to obtain them need to be considered with care.

## 22.2 PROBABILITY

With gambling apparatus, such as dice and well-shuffled cards, the results of a trial can be defined in terms of equally likely possible outcomes. In a lottery the organisers attempt to ensure all entrants are equally likely to be picked. In such cases the probability of an event $A$ is defined as the ratio of the number of equally likely possible outcomes that result in $A$ divided by the total number of equally likely outcomes. In a simple random sample, if the population is of size $N$, and the sample is of size $n$, every member of the population has a probability $n/N$ of being in the sample. For example, if entrants are allowed to buy only one ticket in a 100-ticket raffle and 5 winning tickets are drawn, then each person has a 0.05 probability of winning.

**Sample Bias**

On 3 November 1948, the Chicago Tribune printed the headline "Dewey Defeats Truman" in anticipation of the result of the US presidential election. To the newspaper's embarrassment, Harry Truman was announced president the following day, with an emphatic majority. The reason for this mistake was that the newspaper relied on the results of a phone survey. Phones at the time were not widespread and were more common in affluent households, the same households that were more likely to vote for Dewey.

Another famous example of bias is the experiments of H. Mann on the provision of free milk to British school children in the period 1921–1925. The results were used as a justification that state-sponsored milk should play a greater role in the diet of school children. E. Petty later showed that the results were unreliable because they focused on undernourished boys. While the milk did indeed benefit the undernourished boys (due to 'catch up' growth) the conclusions were erroneously generalised to represent all children (Aitkins, 2005).

The idea of equally likely is far too restrictive for most engineering applications and instead the probability of an event occurring is estimated by the proportion of times it has occurred in a random sample. Thus a sample proportion is an estimate of the population probability. In other cases, the probability is an expert's subjective opinion; for example, the probability of striking oil in a certain location. However probability is defined, it is measured on the same scale between 0 and 1, where 0 corresponds to impossible and 1 corresponds to a certain event. No matter how it is defined, the same rules apply:

1. $0 \leq P(A) \leq 1$.
2. $P(A \text{ or } B) = P(A) + P(B) - P(A \text{ and } B)$ where by mathematical convention "or" includes the possibility of both.
3. $P(A|B) = P(A \text{ and } B)/P(B)$ where the vertical line is read as "given that" or "conditional on". It follows that $P(A \text{ and } B) = P(A)\, P(B|A) = P(B)P(A|B)$.

The rules of probability are largely intuitive and can be verified if Figure 22.1 is considered carefully. In this figure the areas represent the proportions in the different categories and the area of the enclosing box is one. If $A$ and $B$ cannot occur together they are said to be 'mutually exclusive' or 'disjoint'. In this case $P(A \text{ and } B) = P(A) + P(B)$. An important special case is $P(\text{not } A) = 1 - P(A)$. If the fact that A occurs does not affect the probability that $B$ occurs then $A$ and $B$ are independent. Then, $P(B|A) = P(B)$ which implies that $P(A \text{ and } B) = P(A)P(B)$.

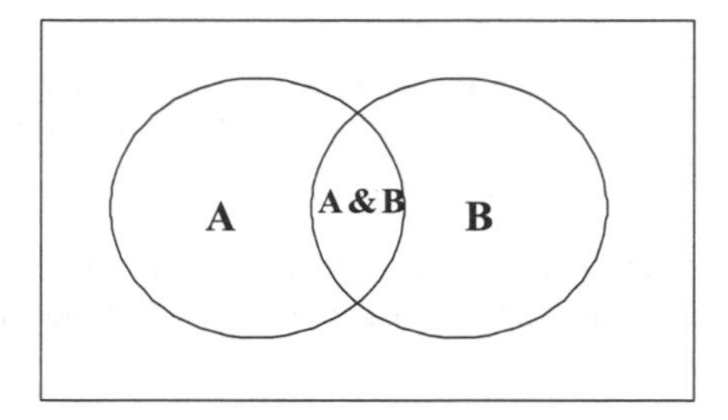


**Figure 22.1** Diagram of two events A, B, where the box represents all possible events.

*Example*

A manufacturer of sports cars finds at final inspection that 70% need timing adjustments, 40% need carburettor adjustments and 30% need both. Calculate the probability that a randomly selected sports car needs (i) a carburettor or timing adjustment, (ii) carburettor or timing adjustment but not both, (iii) not a timing adjustment, (iv) a timing adjustment given that it needs a carburettor adjustment, and (v) a carburettor adjustment given that it needs a timing adjustment.

*Solution*

It is usually very helpful to draw diagrams to assist in the solution of probability questions. Assume event $A$ represents carburettor adjustments and event $B$ represents timing adjustments. It follows that:

(i)

$$P(A \text{ or } B) = P(A) + P(B) - P(A \text{ and } B)$$
$$= 0.4 + 0.7 - 0.3$$
$$= 0.8$$

(ii)

$$P(A \text{ or } B) - P(A \text{ and } B) =$$
$$= 0.8 - 0.3$$
$$= 0.5$$

(iii)

$$P(\text{not } B) = 1 - P(B)$$
$$= 1 - 0.7$$
$$= 0.3$$

(iv)

$$P(B/A) = P(A \text{ and } B)/P(A)$$
$$= 0.3/0.4$$
$$= 0.75$$

(v)

$$P(A/B) = P(A \text{ and } B)/P(B)$$
$$= 0.3/0.7$$
$$= 0.428$$

*Example*

3% of cars are returned under warranty for electrical repairs and 5% are returned for mechanical repairs. Assume that faults occur independently. What proportion of cars are returned for both? What proportion of cars are returned?

*Solution*
Since the faults occur independently $P(A \text{ and } B) = P(A)P(B) = 0.03(0.05)=0.0015$. The proportion of cars returned is $P(A \text{ or } B) = 0.03 + 0.05 - 0.0015 = 0.0785$.

*Example*
If the probability of a flood in any one year is 1 in 100 and flood years occur independently, what is the probability of at least one flood year in the next 100 years?

*Solution*
Denoting $F_y$ as the event of a flood in year $y$, one method is to determine the probability $P(F_1 \text{ or } F_2 \text{ or} \ldots \text{or } F_{100})$, but this soon becomes complicated and tedious. A quicker method is to determine the probability of no floods in 100 years and subtract from one: $1 - P(\text{not } F_1 \text{ and not } F_2 \ldots \text{ and not } F_{100})$. $P(\text{not } F_y) = 0.99$ for all years. As years are independent the probability becomes $1 - P(\text{not } F_1)\, P(\text{not } F_2) \ldots P(\text{not } F_{100}) = 1 - (0.99)^{100} = 0.634$. In fact, the probability of the 1 in $N$ year event occurring in the next $N$ years tends towards 0.632 for large $N$.

## 22.3 ESTIMATION

In Chapter 21 the sample mean $m$, a standard deviation $s$ and a skewness $k$ were calculated. A statistic is a summary number calculated from a sample and some standard statistics are listed in Table 22.1.

**Table 22.1** Standard sample statistics.

| Quantity | Sample statistic | Population parameter |
|---|---|---|
| Size | $n$ | $N$ |
| Mean | $m = \frac{1}{n}\sum_{i=1}^{n} x_i$ | $\mu = \frac{1}{N}\sum_{j=1}^{N} x_j$ |
| Std Dev. | $s = \sqrt{\frac{1}{n-1}\sum_{i=1}^{n}(x_i - m)^2}$ | $\sigma = \sqrt{\frac{1}{N}\sum_{j=1}^{N}(x_j - \mu)^2}$ |
| Variance | $s^2$ | $\sigma^2$ |
| Mean of a function $\phi(x)$ | Sample average $= \frac{1}{n}\sum_{i=1}^{n}\phi(x_i)$ | Expected value = $E[\phi(x)] = \frac{1}{N}\sum_{j=1}^{N}\phi(x_j)$ |

These statistics are now interpreted as estimates of the corresponding quantities in the population which are referred to as population parameters. If the population is infinite, the parameters can be thought of as the limits as $N \rightarrow \infty$. The expected value, $E[\ ]$, is the population average. In particular $\mu = E[X]$ and $\sigma^2 = E[(X-\mu)^2]$.

An estimate is most unlikely to equal the population parameter because of sampling variability. It is useful to distinguish between an estimate and the estimator. The estimator is defined by a formula and will vary from sample to sample, whereas an estimate is the value the formula takes in a specific instance. The distinction is usually clear from the context. The property of estimators can conveniently be described in terms of accuracy and precision. An estimator is accurate if on average it is close to the population parameter. An estimator is precise if estimates do not vary much. In Figure 22.2 suppose the parameter lies at the middle of the target and each of the marks shows an estimate. Target (a) shows a precise but not accurate estimator, Target (b) shows an accurate but not precise estimator and Target (c) shows an estimator that is both precise and accurate.

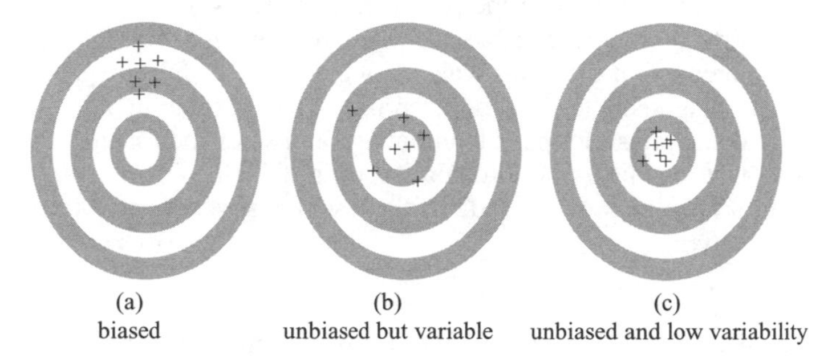

| (a) | (b) | (c) |
|---|---|---|
| biased | unbiased but variable | unbiased and low variability |

**Figure 22.2** Target analogy for bias and variability in estimators.

The formal definition of bias is the difference between the average value of the estimator and the parameter it purports to estimate. Therefore, an accurate estimator is either unbiased or only has a slight bias. With the divisor ($n$–1) the sample variance, $s^2$, is unbiased for the population variance. However, the sample standard deviation, $s$, is slightly biased for the population standard deviation.

**Bayesian Spam Filters**

Many modern spam filters rely on Bayes' rule. Consider that one event is spam and the other event is words in a current e-mail. From Equation (22.1) it is possible to write

P(spam | words) = P(words | spam) P(spam) / P(words)

which literally states that the probability of a certain e-mail being spam given a set of words equals the probability of those words occurring in spam e-mails, multiplied by the probability of a spam e-mail and divided by the probability of finding those same words in any e-mail. The filter does not know these probabilities beforehand, but determines them over time by viewing hundreds of e-mails and having a user confirm if they are spam.

## 22.4 BAYES' RULE AND TOTAL PROBABILITY

Bayes' rule follows in a straightforward manner from the rules of probability and it can be written as:

$$P(B|A)P(A) = P(A/B)P(B) \qquad (22.1)$$

Bayes' rule is useful because it relates the conditional probabilities of events *A* and *B*. It is quite common that an event is not measured in isolation but is measured conditioned on some other event – for example, the probability of failure given different failure modes or the probability of a flood given that the season is summer or that the season is winter.

*Example*
The replacement of a leaking underground water pipeline is an expensive exercise. Researchers have developed an acoustic test that will enable an engineer to assess whether or not there is a leak in the pipeline. This test is, however, subject to two errors: it can indicate that there is a leak when there wasn't one (false-positive), or it can fail to indicate that there is a leak when there is one (false-negative). Assume that before any diagnostic test an engineer assesses the probability of a leak is 0.01 (based on breakage rates of pipes of similar age and size) and assume the makers of the acoustic test estimate the probability of a false-positive as 0.05 and the probability of a false-negative as 0.02. If the test is positive, what is the chance that the pipe is defective?

*Solution*
In this example, Bayes' rule enables the use of the diagnostic test to improve the engineer's estimate. Let *G* represent the event that a pipe is in good condition (no leak) and *D* represent a defective pipe that has a leak. Let *A* be the event that a test is affirmative and *N* be the event that the test is negative. The probability of a defective pipe given a positive test result, $P(D/A)$, is $P(D/A) = P(D)P(A/D) / P(A)$.

From Equation (22.1) the probability of A can be expanded using the theorem of total probability. As shown in Figure 22.3 the events *G* and *D* are mutually exclusive and mutually exhaustive because they cannot occur together and one or the other must occur. It follows that the event *A* is equivalent to (*A and D*) or (*A and G*), and these two composite events are themselves mutually exclusive. The probability of (*A and D*) is equal to $P(A|D)\,P(D)$, and likewise $P(A \text{ and } G) = P(A|G)P(G)$. The theorem of total probability gives the following expression: $P(A) = P(A|G)\,P(G) + P(A|D)\,P(A)$. This theorem is useful as quite often the event A is not observed in isolation and must be expressed it in terms of conditional probabilities.

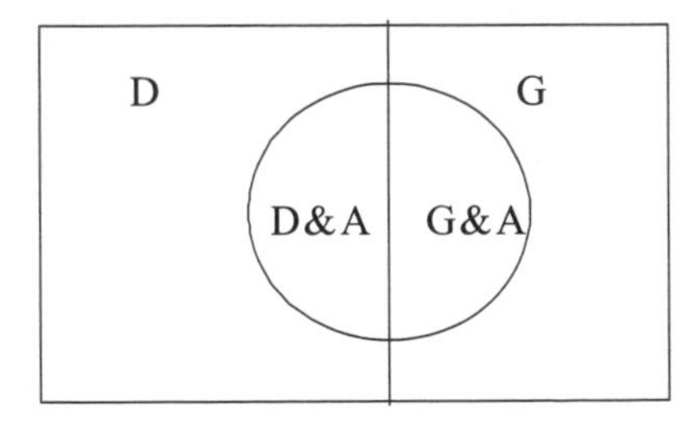


**Figure 22.3** Schematic of mutually exclusive events to explain the theorem of total probability.

Incorporating the theorem of total probability with Bayes' rule gives:

$$P(D|A) = \frac{P(D)P(A \mid D)}{P(A)} = \frac{P(D)P(A \mid D)}{P(D)P(A \mid D) + P(G)P(A \mid G)} \tag{22.2}$$

Figure 22.3 outlines the mutually exclusive probabilities in a tree-diagram, and from Equation (22.2) P(D|A) = 0.01*0.98/(0.01*0.98+0.99*0.05) = 0.16. Also, note that P(G|A) = 1–P(D|A). Consequently, the probability of being defective given the test results is much higher than the original assessment and might be sufficient to justify the expense of replacing the pipe.

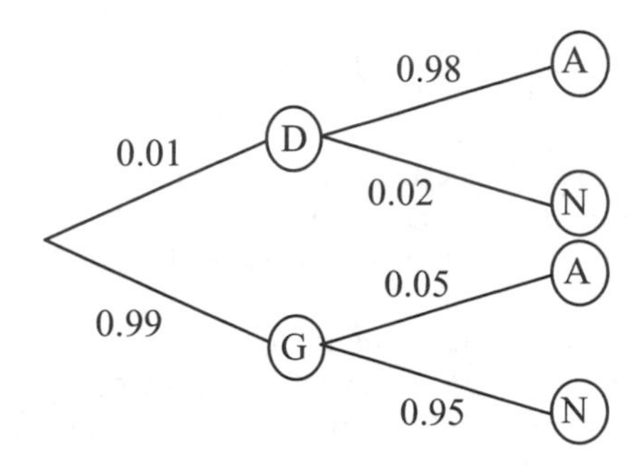


**Figure 22.4** Tree-diagram of probabilities from mutually exclusive events.

## PROBLEMS

**22.1** A gas company operates in two divisions, *A* and *B*, which contain $N_A = 40\ 000$ and $N_B = 80\ 000$ properties respectively. The company is considering replacing pre-1975 meters with a modern design, and wishes to estimate the cost. Let $p_A$ and $p_B$ be proportions of all properties in *A* and *B* with pre-1975 meters and $p$ be the overall proportion.

a) Express $p$ in terms of $p_A$ and $p_B$.
b) Assume two random samples of size $n_A$=400, and $n_B$=800. If 240 old meters were found in *A* and 200 were found in *B*, estimate $\hat{p}_A$, $\hat{p}_B$ and $\hat{p}$.
c) As for (b) except $n_A = n_B = 400$. 240 and 100 old meters were found in the respective samples. For each division, what is the chance a property will be selected? Estimate the overall proportion of pre-1975 meters.

**22.2** A water company for a city in the Great Britain has a policy of replacing lead pipes. The company recently took over management of two jurisdictions that have had this policy in place for some time. The company surveyed three divisions with respective populations 220,000, 149,000 and 413,000 properties. The estimated proportion of properties having a lead pipe connection was 0.23, 0.21 and 0.46 respectively. What is the weighted proportion of lead pipes?

**22.3** A company having data acquisition products decides to open a trade booth at an engineering conference. The conference has the following attendees: mining (27%), civil (45%), environmental (16%), mechanical (7%) and other (5%). For each group the company estimates the respective chances of making one sale as 0.09, 0.17, 0.4, 0.1 and 0.01. Of the civil engineers 27% are academics and 14% are students. Of the environmental engineers 66% are government employees. All other attendees are industry consultants. The company estimates the following breakdown of their previous sales: 40% (government), 53% (industry), 7% (academic) and 0% (students). What is the probability of making a sale?

**22.4** An engineering company estimates that customers visit their webpage in the following proportion: prior clientele (0.34), word of mouth (0.31), search engines (0.22) and online advertisements (0.13). They estimate that from prior clientele and word-of-mouth customers, there is 1 sale for every 10,000 visits to their webpage. From search engines they estimate 1 sale for every 1,000 visits and for advertisements they estimate 1 sale for every 100 visits. For a given webpage visit, what is the probability of making a sale? Given that there was a sale, what is the probability that it was due to an online advertisement?

**22.5** For some value x consider the function $\phi(x) = Profit(x)P(x)$ where $P(x)$ is the probability of $x$ occurring and $Profit(x)$ is the profit of that event. A wind-farm operator estimates the probability of occurrence of four different wind categories (x): no wind (0.11), infrequent gusts (0.48), consistent gale (0.39) and extreme winds (0.02). For a day classified according to the respective categories, the estimated profit is: –$10,000, $50,000, $160,000 and –$10,000. What is the expected daily profit?

**22.6** One raffle sells 1000 tickets at $100 each for the chance to win $50,000. Another raffle sells 100,000 tickets for $1 each for the chance to win $50,000. What is the probability of winning each raffle? Which raffle is the more worthwhile to enter? Hypothetically, if 100 tickets are bought in the 1000-ticket raffle, what are the expected winnings? How is this different from the actual winnings if the lottery is entered?

CHAPTER TWENTY THREE

# Probability and Statistics (Linear Combination of Random Variables)

## 23.1 INTRODUCTION

In many engineering applications it is necessary to consider the distribution of some linear combination of random variables. For example, in the manufacture of single-cylinder motor cycle engines, the cylinders and pistons are selected at random. The clearance is the cylinder diameter minus the piston diameter and will clearly depend on variations in both sizes. Another example is the sample mean, which is the sum of the sample values divided by the sample size.

As a more detailed example, consider a student who has 5 assessments, each worth 25 marks. The student estimates that they will obtain a mark of $20 \pm 2$ for each assessment. The final grade is a linear combination of the 5 assessment marks: $80 \pm 10$. The estimated range of the final grade is larger than for the individual assessments since it is possible that they may achieve 18 for all assessments, or may achieve 22 for all assessments. Note also that the variability in their estimated grade will depend on whether the assessments were related in some way. It is possible that the assessments test the same piece of knowledge; for example, a course may have an assignment, essay, project, experiment, exam all related to one topic. Depending on how well the student grasps this one topic they will be more likely to get all 18 or all 22 than if the assessments were on distinct separate topics within the course.

## 23.2 MATHEMATICAL MODEL

Let $X$ and $Y$ be variables with means $\mu_X$, $\mu_Y$ and variances $\sigma_X^2$, $\sigma_Y^2$ respectively. For constants $a$ and $b$, a variable which is the linear combination, $W = aX + bY$, will have the following properties:

$$\mu_W = a\mu_X + b\mu_Y \tag{23.1}$$

$$\sigma_W^2 = a^2\sigma_X^2 + b^2\sigma_Y^2 + 2ab\ \mathrm{cov}(X, Y) \tag{23.2}$$

To prove this result recall that the expected value $E[\ ]$ is the average value of the population. It follows that:

$$E[a\phi(x)+b\psi(x)]=\frac{1}{N}\sum_{i=1}^{N}(a\phi(x)+b\psi(x))$$

$$=\frac{a}{N}\sum_{i=1}^{N}\phi(x)+\frac{b}{N}\sum_{i=1}^{N}\psi(x) \qquad (23.3)$$

$$=aE[\phi(x)]+bE[\psi(x)]$$

Using this result the mean is:

$$\mu_W = E[W] = E[aX+bY] = aE[X] + bE[Y] = a\mu_X + b\mu_Y \qquad (23.4)$$

This process can be repeated for the expression $\sigma_W^2 = E[(W - \mu_W)^2]$ by substituting the linear combination $W = aX + bY$, collecting like terms, applying the result from Equation (23.3) and noting that the covariance is defined as $cov(X,Y) = E[(X-\mu_X)(Y-\mu_Y)]$. The covariance term will be equal to zero if the two random variables are independent.

The properties outlined in Equation (23.3) are useful largely because the linear combination occurs frequently in practice. This is illustrated with a variety of examples.

*Example*

A wind turbine produces 600kW of electricity when it operates, but due to the variability in wind conditions it has output $\mu_X = 410$ kW and $\sigma_X = 30$ kW. For any two turbines the covariance is of the form $cov(X,Y) = \rho_{XY}\sigma_X\sigma_X$ where $\rho_{XY}$ is the correlation coefficient. What is the output from three turbines if (i) they are uncorrelated, and (ii) $\rho_{XY} = 0.8$ for all pairs of turbines.

*Solution*

Denote the combined output from three respective turbines as $W = X + Y + Z$. The mean output is $\mu_W = \mu_X + \mu_Y + \mu_Z = 1230$ kW. If the turbine output is uncorrelated the variance is $\sigma_W^2 = \sigma_X^2 + \sigma_Y^2 + \sigma_Z^2 = 2700$ giving a standard deviation of 51.96 kW. If the turbines are correlated the variance becomes 7020 since $\sigma_W^2 = \sigma_X^2 + \sigma_Y^2 + \sigma_Z^2 + 2\rho_{XY}\sigma_X\sigma_Y + 2\rho_{XZ}\sigma_X\sigma_Z + 2\rho_{YZ}\sigma_Y\sigma_Z$ and the standard deviation is 83.79 kW.

*Example*

An oil company has a survey vessel which measures the depth of the sea bed with two instruments which give readings $X$ and $Y$. Both instruments give unbiased estimates of depth. That is, if the depth is $\theta$, $E[X] = E[Y] = \theta$. However, the second instrument is more precise and $\sigma_X^2 = 2\sigma_Y^2$. A surveyor intends on averaging the two results, but she thinks some weighted average will be better than W=(X+Y)/2.

(i) Find the mean and variance of the average.
(ii) Write the expressions for mean and variance for a general weighted average.
(iii) Find the best weighted average that minimises the combined variance.

*Solution*

Assume the errors are independent. In the case of $W=0.5X+0.5Y$ the mean is $\mu_W = 0.5\theta + 0.5\,\theta = \theta$. The variance is $\sigma_W^2 = 0.25(2\sigma_Y^2) + 0.25\sigma_Y^2 = 0.75\sigma_Y^2$. A general weighted average will have the form $W = aX + (1-a)Y$ where $0 < a < 1$. Thus the variance will be $\sigma_W^2 = a^2\sigma_X^2 + (1-a)^2\sigma_Y^2 = a^2 2\sigma_Y^2 + (1-a)^2\sigma_Y^2$ and the mean will be $\mu_W = a\theta + (1-a)\theta = \theta$. To find the best weighted average differentiate the variance with respect to the weight parameter $a$:

$$\frac{d\sigma_W^2}{da} = 4a\sigma_Y^2 - 2(1-a)\sigma_Y^2 = 0 \tag{23.5}$$

Solving Equation (23.5) gives $a = 1/3$ and for this combination the variance is $\sigma_W^2 = (2/3)\sigma_Y^2$ which is less than the variance obtained for $W=0.5X+0.5Y$.

## 23.3 DISTRIBUTION OF THE SAMPLE MEAN

Let $\{X_i\}$ be a random sample of size $n$ from a distribution with mean $\mu$ and finite variance $\sigma^2$. Define the sample total, $T$, by $T=X_1+\ldots+X_n$. The results for a linear combination of variables and the fact that randomisation makes an assumption of independence valid gives:

$$\mu_T = \mu + \ldots + \mu = n\mu \tag{23.6}$$

$$\sigma_T^2 = \sigma^2 + \ldots + \sigma^2 = n\sigma^2 \tag{23.7}$$

$$\sigma_T = \sigma\sqrt{n} \tag{23.8}$$

Now the sample mean $m$ is, by definition, $m=T/n$. Therefore, $\mu_m = \mu_T/n = \mu$ and $\sigma_m^2 = \sigma_T^2/n^2 = n\sigma^2/n^2 = \sigma^2/n$. The standard deviation is obtained from the variance as $\sigma_m = \sigma/\sqrt{n}$. These results give the mean and standard deviation of the distribution of the sample mean, $m$. The central limit theorem states that the shape of this distribution is Gaussian. In other words:

$$\frac{m-\mu}{\sigma/\sqrt{n}} \text{ is approximately distributed as a } N(0,1) \tag{23.9}$$

The approximation improves as $n$ increases, and is usually good for $n$ above 30. If the population itself is nearly normal the approximation is excellent for any value of $n$. The distribution of $m$ is an example of a *sampling distribution*. Imagine taking millions of samples of size $n$ from a population, replacing after each selection (unless it is an infinite population), and recording the means. Plotting a histogram of all these means would look like a normal distribution. The mean of the means would be $\mu$ and the standard deviation of the means would be $\sigma/\sqrt{n}$. The term *standard error* is often used for the standard deviation of the mean.

## PROBLEMS

**23.1** Find the respective mean and standard deviation totals from one, two and three dice. What will be the mean and standard deviation of 100 dice?

**23.2** A rainfall model estimates the annual total of rainfall by independently sampling monthly rainfall amounts and adding them. Would this model correctly match the mean and standard deviation of annual totals?

**23.3** A lift can carry ten passengers. Suppose passengers' weights are normally distributed with a mean of 80 kg and a standard deviation of 14 kg. Let $X$ represent passenger weight. Then $X \sim N(80,(14)^2)$. Find the following probabilities:

(i) Pr (a randomly selected passenger exceeds 90 kg);
(ii) Pr (total weight of 10 passengers selected at random exceeds 900 kg);
(iii) Pr (mean weight of 10 passengers selected at random exceeds 90 kg).

**23.4** A workshop cuts plates for tankers. The distributions of lengths of plates are known. A particular unit has a fore (forward) and aft plate on either side. A designer needs to know the likely difference in lengths between the port and starboard sides, before units are built. If lengths of forward plates $X_1, X_2$ are distributed $N(3.0,(0.020)^2)$ and lengths of aft plates $Y_1, Y_2$ are distributed $N(2.5,(0.016)^2)$ find the probability that the difference in lengths, $D$, exceeds 0.05 m when plates are selected at random. This involves the linear combination $D = X_1 + Y_1 - X_2 - Y_2$ and requires the result that a linear combination of normally distributed random variables has a normal distribution.

CHAPTER TWENTY FOUR

# Probability and Statistics (Correlation and Regression)

## 24.1 INTRODUCTION

There are numerous situations where an engineer may be interested in two variables for each item of some population, and the relationship between them. Some examples are:

(i) Rainfall on a catchment and flow in a river. In this case, the flow depends on the rainfall but rainfall does not depend on flow.

(ii) The stopping distance of cars and their speed when the test driver is asked to brake.

(iii) Breathalyser readings and the blood-alcohol level of drivers stopped for random breath testing. In this case, the breathalyser provides an approximate measure of the blood alcohol level.

(iv) The concentrations of carbon monoxide and benzoa pyrene in air samples. In this case, neither chemical causes the other, but it is possible that both are a result of automobile engines.

(v) The normal strength and the shear strength of soil samples. These are two characteristics of soil that are generally related, but neither can be said to cause the other.

Assume that $n$ data pairs $(x_i, y_i)$ are available and the first step in the analysis is to plot the data. It is usual to denote the variable to be predicted by $y$ and the predictor variable by $x$, but common alternative names for $y$ include the 'response' or 'dependent' variable and 'explanatory' variable is often used for $x$. In (i) take flow as $y$, in (ii) take stopping distance as $y$, and in (iv) and (v) the choice is arbitrary. In (iii) the choice is less clear because although the objective is to predict blood alcohol from the breathalyser test, the standard statistical model is more appropriate for the breathalyser being the response variable.

## 24.2 CORRELATION

Correlation is a measure of linear association between two variables. As an example, consider 16 air samples from Herald Square in Manhattan (Colucci and Begeman, 1971). The air samples were analysed for carbon monoxide (denoted $x$ in ppm) and benzoa pyrene (denoted $y$ in $\mu g/10^4 m^3$). The measurements are provided in Table 24.1 and Figure 24.1 shows a scatter plot. It appears the two may be associated.

**Table 24.1** Carbon monoxide (denoted $x$ in ppm) and benzoa pyrene (denoted $y$ in $\mu g/10^4 m^3$).

| $i$ | 1 | 2 | 3 | 4 | 5 | 6 | 7 | 8 | 9 | 10 | 11 | 12 | 13 | 14 | 15 | 16 |
|---|---|---|---|---|---|---|---|---|---|---|---|---|---|---|---|---|
| $x_i$ | 3 | 15 | 19 | 7 | 5 | 6 | 10 | 13 | 5 | 12 | 6 | 20 | 11 | 13 | 5 | 10 |
| $y_i$ | 5 | 1 | 8 | 9 | 10 | 16 | 39 | 40 | 13 | 57 | 15 | 60 | 73 | 81 | 22 | 95 |

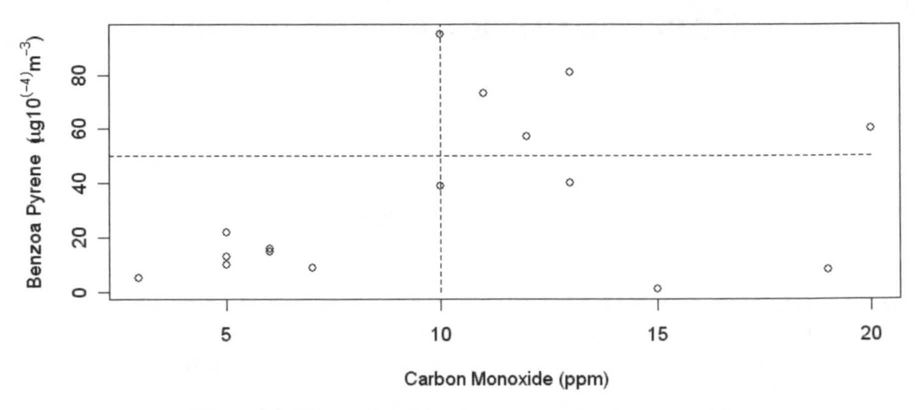


**Figure 24.1** Scatterplot of benzoa pyrene and carbon monoxide.

A measure of linear association can be obtained from the sample covariance. This is defined by:

$$\hat{cov} = \frac{\sum (x_i - m_x)(y_i - m_y)}{n-1} \tag{24.1}$$

where $m_x$ and $m_y$ are the means of the respective variables. The points in the scatterplot (Figure 24.1) are divided into four quadrants by drawing the lines $x = m_x$ and $y = m_y$. For a typical point $(x_i, y_i)$, in the lower right quadrant $x_i - m_x > 0$ and $y_i - m_y < 0$ so the product $(x_i - m_x)(y_i - m_y)$ is negative. Similarly for all other points in the lower right quadrant and the upper left quadrant.

Points in the other two quadrants will make positive contributions to the sum in the numerator of $\hat{cov}$. If there is no association between the two variables, the positive and negative contributions will tend to balance out and, after division by $(n-1)$, $\hat{cov}$ will be negligible. In contrast to this, if there is a tendency for $y$ to increase as $x$ increases, most of the contributions will be positive, and $\hat{cov}$ will be large and positive. Similarly, if there is a tendency for $y$ to decrease as $x$ increases, $\hat{cov}$ will be large and negative. The unit of $\hat{cov}$ is the product of the unit of $x$ with the unit of $y$, so the interpretation of 'large' depends on the choice of units. The sample correlation, $r$, is a dimensionless quantity obtained from the covariance by:

$$r = \frac{\hat{cov}}{s_x s_y} \tag{24.2}$$

where $s_x$ and $s_y$ are the standard deviation of the respective variables. Notice that the covariance of $x$ with itself is $s_x^2$. The value of $r$ for these 16 data is 0.35. This corresponds to a discernable increasing relationship between the two variables, but there is considerable scatter. It can be shown that $-1 \leq r \leq 1$.

The correlation will take its extreme value of $-1$ if the points lie on a straight line with a negative slope, and 1 if they lie on a line with a positive slope (see Figure 24.2). If points are scattered equally over all four quadrants, the correlation will be close to zero. However, correlation is only a measure of linear association and points displaying a clear pattern can have correlations close to zero. So, zero correlation does not always imply random scatter. Also be aware that correlation does not imply causation.

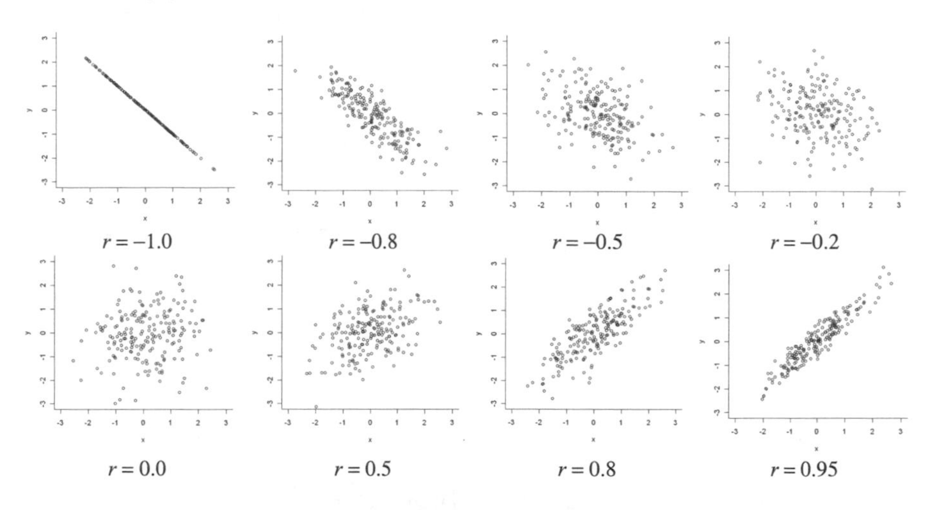


**Figure 24.2** Sample of 200 random numbers with various correlations, $r$.

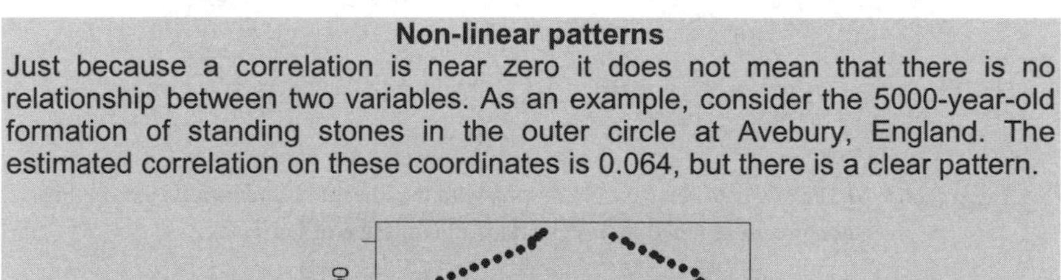

**Non-linear patterns**

Just because a correlation is near zero it does not mean that there is no relationship between two variables. As an example, consider the 5000-year-old formation of standing stones in the outer circle at Avebury, England. The estimated correlation on these coordinates is 0.064, but there is a clear pattern.

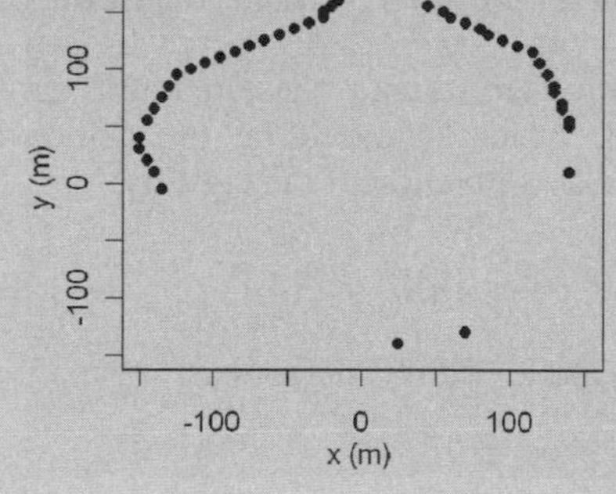

## 24.3 CORRELATION AND TIME SERIES

A time series is a sequence of values from one random variable at equally spaced time steps. If there are no trends in the data there may still be correlations between values separated by some fixed time increment known as the lag. For example, Gy (2004) points out the importance of correlated patterns when trying to determine propcrtics of a stream of minerals on a conveyor belt heading for processing. If there are non-random fluctuations then this may compromise the sampling and lead to biased results. Turner et al. (2006) used the autocorrelation function to investigate and identify long-term changes in a multi-year beach volume plot. After accounting for a long-term trend the analysis showed a strong annual cycle of erosion and accretion on the Gold Coast beach under study. The results, created in Matlab, are shown in Figure 24.3.

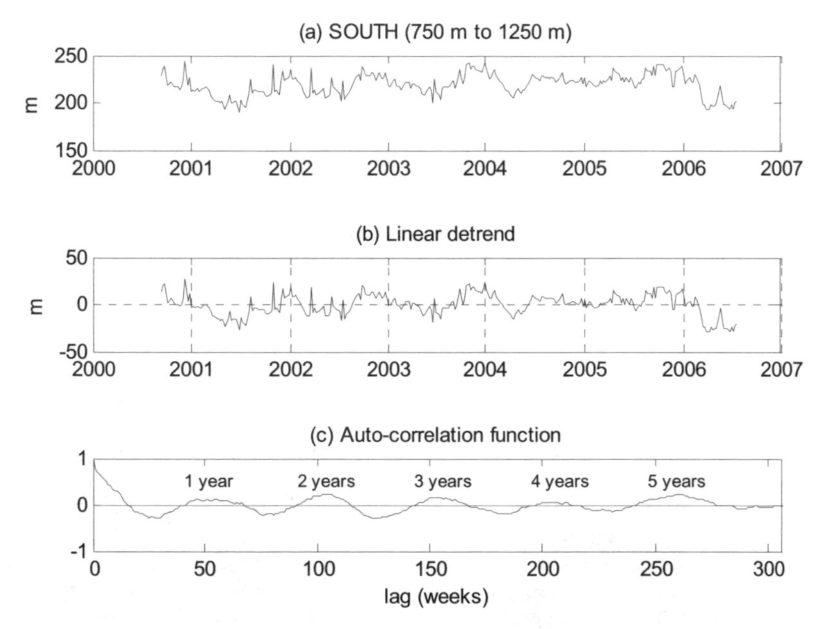


**Figure 24.3** (a) The raw data set, (b) the raw data with the linear trend removed, and (c) the autocorrelation function. Used with the permission of Ian Turner.

The analysis required to detect these patterns involves calculating the autocorrelation function, which is based on the autocovariance function. The autocovariance is defined as a function of the lag $k$ by:

$$\gamma(k) = \mathrm{E}[(x(t)-\mu_x)(x(t+k)-\mu_x)] \tag{24.3}$$

and the autocorrelation function can be defined as:

$$\rho(k) = \frac{\gamma(k)}{\gamma(0)} \tag{24.4}$$

which can be estimated by:

$$r(k)=\frac{\sum_{t=1}^{n-k}(x(t)-m_x)(x(t+k)-m_x)}{\sum_{t=1}^{n}(x(t)-m_x)^2} \tag{24.5}$$

where $r(k)$ is the $k^{th}$ element of an array of correlation coefficients, the data are stored in $x(t)$ where $t$ goes from 1 to $n$, $n$ is the total number of data, $k$ is the lag number, and $m_x$ is the estimated mean. Note that the denominator does not depend on $k$ and so can be calculated once and then used for all $k$.

When the $r(k)$ are plotted the result is called a correlogram. It can be shown that if the data are serially uncorrelated then 95% of values of the correlation function will lie between $-1/n \pm 1.96/\sqrt{n}$. Autocorrelations outside these limits are statistically significant at the 5% level.

Figure 24.4 shows 100 data from a set of raw wave data collected at Seacliff, Adelaide. The data were collected at 0.5-second intervals. The autocorrelation function was calculated and plotted in Figure 24.5. The effect of the marked periodicity in the input data is evident with the pattern of correlation coefficients.

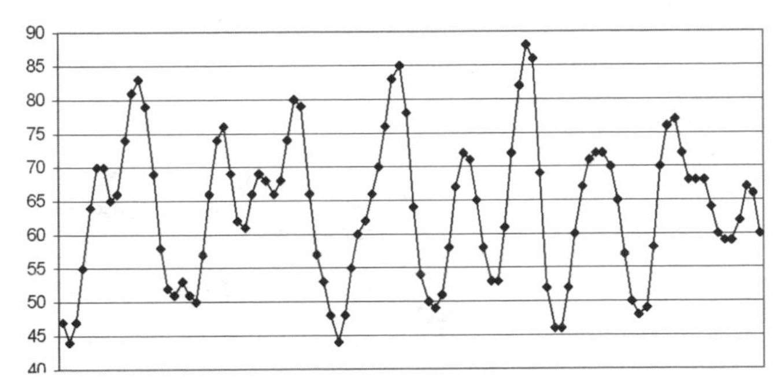


**Figure 24.4** Raw wave data from Seacliff, Adelaide.

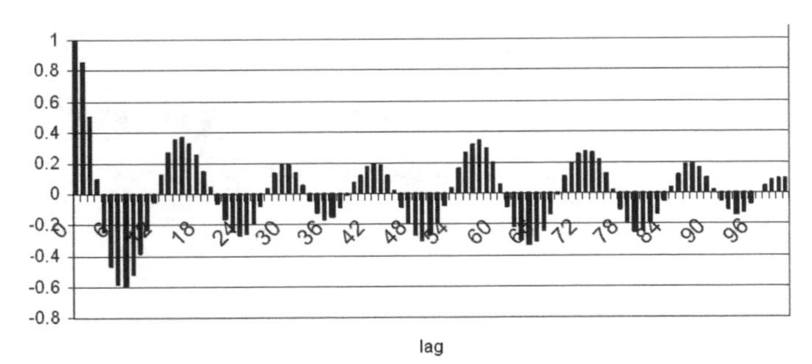


**Figure 24.5** Autocorrelation function. The first 101 lags (from 0 to 100) are shown.

## 24.4 TESTING A RANDOM NUMBER GENERATOR

For a random number generator to be useful it must (i) produce numbers with the specified probability distribution; and (ii) produce numbers with no discernable pattern. In other words, each generated number should be independent of preceding numbers. The autocorrelation function can be used to test the latter.

As a test case, 1000 numbers were generated in Fortran using the inbuilt random number generator, `random_number( )`, as shown in Figure 24.6. The autocorrelation function was calculated for the first 20 lags ($k$=20 in Equation 24.5) and these were then plotted, as shown in Figure 24.7. The 95% limits were calculated as $-1/n \pm 1.96/\sqrt{n}$ and were plotted also. The plot shows that only one correlation (lag 6) lies outside the limits, which is acceptable since 95% limits imply that 1 in 20 numbers will lie outside these lines on average for a random sample. This indicates that the random number generator produces uncorrelated numbers. Note, however, that the requirement of independent random numbers is more rigorous than this as there may be other patterns in the random number generator that need to be checked. For example, a transformation of the random numbers might show significant correlations at some lag – for example, if the numbers were squared or cubed.

```
program autocor
! Autocorrelation function for random numbers
! written m.leonard, april 2008

implicit none
integer,parameter :: n=1000  ! no of data
integer,parameter :: nlag=20     ! no of lags
real :: r(0:nlag)                ! autocorrelations
real :: x(1:n)                   ! data array
real :: mx,sx2                   ! mean and sum of squares
integer :: t,k                   ! index counters

call random_number(x)            ! get array of random numbers
mx = sum(x)/real(n)          ! calculate mean
sx2 = sum((x(:)-mx)**2.0)        ! sum x**2 - note: array syntax
                             ! calculate correlations from 0 to nlag
do k=0,nlag
 r(k)=0.0
 do t=1,n-k
  r(k)=r(k)+(x(t)-mx)*(x(t+k)-mx)
 end do
 r(k) = r(k)/sx2             ! standardise to get correlation
end do
                             ! output results
open(unit=98,file='output.txt')
do k=0,nlag
 write(*,*)k,',',r(k)            ! screen output
 write(98,*) k,',',r(k)      ! file output
end do
close(98)
write(*,*)'Press ENTER to exit'
read(*,*)
end program
```

**Figure 24.6** Fortran code to produce correlogram of 1000 random numbers.

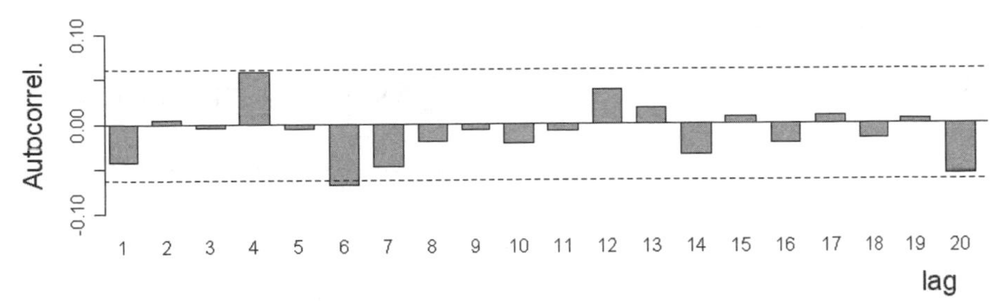


**Figure 24.7** Autocorrelation function for Fortran random number generator. The 95% confidence limits are also shown.

## 24.5 SIMPLE LINEAR REGRESSION

In some cases the scatter-plot of the data pairs will indicate that a linear relationship is a reasonable model over the range of values covered by the sample. Simple linear regression involves fitting a straight line through a set of data. The basic idea is to minimise the error where that error is measured parallel to the $y$-axis between the line and the individual data points. This is shown in Figure 24.8 and called the regression of $y$ on $x$.

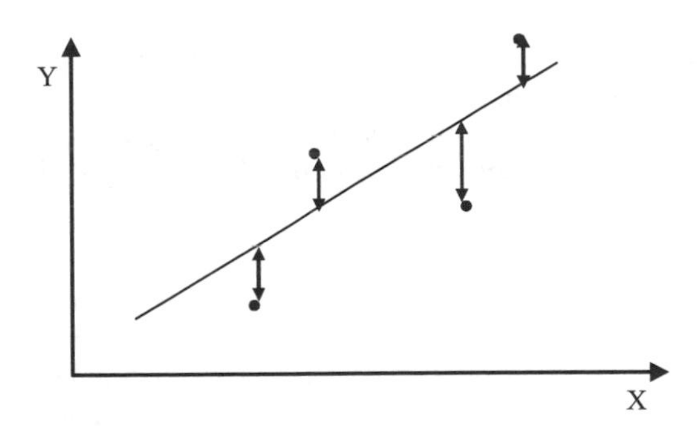


**Figure 24.8** A curve-fitting exercise, where the errors (shown with the vertical arrows) are used to derive a straight line such that the sum of the errors squared is a minimum.

*Example*

Tungsten steel erosion shields are fitted to the low-pressure blading in steam turbines. The manufacturer wishes to assess the quality of each batch of erosion shields on the basis of a random sample from each batch. The direct measurement of abrasion loss requires that the test pieces are subjected to steam erosion over a long period. This is an expensive test and would result in considerable delays before a batch could be used. It would be useful if a related measurement that is quick and inexpensive can be used as a surrogate. The Vickers hardness of the erosion blades is a possible surrogate, and the aim is to investigate the relationship between abrasion loss and Vickers hardness measurements. An experiment was performed with 13 shields and the results are given in Table 24.2 and plotted in Figure 24.9.

**Table 24.2** Vickers hardness and abrasion loss of erosion shields in pressure turbines.

| Vickers hardness ($x$, $10\text{Nmm}^{-2}$) | 665 | 719 | 659 | 756 | 711 | 671 | 709 | 722 | 718 | 714 | 701 | 683 | 731 |
|---|---|---|---|---|---|---|---|---|---|---|---|---|---|
| Abrasion loss ($y$, mg) | 597 | 436 | 602 | 297 | 393 | 561 | 385 | 380 | 340 | 513 | 499 | 553 | 416 |

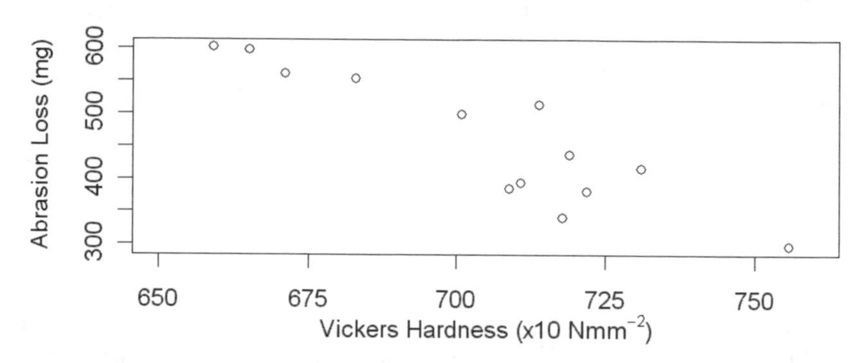


**Figure 24.9** Scatterplot of Vickers hardness and abrasion loss.

There is a tendency for the abrasion loss to decrease as the Vickers hardness increases, but the points do not all lie on a straight line: *abrasion loss* = $\alpha+\beta$(*Vickers hardness*). By including random variation in the model, the points lying off the line can be accommodated. This is given as, *abrasion loss* = $\alpha+\beta$(*Vickers hardness*) + *random variation*. The general form of a linear regression is:

$$Y_i = \alpha + \beta x_i + \varepsilon_i\,, \quad i = 1,\ldots,n \tag{24.6}$$

where $Y$ is the variable to be predicted, $x$ is the predictor variable, there are $n$ observations, $\alpha$ is a parameter for the intercept and $\beta$ is a parameter for the slope. The $\varepsilon_i$ represents the random variation and it is convenient to refer to the random variation as 'error'. The errors are not necessarily measurement error and often represent inherent variability in the population. The random variation has the following assumptions:

*Assumption 1* $E[\varepsilon_i]=0$
*Assumption 2* The $\varepsilon_i$ are uncorrelated with the $x_i$ .
*Assumption 3* The $\varepsilon_i$ are uncorrelated with each other.
*Assumption 4* The $\varepsilon_i$ all have the same variance $\sigma^2$.
*Assumption 5* The $\varepsilon_i$ are normally distributed.

Assumption 1 is crucial and if it is not satisfied the parameter estimate of the intercept will be biased. Assumption 2 is also crucial: if it is not satisfied the estimate of the slope will be biased. Neither Assumption 1 nor Assumption 2 can be checked from the data. Correct calibration of measuring equipment is therefore critical. Assumptions 3, 4 and 5 can be relaxed as is done using generalised least squares.

A consequence of Equation (24.6) is that the model defines the conditional distribution of $Y$ given $x$ and the $x_i$ in the model are treated as known constants. Therefore, $E[Y/x] = \alpha + \beta x$ and $Var[Y \mid x] = \sigma^2$.

The unknown population parameters are $\alpha$, $\beta$ and $\sigma^2$. The least squares estimators of $\alpha$ and $\beta$ are the values $\hat{\alpha}$ and $\hat{\beta}$ which minimise the sum of squared errors:

$$SSE = \sum_{i=1}^{n} \varepsilon_i^2 = \sum_{i=1}^{n} (Y_i - (\alpha + \beta x_i))^2 \tag{24.7}$$

and are obtained by solving:

$$\frac{\partial SSE}{\partial \alpha} = 0, \quad \frac{\partial SSE}{\partial \beta} = 0 \tag{24.8}$$

Since these equations are linear in $\alpha$ and $\beta$, there is an analytic solution:

$$\hat{\beta} = \frac{\sum (x_i - m_x)(y_i - m_y)}{\sum (x_i - m_x)^2}, \quad \hat{\alpha} = m_y - \hat{\beta} m_x \tag{24.9}$$

The fitted regression line is $y = \hat{\alpha} + \hat{\beta} x$, which can also be written as $y = m_y + \hat{\beta}(x - m_x)$. This line is the line such that the sum of the squared distances, parallel to the $y$-axis, from the points to the line is a minimum. The fitted values of $y$ are defined as $\hat{y}_i = \hat{\alpha} + \hat{\beta} x_i$ and the residuals, which are estimates of the errors, are defined as $r_i = y_i - \hat{y}_i$. The estimate of $\sigma^2$ is $s^2 = \sum r_i^2 / (n-2)$.

Notice that the denominator is $(n–2)$ rather than $n$. It is said that two degrees of freedom are lost because two parameters, $\alpha$ and $\beta$, have been estimated from the data. One way of justifying the division by $(n–2)$ is that if a straight line is drawn between two points, the estimated standard deviation of the errors is then an undefined quantity rather than zero.

These steps can be implemented easily in a spreadsheet. Consider Table 24.3, where the Vickers hardness and abrasion loss data are entered in the first few columns. From columns B and C the respective means are $m_x = 704.5$ and $m_y = 459.4$. These are used to obtain the standardised value in columns D and E. Columns D and E are used to obtain the products in columns F and G. From Equation (24.9) the slope is estimated from the sum of values in column F divided by the sum of column G. This gives $\hat{\beta} = –3.24$. The intercept is then $\hat{\alpha} = 459.4 – (–3.24)(704.5) = 2744.0$. Column H uses these estimates to determine the fitted values $\hat{y}_i$. The residuals are the discrepancy between the observed and fitted values and are shown in column I. The squared residuals in column J are summed to obtain an estimate of their variability, $s^2$=2082.6.

Suppose that the Vickers hardness of an erosion shield is 670. The predicted hardness measurement is 575. An approximate prediction interval for an individual erosion shield with a Vickers hardness of 670 is 575 ± 2s, which is 575 ± 84. This is based on an assumption that the errors are normally distributed. It would not be a good idea to use the fitted relationship to predict abrasion loss if the Vicker's hardness is outside of the range of values used to fit the regression. For example, a Vicker's hardness of 900 would correspond to a negative abrasion loss.

**Table 24.3** Calculation steps to estimate regression parameters.

| A | B | C | D | E | F | G | H | I | J |
|---|---|---|---|---|---|---|---|---|---|
| $i$ | $x$ | $y$ | $(x-m_x)$ | $(y-m_y)$ | $(x-m_x)$ $(y-m_y)$ | $(x-m_x)^2$ | $\hat{y}_i$ $\hat{\alpha}+\hat{\beta}x_i$ | $r_i$ $y_i-\hat{y}_i$ | $r_i^2$ |
| 1 | 665 | 597 | −39.5 | 137.6 | −5441.1 | 1563.3 | 587.6 | 9.4 | 88.4 |
| 2 | 719 | 436 | 14.5 | −23.4 | -338.2 | 209.1 | 412.5 | 23.5 | 552.8 |
| 3 | 659 | 602 | −45.5 | 142.6 | −6494.5 | 2073.8 | 607.1 | −5.1 | 25.6 |
| 4 | 756 | 297 | 51.5 | −162.4 | −8356.6 | 2648.3 | 292.5 | 4.5 | 20.2 |
| 5 | 711 | 393 | 6.5 | -66.4 | −428.9 | 41.8 | 438.4 | −45.4 | 2064.0 |
| 6 | 671 | 561 | −33.5 | 101.6 | −3408.0 | 1124.8 | 568.1 | −7.1 | 51.0 |
| 7 | 709 | 385 | 4.5 | −74.4 | -331.9 | 19.9 | 444.9 | −59.9 | 3590.0 |
| 8 | 722 | 380 | 17.5 | −79.4 | −1386.2 | 304.9 | 402.8 | −22.8 | 518.1 |
| 9 | 718 | 340 | 13.5 | −119.4 | −1607.1 | 181.2 | 415.7 | −75.7 | 5735.3 |
| 10 | 714 | 513 | 9.5 | 53.6 | 507.3 | 89.5 | 428.7 | 84.3 | 7106.0 |
| 11 | 701 | 499 | −3.5 | 39.6 | −140.2 | 12.5 | 470.9 | 28.1 | 791.9 |
| 12 | 683 | 553 | −21.5 | 93.6 | −2016.3 | 463.9 | 529.2 | 23.8 | 565.1 |
| 13 | 731 | 416 | 26.5 | −43.4 | −1148.0 | 700.2 | 373.6 | 42.4 | 1799.8 |

In general a linear regression is an empirical approximation to some more complex underlying relationship and it is only applicable within the range of the data. Extrapolation should be avoided, or if this is not feasible it should be at least kept to a minimum.

## 24.6 REGRESSION TOWARDS THE MEAN

In a regression of $Y$ on $x$, $Y$ is the random variable and the values of $x$ are treated as fixed. In a designed experiment, the investigator chooses the values of $x$ at which $Y$ is observed. In a correlation analysis, the pairs $(x_i, y_i)$ are considered as a random sample from some population. Provided that the inherent variability in this population is considerably larger than any measurement error, it is also valid to carry out a regression of $Y$ on $x$ or $X$ on $y$. However, these two regression lines will be different.

Consider the benzoa pyrene and carbon monoxide concentrations of air samples in Figure 24.10. The samples have been standardised using $x = (x' - m_x)/s_x$ where $x'$ is the original data, $m_x$ is the mean and $s_x$ is the standard deviation. A similar expression is used for $y$. The regression line $E[Y \mid x]$ is obtained using the values of $x$ to predict $Y$ and the regression line $E[X \mid y]$ is obtained taking the values of $y$ to predict $X$. The parameters for the two lines are different and this is also clear from Figure 24.10. The lines cross at the mean values $(m_x, m_y)$.

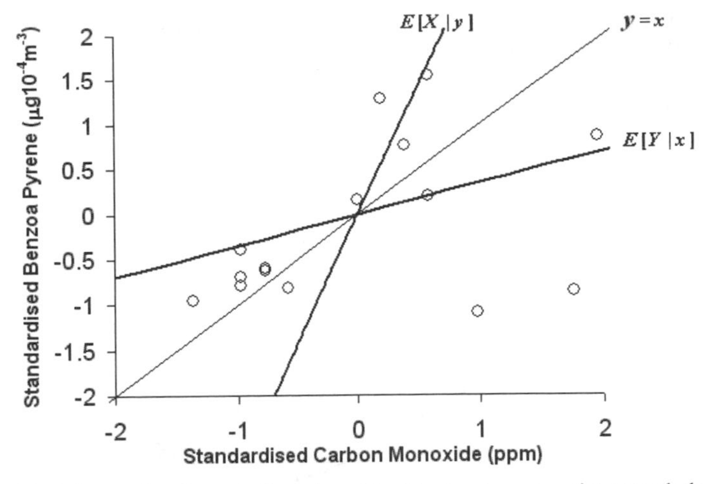


**Figure 24.10** Scatterplot of standardised air quality data showing regression towards the mean.

Figure 24.10 can be used to explain the phenomenon of 'regression towards the mean'. If, for example, carbon monoxide is 2 standard deviations above its mean, from the line $E[Y \mid x]$ in Figure 24.10, the expected value of benzoa pyrene is about 0.6 standard deviations above the mean. That is, the average prediction of benzoa pyrene is not also 2 standard deviation above the mean. This is because the benzoa pyrene is not solely predicted by the carbon monoxide concentration, and the other unknown predicting variables are unlikely to be at similar above average levels. The phenomenon of regression toward the mean was first described by Sir Francis Galton, who studied the heights of parents and their children. In this context, if a parent is extremely tall then it is more likely for their child to be shorter than them. A similar argument applies for an extremely short parent. This tendency to 'return to' the mean is the reason for the terminology 'regression' rather than a more prosaic term such as 'line-fitting'.

## PROBLEMS

**24.1** The following concentrations of nickel and chromium were obtained from mineral samples:

| Nickel | (ppm) | 130 | 165 | 95 | 140 | 135 | 145 |
|---|---|---|---|---|---|---|---|
| Chromium | (ppm) | 400 | 490 | 380 | 440 | 460 | 470 |

a) Plot the data.
b) Determine the correlation between nickel and chromium.
c) Obtain the parameter estimates for the regression of nickel on chromium.
d) Obtain the parameter estimates for the regression of chromium on nickel.
e) Plot the two regression lines.
f) Give reasons why these regression relationships might be useful in practice.

g) If a further sample is taken and the Chromium levels are estimated to be 520 ppm, what is the expected Nickel concentration?

**24.2** Chatfield (1984) derives the form of the autocorrelation function for a Markov process shown in Figure 24.11. The form of the data that make up a Markov process can be written $x_n = \alpha x_{n-1} + \varepsilon_n$ where ε is a Gaussian noise series with a mean of zero and a standard deviation of 1.0 and $x_0 = 0.0$.

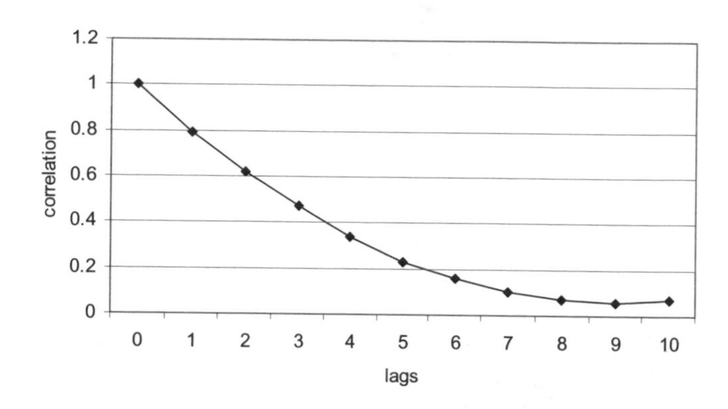


**Figure 24.11** Autocorrelation function for Markov process, α = 0.8.

Write a Fortran program to generate 1000 data points according to the Markov equation and then calculate the autocorrelation function. With other values of α the behaviour should vary. With α = 0.3 the drop-off is much more sudden and with α = –0.8 the values oscillate positive and negative with the absolute magnitude gradually reducing.

**24.3** An experiment was conducted to measure the blood-alcohol reading of 19 individuals from a blood sample compared to a reading from a breathalyser.

| | | | | | | | | | | |
|---|---|---|---|---|---|---|---|---|---|---|
| Blood alcohol (mm alcohol /100 g blood) | 0 | 14 | 20 | 46 | 71 | 74 | 86 | 79 | 96 | 114 |
| Breathalyser | 40 | 21 | 35 | 34 | 59 | 44 | 47 | 51 | 76 | 70 |

| | | | | | | | | | |
|---|---|---|---|---|---|---|---|---|---|
| Blood alcohol (mm alcohol /100 g blood) | 112 | 152 | 150 | 165 | 190 | 192 | 202 | 226 | 229 |
| Breathalyser | 52 | 73 | 71 | 88 | 67 | 90 | 103 | 88 | 101 |

Estimate the standard deviation of the residuals. Is there significant variability in the data? For a blood-alcohol level of 170 mm alcohol per 100 g of blood, what is the expected breathalyser reading?

## CHAPTER TWENTY FIVE

# Probability and Statistics (Multiple Regression)

### 25.1 INTRODUCTION

Multiple regression is a means of predicting one variable from measurements of other variables that are associated with it. These other variables are either easier to measure than the variable to be predicted or they are antecedent to it. For example, a crucial measure of the quality of concrete used in structures is the compressive strength after 28 days (28-day strength). However, a contractor needs to detect any sub-standard concrete when it is delivered and will test random samples for slump and density that are used to predict the eventual 28-day strength.

### 25.2 MULTIPLE REGRESSION MODEL

The general model relates the variable to be predicted to a linear combination of a set of predictor variables. The variable to be predicted is also commonly referred to as the response or the dependent variable, and the predictor variables are often called explanatory variables. The model is proposed for a sample of $n$ observations, and the objective is to estimate the coefficients in the linear combination from these observations. Once the coefficients have been estimated the model can be used for predictions.

Let $y_i$, for $i = 1,...n$, be the response values and $x_{1i},...x_{ki}$ be the corresponding values of the $k$ predictor variables. The sample is $(x_{1i},...x_{ki}, y_i)$, for $i = 1,\ldots,n$ and the model for the sample observations is:

$$y_i = \beta_0 + \beta_1 x_{1i} + \cdots + \beta_k x_{ki} + e_i \qquad \text{for } i = 1, ..., n \qquad (25.1)$$

where the $e_i$ are independent random errors with a mean of 0. The errors represent inherent variation in the population, which will be in part, at least, due to the effects of factors that are not included in the model, or errors in measuring the response, or both. The objective is to estimate the coefficients, $\beta_j$, from the sample. The method can be expressed far more succinctly if matrix algebra is used. To begin with, the model can be expressed as:

$$Y = XB + E \qquad (25.2)$$

where

$$Y = \begin{pmatrix} y_1 \\ y_2 \\ \vdots \\ y_{n-1} \\ y_n \end{pmatrix} \quad X = \begin{pmatrix} 1 & x_{11} & \cdots & x_{k1} \\ 1 & x_{12} & \cdots & x_{k2} \\ \vdots & \vdots & \cdots & \vdots \\ 1 & x_{1,n-1} & \cdots & x_{k,n-1} \\ 1 & x_{1,n} & \cdots & x_{k,n} \end{pmatrix} \quad B = \begin{pmatrix} \beta_0 \\ \beta_1 \\ \vdots \\ \beta_k \end{pmatrix} \quad E = \begin{pmatrix} e_1 \\ e_2 \\ \vdots \\ e_{n-1} \\ e_n \end{pmatrix} \tag{25.3}$$

The principle of least squares is used to estimate the coefficients and was first published in 1805 by A. M. Legendre in his work on the estimation of the orbits of comets. The regression model can be rewritten with the errors on the left-hand side:

$$e_i = y_i - (\beta_0 + \beta_1 x_{1i} + \cdots + \beta_k x_{ki}) \qquad for \quad i = 1,\ldots,n \tag{25.4}$$

and by considering all of the data points, the sum of squared errors is:

$$\sum_{i=1}^{n} e_i^2 = \sum_{i=1}^{n} \left(y_i - (\beta_0 + \beta_1 x_{1i} + \cdots + \beta_k x_{ki})\right)^2 \tag{25.5}$$

If the known numerical values obtained from the sample are substituted into Equation (25.5), the principle of least squares is to estimate the $\beta_j$ by the values which minimise the sum of squared errors, denoted by $\hat{\beta}_j$. It can be shown (see Problems 25.1 and 25.2) that the unknown coefficients can be obtained from:

$$\hat{B} = (X'X)^{-1} X'Y \tag{25.6}$$

where the fitted values $\hat{y}_i$ of the $y_i$ are their predicted values from the fitted model. In matrix terms:

$$\hat{Y} = X\hat{B} \tag{25.7}$$

The residuals, $r_i$, are the differences between the observed and fitted values and are estimates of the errors:

$$r_i = y_i - \left(\hat{\beta}_0 + \hat{\beta}_1 x_{1i} + \cdots + \hat{\beta}_k x_{ki}\right) \tag{25.8}$$

If the column of residuals is denoted R, it is given by:

$$R = Y - \hat{Y} \tag{25.9}$$

the sum of the residuals is zero. The estimate of the standard deviation of the errors in the model is given by:

$$s = \sqrt{\frac{1}{n-k-1} \sum_{i=1}^{n} r_i^2} \tag{25.10}$$

Since the estimates of the coefficients are found by minimising the sum of squared errors, and the sum of squared residuals equals this minimum, the sum of squared residuals will be less than the unknown sum of squared errors. The division by $(n - k - 1)$ compensates for the sum of squared residuals being less than the unknown sum of squared errors.

## 25.3 WORKED EXAMPLE

The following example is based on real data from a retail company that has been successful and wishes to expand by purchasing a new shop. Two suitable shops have been identified, but the company only has the resources to buy one of these. One shop has a larger floor area, but the other is in a busier area. Which should it buy? There is little difference between them in terms of outlay and running costs, so the company's priority is to select the shop which is likely to generate more sales. The company has records of sales from its existing shops and knows their floor areas. It also has a measure of the pedestrians per hour, during the working day, using the streets in which these shops are located, from local councils. These data are summarised in Table 25.1. Data should always be plotted before attempting any further analysis and a 3D plot is shown in Figure 25.1.

**Table 25.1** Sales and explanatory data for 10 shops.

| Pedestrians | 564 | 1072 | 326 | 1172 | 798 | 584 | 280 | 970 | 802 | 650 |
|---|---|---|---|---|---|---|---|---|---|---|
| Area | 650 | 700 | 450 | 500 | 550 | 650 | 675 | 750 | 625 | 500 |
| Sales | 980 | 1160 | 800 | 1130 | 1040 | 1000 | 740 | 1250 | 1080 | 876 |

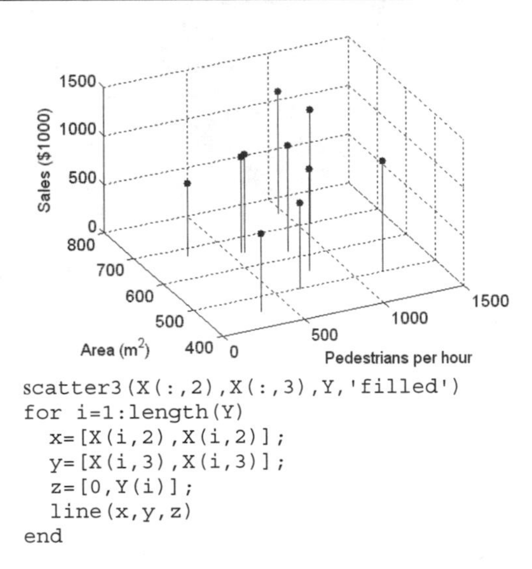


```
scatter3(X(:,2),X(:,3),Y,'filled')
for i=1:length(Y)
  x=[X(i,2),X(i,2)];
  y=[X(i,3),X(i,3)];
  z=[0,Y(i)];
  line(x,y,z)
end
```

**Figure 25.1** 3D scatterplot of sales with respect to pedestrian numbers and floor area.

The Matlab commands to input the data and perform the regression are shown in Figure 25.2 along with the output regression coefficients, observed values, fitted values and residuals.

```
% Input
Y=[980 1160 800 1130 1040 1000 740 1250 1080 876]';
X=[1   1    1   1    1   1   1   1    1   1
   564 1072 326 1172 798 584 280 970 802 650
   650 700  450 500  550 650 675 750 625 500]';
% Multiple regression
B=inv(X'*X)* X'*Y
Yhat=X*B
R=Y-Yhat
%%%%%%%%%%%%%%output
```

$$B = (385.7, 0.46424, 0.4708)$$

$$Y = (980, 1160, 800, 1130, 1040, 1000, 740, 1250, 1080, 876)$$

$$\hat{Y} = (853.53, 1212.90, 748.88, 1165.17, 1015.08, 962.81, 833.45, 1189.09, 1052.25, 922.83)$$

$$R = (26.47, -52.90, 51.12, -35.17, 24.92, 37.19, -93.45, 60.91, 27.75, -46.83)$$

**Figure 25.2** Matlab commands and output from multiple regression.

Using Equation (25.9) the estimated standard deviation of the errors, $s$, is 59.47. Finally, the fitted plane from this regression is shown in Figure 25.3.

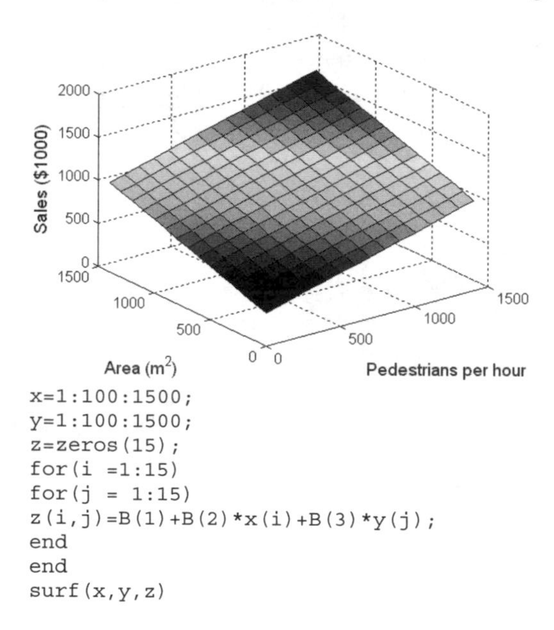


**Figure 25.3** Surface plot of fitted multiple regression.

The company must decide between shop A and shop B. The company knows their sizes and the pedestrian traffic but there is no sales information for the proposed new shops. Even if they were established shops it is most unlikely that they would have been selling the same products. The relevant data for the new shops are listed in Table 25.2.

**Table 25.2** Data for choosing shop to buy.

| Shop | Pedestrians | Floor area |
|---|---|---|
| A | 475 | 1000 |
| B | 880 | 550 |

Using the fitted regression model, the predicted sales are:

Shop A: $385.7 + 0.46424 \times 475 + 0.4708 \times 1000 = 1077$
Shop B: $385.7 + 0.46424 \times 880 + 0.4708 \times 550 = 1053$

At a glance it appears that Shop A is the better purchase. But this would be a risky decision. A regression is an empirical approximation that is valid over the range of values of the predictor variables used to fit it, and predictions should be based on interpolation. Generally, extrapolation should be avoided. The regression plane for sales is based on data from 10 shops which have pedestrians per hour ranging from 280 to 1172, and floor areas ranging from 450 to 750. The floor area for Shop A is 1000, which is substantially greater than any of the 10 shops used to fit the regression. It is possible that there isn't sufficient variety of stock to fill the display space, and expanses of open floor may deter customers.

The predicted sales for Shop A are only moderately higher, 24, than for Shop B when compared with the estimated standard deviation of the errors, which is 59. Approximate, and rather too narrow, 67% prediction limits for sales from a single shop are the predicted sales plus or minus 59. Unless the company has specific plans for taking advantage of the extra floor space, Shop B would be the more prudent choice.

## 25.4 REGRESSION IN EXCEL

Excel is a convenient tool for multiple regression. The only slight limitation is that the predictor variables must be arranged into contiguous columns. The Analysis ToolPak Add-in must be enabled, which can be checked on the submenu Tools – Add-ins – Analysis ToolPak – Analysis ToolPak VBA. The multiple regression can be performed using the form located on the submenu Tools – Data Analysis – Regression.

The input of the pedestrian and floor area data ($x$) along with the sales data ($y$) is shown in Figure 25.4. The results appear in another sheet, as shown in Figure 25.5. The main interest from the Excel output is in the regression coefficients and the standard error, which is the estimated standard deviation of the errors. The $R^2$ (R-square) value is the proportion of the variance explained by the regression and is often expressed as a percentage:

$$R^2 = 1 - \frac{\sum r_i^2}{\sum (y_i - \bar{y})^2} \tag{25.11}$$

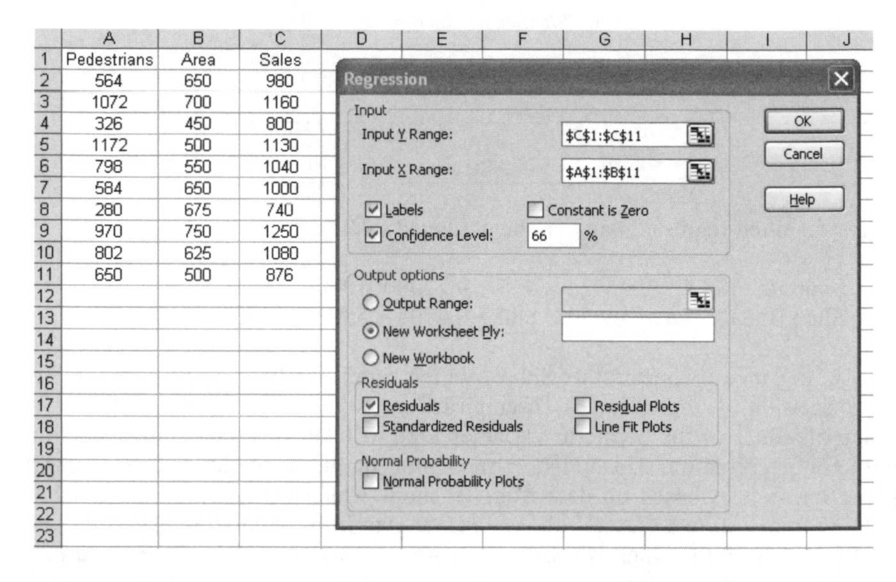


**Figure 25.4** Input data and user form for multiple regression in Excel.

| | A | B | C | D | E | F | G | H | I |
|---|---|---|---|---|---|---|---|---|---|
| 3 | *Regression Statistics* | | | | | | | | |
| 4 | Multiple R | 0.95 | | | | | | | |
| 5 | R Square | 0.90 | | | | | | | |
| 6 | Adjusted R Square | 0.87 | | | | | | | |
| 7 | Standard Error | 59.47 | | | | | | | |
| 8 | Observations | 10 | | | | | | | |
| 9 | | | | | | | | | |
| 10 | ANOVA | | | | | | | | |
| 11 | | *df* | *SS* | *MS* | *F* | *Significance F* | | | |
| 12 | Regression | 2 | 211301.9 | 105650.9 | 29.9 | 0.000374 | | | |
| 13 | Residual | 7 | 24760.5 | 3537.2 | | | | | |
| 14 | Total | 9 | 236062.4 | | | | | | |
| 15 | | | | | | | | | |
| 16 | | *Coefficients* | *Standard Error* | *t Stat* | *P-value* | *Lower 95%* | *Upper 95%* | *Lower 66.0%* | *Upper 66.0%* |
| 17 | Intercept | 385.68 | 125.65 | 3.07 | 0.02 | 88.57 | 682.79 | 257.04 | 514.32 |
| 18 | Pedestrians | 0.46 | 0.067 | 6.885 | 0.0002 | 0.305 | 0.624 | 0.395 | 0.533 |
| 19 | Area | 0.47 | 0.203 | 2.322 | 0.053 | -0.009 | 0.950 | 0.263 | 0.678 |
| 20 | | | | | | | | | |
| 21 | RESIDUAL OUTPUT | | | | | | | | |
| 22 | | | | | | | | | |
| 23 | *Observation* | *Predicted Sales* | *Residuals* | | | | | | |
| 24 | 1 | 953.53 | 26.47 | | | | | | |
| 25 | 2 | 1212.90 | -52.90 | | | | | | |
| 26 | 3 | 748.88 | 51.12 | | | | | | |
| 27 | 4 | 1165.17 | -35.17 | | | | | | |
| 28 | 5 | 1015.08 | 24.92 | | | | | | |
| 29 | 6 | 962.81 | 37.19 | | | | | | |
| 30 | 7 | 833.45 | -93.45 | | | | | | |
| 31 | 8 | 1189.09 | 60.91 | | | | | | |
| 32 | 9 | 1052.25 | 27.75 | | | | | | |
| 33 | 10 | 922.83 | -46.83 | | | | | | |

**Figure 25.5** Analysis of regression outputs in Excel.

From Figure 25.5 multiple-R is the square root of $R^2$. The $R^2$ value can range from 0, if the estimates of all the coefficients with the exception of the mean, $\beta_0$, are 0, to 1 if the fit is exact. An exact fit is not necessarily impressive; an exact fit to 10 data can be obtained using any 9 linearly independent columns of predictor variables. An adjusted $R^2$ discounts such over-fitting by dividing the numerator and denominator by the corresponding degrees of freedom, so adjusted $R^2$ increases or decreases as $s$ decreases or increases:

$$R^2_{adj} = 1 - \frac{\sum r_i^2 /(n-k-1)}{\sum (y_i - \bar{y})^2 /(n-1)} = 1 - \frac{s^2}{s_y^2} \qquad (25.12)$$

In the analysis of variance (ANOVA) the significance F can be useful. It is the probability of such a high $R^2$ value if there is no association between the predictors and response. If the model is to be plausible, significance F should be a low probability, generally less than 0.1. However, if the data set is large it is possible to get small Significance F associated with models that do not explain much of the variability in the data despite the evidence of some association between predictors and response. In other words, the model would be valid but of little use.

Turning to the next table in Figure 25.5, the coefficients and their labels are given in the first two columns. The third column, standard error, gives the estimated standard deviation of the estimators of the coefficients. To understand the concept of the standard deviation of an estimator it is necessary to imagine the investigation being repeated many times, each with a different realisation of random errors. This will result in many estimates $\beta_j$ for each $j$. The mean of this distribution, which is called a sampling distribution, of estimates $\hat{\beta}_j$ is $\beta_j$ and the standard deviation of the distribution is the standard deviation of the estimator, commonly referred to as its standard error. Although this is an imaginary exercise and there is only the single estimate $\hat{\beta}_j$ the sampling distribution of the $\hat{\beta}_j$ can be deduced from the model. The mean and variance of the sampling distribution of the coefficients in a multiple regression are given by:

$$mean[\hat{B}] = B \qquad cov[\hat{B}] = (X'X)^{-1}\sigma^2 \qquad (25.13)$$

The $cov[\,]$ matrix in Equation (25.13) is square and has variances of the estimators along the leading diagonal, and covariances of estimator pairs in the off-diagonal cells. The covariances show any correlations between estimated parameters and can be ignored for now. The standard error column in Figure 25.5 is calculated by replacing the unknown variance of the errors in Equation (25.13), $\sigma^2$, with $s^2$ and then taking square root of the terms on the leading diagonal. The t-Stat column in Figure 25.5 is the ratio of the coefficient to its standard error, and the P-value is the probability of such a large t-Stat, in absolute value, if the hypothetical true value of the coefficient is 0. If the P-value is small – less than 0.05, for example – empirical evidence is claimed of an association between predictor and response at the 5% level.

The remaining columns of this coefficients table are confidence limits. An interpretation is that the lower/upper 95% values are limits within which there is a 95% chance that the hypothetical true value of the coefficient lies. The interval needs to be all positive, or all negative, to be able to claim empirical evidence of an association between predictor and response at the 5% level. The lower/upper 66% values can be interpreted as limits within which there is a 66% chance that the hypothetical true value of the coefficient lies, and these are approximately equal to the estimate of coefficient plus or minus the estimated standard error.

For the worked example the 95% confidence interval for the coefficient of floor area is [–0.01, 0.95], so its value has not been identified very precisely. This is partly due to the small sample and partly due to sales being influenced by many things other than the pedestrian traffic and floor area. Nevertheless, it seems reasonable to suppose that floor area has some effect on sales and the estimated coefficient, 0.4708, is the best estimate of this effect within the range of floor areas in the sample.

Excel provides a bubble plot feature, which can be a useful alternative to a 3D plot. In Figure 25.6 the diameters of the bubbles are equal to the floor area less 400. The reason for subtracting 400 is to emphasise the differences in size, rather than their absolute value. The data must be in three contiguous columns: $X$, $Y$ and bubble diameter.

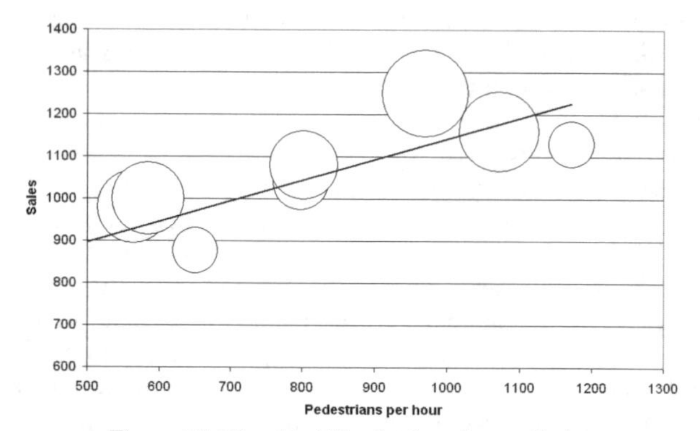


**Figure 25.6** Excel bubble plot for sales prediction.

The bubbles are opaque and two are obscured. However, there is a tendency for the larger bubbles to lie above the regression line of sales on *pedest*, suggesting that floor area may help explain sales.

## 25.5 FITTING QUADRATIC SURFACES

A regression is an empirical approximation to some unknown underlying relationship between predictor variables and the response. The underlying relationship will typically be complex, but a model which is linear in the predictor variable or predictor variables and is geometrically represented by a line, plane (2 predictors) or hyper-plane may be an adequate approximation. However, a Taylor series can be emulated by introducing squared and cross-product terms in the hope of obtaining a better approximation. The multiple regression model is still referred to as a linear model by statisticians because it remains linear in the unknown parameters. As an example, a general quadratic surface has the form:

$$y_i = \beta_0 + \beta_1 x_{1i} + \beta_2 x_{2i} + \beta_3 x_{1i}^2 + \beta_4 x_{1i} x_{2i} + \beta_5 x_{2i}^2 + e_i \quad for \quad i = 1, \ldots, n \tag{25.14}$$

Rebeiz et al. (1996) performed an experiment to investigate the effect of sand and fly ash filler on the tensile strength of a polyester mortar. This was a designed experiment and they were free to choose values for the percentage of sand and ash filler. The results are given in Table 25.3.

**Table 25.3** Tensile strength of 34 preparations of a polyester mortar.

| Sand | Ash | Strength |
|---|---|---|
| 0 | 0 | 38.9 |
| 0 | 0 | 37.8 |
| 10 | 0 | 26.4 |
| 10 | 0 | 25.8 |
| 20 | 0 | 23.1 |
| 20 | 0 | 25.3 |
| 30 | 0 | 18.2 |
| 30 | 0 | 17.1 |
| 40 | 0 | 24.2 |
| 40 | 0 | 26.4 |
| 0 | 0 | 39.5 |
| 0 | 0 | 37.8 |
| 0 | 0 | 36.7 |
| 0 | 10 | 31.8 |
| 0 | 10 | 31.3 |
| 0 | 10 | 28 |
| 0 | 20 | 24.7 |
| 0 | 20 | 23.6 |
| 0 | 30 | 20.9 |
| 0 | 30 | 17.6 |
| 0 | 30 | 13.8 |
| 0 | 40 | 18.2 |
| 0 | 40 | 17.6 |
| 0 | 40 | 14.4 |
| 0 | 40 | 12.7 |
| 40 | 0 | 22.5 |
| 40 | 0 | 21.5 |
| 40 | 0 | 20.4 |
| 30 | 10 | 23.1 |
| 30 | 10 | 25.3 |
| 20 | 20 | 31.3 |
| 20 | 20 | 26.9 |
| 10 | 30 | 21.5 |
| 10 | 30 | 14.9 |

A regression on sand and ash has an estimated standard deviation of the errors of 4.59 with an $R^2$ of 64.5%, whereas the quadratic surface has an estimated standard deviation of the errors of 2.76 with an $R^2$ of 88.4%. The quadratic surface is a substantially better fit. When fitting a quadratic surface it is good practice to mean-adjust the original predictor variables. If this is not done the correlations between sand and sand-squared and between ash and ash-squared are 0.964 and 0.961 respectively. This can cause numerical instability and lead to inconveniently large numbers for some of the predictors. It is also common practice to scale the mean-adjusted predictor variables to avoid very small or very large estimates of the coefficients. Define:

$$x_1 = \frac{sand - 20}{10} \qquad x_2 = \frac{ash - 20}{10} \tag{25.15}$$

The fitted regression surface is then:

$$y = 25.3 + 5.44 x_{1i} + 3.48 x_{2i} + 2.24 x_{1i}^2 + 4.65 x_{1i} x_{2i} + 0.77 x_{2i}^2 \tag{25.16}$$

Notice that the cross-product term has a substantial effect on the strength. The cross-product represents an interaction between sand and ash, and so the effect of sand will depend on the percentage of ash and vice versa. A contour plot of the fitted surface is shown in Figure 25.7, but notice that only the region corresponding to less than 40% filler, $x_1+x_2 \le 2$, is applicable. The tensile strength of polyester mortar decreases as filler is added, but other important properties such as elasticity increase with filler so a compromise is needed. Rebeiz et al. (1996) recommend 20% sand and 20% ash, with a predicted tensile strength of 25.3 MPa.

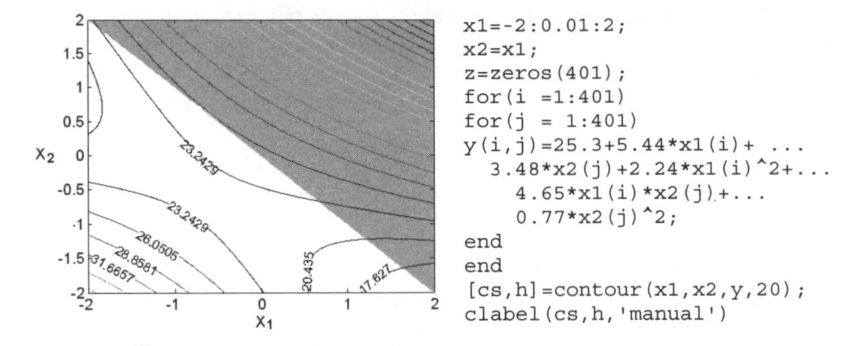


**Figure 25.7** Contours of strength of a polyester mortar.

## 25.6 INDICATOR VARIABLES

The predictor variables that have been considered thus far have all been measured on some continuous scale. However, categorical variables can be included in a multiple regression by the use of indicator variables. Day (1999) investigated the relationship between slump, temperature at time of pouring, density, grade, and 7-day and 28-day strengths for test cubes of concrete made from the same batch of cement. There were three grades: 320LWTB, 320LWTA, and 400. One grade has to be chosen as a base category for the comparisons. It is usual to take the base category as that with the most data, so that comparisons have the highest precision, but the estimates of effects of categories are unaffected by the choice. Two indicator variables need to be defined to distinguish 320LWTA and 400 from 320LWTB. With the coding shown in Table 25.4, the coefficient of $x_1$ represents the difference in strength of 320LWTA and 320LWTB. If the estimate of the coefficient is positive the estimated strength of 320LWTA is higher. Similarly, the coefficient of $x_2$ represents the difference in strength between 400 and 320LWTB.

The data and results of a regression of 28-day strength on the mean adjusted slump, mean adjusted temperature, the squares and product of these two variables, and the two indicator variables are shown in the Figures 25.8 and 25.9.

**Table 25.4** Indicator variables for concrete grade.

| Grade | $x_1$ | $x_2$ |
|---|---|---|
| 320LWTB | 0 | 0 |
| 320LWTA | 1 | 0 |
| 400 | 0 | 1 |

The product of slump and temperature is an example of an interaction term. The practical interpretation of an interaction is that the effect of one predictor variable depends on the value taken by the other. In this case the interaction is positive and indicates that the slump should be higher at higher temperatures and lower at lower temperatures, in order to increase the strength. There is strong evidence that grade 400 has a higher compressive strength than 320LWTB and some weaker evidence that 320LWTA is slightly stronger than 320LWTB. The fitted model can be written as:

$$strength = 40.97 - 0.01S + 0.12T + 0.03S^2 + 0.32T^2 + 0.30ST + \begin{cases} +3.79 & if\,320LWTA \\ +13.64 & if\,400 \end{cases}$$

| | A | B | C | D | E | F | G | H | I | J | K | L | M | N | O | P |
|---|---|---|---|---|---|---|---|---|---|---|---|---|---|---|---|---|
| 1 | Original Data | | | | | | | | | Regression Inputs | | | | | | |
| 2 | grade | slump | temp | density | x1 | x2 | strng_7d | strng_28d | | slump | temp | slump^2 | temp^2 | slump*temp | x1 | x2 |
| 3 | 320LWTA | 172 | 25 | 2058 | 1 | 0 | 40.3 | 46.2 | | -5 | 0.36 | 25 | 0.128 | -1.786 | 1 | 0 |
| 4 | 320LWTA | 180 | 24 | 2074 | 1 | 0 | 38.1 | 43.2 | | 3 | -0.64 | 9 | 0.413 | -1.929 | 1 | 0 |
| 5 | 320LWTA | 180 | 23 | 2045 | 1 | 0 | 36 | 42.5 | | 3 | -1.64 | 9 | 2.699 | -4.929 | 1 | 0 |
| 6 | 320LWTA | 173 | 24 | 2085 | 1 | 0 | 40.5 | 47.8 | | -4 | -0.64 | 16 | 0.413 | 2.571 | 1 | 0 |
| 7 | 320LWTB | 197 | 23 | 2073 | 0 | 0 | 34.9 | 42.5 | | 20 | -1.64 | 400 | 2.699 | -32.857 | 0 | 0 |
| 8 | 320LWTB | 170 | 24 | 2087 | 0 | 0 | 33.2 | 42.5 | | -7 | -0.64 | 49 | 0.413 | 4.500 | 0 | 0 |
| 9 | 320LWTB | 176 | 24 | 2069 | 0 | 0 | 34.2 | 42 | | -1 | -0.64 | 1 | 0.413 | 0.643 | 0 | 0 |
| 10 | 320LWTB | 169 | 28 | 2055 | 0 | 0 | 32.5 | 39.2 | | -8 | 3.36 | 64 | 11.270 | -26.857 | 0 | 0 |
| 11 | 320LWTB | 182 | 26 | 2087 | 0 | 0 | 38.8 | 44.3 | | 5 | 1.36 | 25 | 1.842 | 6.786 | 0 | 0 |
| 12 | 400 | 184 | 25 | 2412 | 0 | 1 | 44.4 | 56.7 | | 7 | 0.36 | 49 | 0.128 | 2.500 | 0 | 1 |
| 13 | 400 | 176 | 23 | 2409 | 0 | 1 | 45.2 | 56.8 | | -1 | -1.64 | 1 | 2.699 | 1.643 | 0 | 1 |
| 14 | 400 | 174 | 26 | 2418 | 0 | 1 | 37.6 | 52.1 | | -3 | 1.36 | 9 | 1.842 | -4.071 | 0 | 1 |
| 15 | 400 | 168 | 25 | 2406 | 0 | 1 | 42.7 | 55.3 | | -9 | 0.36 | 81 | 0.128 | -3.214 | 0 | 1 |
| 16 | 400 | 177 | 25 | 2415 | 0 | 1 | 43.8 | 56.8 | | 0 | 0.36 | 0 | 0.128 | 0.000 | 0 | 1 |

**Figure 25.8** Regression input. Right-hand side uses mean corrected slump and temperature.

| | A | B | C | D | E | F | G |
|---|---|---|---|---|---|---|---|
| 1 | SUMMARY OUTPUT | | | | | | |
| 2 | | | | | | | |
| 3 | *Regression Statistics* | | | | | | |
| 4 | Multiple R | 0.9792 | | | | | |
| 5 | R Square | 0.9588 | | | | | |
| 6 | Adjusted R Square | 0.9108 | | | | | |
| 7 | Standard Error | 1.9313 | | | | | |
| 8 | Observations | 14 | | | | | |
| 9 | | | | | | | |
| 10 | | *Coefficients* | *Standard Error* | *t Stat* | *P-value* | *Lower 95%* | *Upper 95%* |
| 11 | Intercept | 40.968 | 1.821 | 22.56 | 5.05E-07 | 36.51 | 45.54 |
| 12 | slump | -0.014 | 0.104 | -0.13 | 9.01E-01 | -0.27 | 0.24 |
| 13 | temp | 0.123 | 0.544 | 0.23 | 8.29E-01 | -1.21 | 1.45 |
| 14 | slump^2 | 0.027 | 0.017 | 1.62 | 1.56E-01 | -0.01 | 0.04 |
| 15 | temp^2 | 0.325 | 0.460 | 0.71 | 5.07E-01 | -0.80 | 1.45 |
| 16 | slump*temp | 0.304 | 0.167 | 1.82 | 1.19E-01 | -0.11 | 0.41 |
| 17 | x1 | 3.788 | 1.848 | 2.05 | 8.62E-02 | -0.73 | 8.31 |
| 18 | x2 | 13.644 | 1.504 | 9.07 | 1.01E-04 | 9.96 | 17.32 |

**Figure 25.9** Regression outputs for concrete slump, temperature and grade indicators.

A 95% confidence interval for the increase in strength obtained by using grade 400 rather than grade 320LWTB is [10.0, 17.3], and a 95% confidence interval for the increase in strength obtained by using grade 320LWTA rather than 320LWTB is [–0.7, 8.3]. A larger experiment would lead to narrower confidence intervals. Other regression models for this data set are explored in the sample problems.

If the temperature and desired strength are known the optimum slump can be calculated and compared to slump of the delivered concrete. The slump can be controlled by adding or evaporating water from the concrete.

## PROBLEMS

**25.1** Let $\psi$ be a scalar function of an array $B$ where $B = (B_0, B_1)'$. The derivative of $\psi$ with respect to $B$ is defined as

$$\frac{\partial\psi}{\partial B} = \left(\frac{\partial\psi}{\partial B_o}, \frac{\partial\psi}{\partial B_1}\right)'$$

Now let C and M be 2 x 1 and 2 x 2 constant matrices respectively. Show from the definition that:

$$\frac{\partial}{\partial B}(B'C) = C \quad \text{and} \quad \frac{\partial}{\partial B}(B'MB) = MB + M'B$$

**25.2** The sum of squared errors is $\psi = E'E = (Y - XB)'(Y - XB)$. A necessary condition for Y to have a minimum is dY/dB=0. Use the results of Problem 25.1 to show that the solution of this equation is $\hat{B} = (X'X)^{-1}X'Y$. Explain why this is a minimum.

**25.3** Consider the leg strengths of 13 American footballers, shown in Table 25.5, which can be used to predict the distance they punt a football using their right leg (Dunn, 2008; sourced from Myers, 1990). The strengths are measured using a weight-lifting test. Determine the simple regression using left and right legs respectively. Are both relationships significant? Now apply multiple regression. Are both predictors significant?

**Table 25.5** Left- and right-leg strength with distance football kicked.

| Left | Right | Distance |
|---|---|---|
| 170 | 170 | 162.5 |
| 130 | 140 | 144 |
| 170 | 180 | 174.5 |
| 160 | 160 | 163.5 |
| 150 | 170 | 192 |
| 150 | 150 | 171.75 |
| 180 | 170 | 162 |

| Left | Right | Distance |
|---|---|---|
| 110 | 110 | 104.83 |
| 110 | 120 | 105.67 |
| 120 | 130 | 117.58 |
| 140 | 120 | 140.25 |
| 130 | 140 | 150.17 |
| 150 | 160 | 165.17 |

**25.4** The residuals are in the $n \times 1$ matrix $R = Y - \hat{Y} = Y - X\hat{B}$. Show that $X'R = 0$. Explain why this implies that the assumption that the errors have a mean of zero, and the assumption that the errors are uncorrelated with the predictor variables, cannot be checked from the data.

**25.5** Repeat the regression of strength on ash and sand without centring and scaling the predictor variables. Check that identical fitted values are obtained.

**25.6** Information on a number of chocolates commonly available in Brisbane city stores was gathered and listed in Table 25.6 (Dunn, 2008). The price, weight and nutritional information was gathered for 16 chocolate bars. Develop a multiple regression for:

(i) the price of the chocolate,
(ii) the energy provided by the chocolate.

Are all predictor variables useful?

**Table 25.6** Weight, price and nutrition of various chocolate bars in Brisbane, 2005.

| Chocolate | Size | Price | Unit price | Energy | Protein | Fat | Carbs. | Sodium |
|---|---|---|---|---|---|---|---|---|
| Dark Bounty | 50 | 0.88 | 1.76 | 1970 | 3.1 | 27.2 | 53.2 | 75 |
| Bounty | 50 | 0.88 | 1.76 | 2003 | 4.6 | 26.5 | 59 | 115 |
| Milko Bar | 40 | 1.15 | 2.88 | 2057 | 9.9 | 23 | 60.9 | 116 |
| Viking | 80 | 1.54 | 1.93 | 1920 | 5.1 | 18.4 | 67.5 | 220 |
| KitKat White | 45 | 1.15 | 2.56 | 2250 | 7.2 | 30.1 | 59.4 | 110 |
| KitKat Chunky | 78 | 1.4 | 1.79 | 2186 | 7 | 28.4 | 59.7 | 93 |
| Cherry Ripe | 55 | 1.28 | 2.33 | 1930 | 3.5 | 24.5 | 56.4 | 40 |
| Snickers | 60 | 0.97 | 1.62 | 1980 | 10.2 | 22.9 | 59.9 | 190 |
| Mars | 60 | 0.97 | 1.62 | 1890 | 4.7 | 19.5 | 67.9 | 160 |
| Crunchie | 50 | 1.28 | 2.56 | 2030 | 5.6 | 20.4 | 67.4 | 250 |
| Tim Tam | 40 | 1.1 | 2.75 | 2180 | 5.5 | 26.8 | 67.3 | 160 |
| Turkish Delight | 55 | 1.28 | 2.33 | 1623 | 2.2 | 9.2 | 73.3 | 90 |
| Mars Lite | 44.5 | 0.97 | 2.18 | 1640 | 3.7 | 12 | 77.9 | 220 |
| Dairy Milk King | 75 | 1.58 | 2.11 | 2210 | 8.2 | 29.8 | 57 | 110 |
| Maltesers | 60 | 1.55 | 2.58 | 1980 | 8.5 | 20.6 | 63.3 | 130 |
| M & Ms | 42.5 | 1.18 | 2.78 | 1970 | 5 | 20 | 69 | 148 |

**25.7** Develop a model for the cost of finished steel from 1920 to 1937 as related to the year of interest, the operating capacity of the steel mill and the average hourly earnings of employees at the mill (Dunn, 2008, sourced from Ezekiel and Fox, 1940). The data are shown in Table 25.7.

**Table 25.7** Cost of finished steel along with mill capacity and employee earnings.

| Year | Cost | Capacity | Earnings | Year | Cost | Capacity | Earnings |
|---|---|---|---|---|---|---|---|
| 1920 | 72.3 | 88.3 | 77.5 | 1929 | 51.5 | 89.2 | 72.5 |
| 1921 | 78.5 | 47.5 | 60.2 | 1930 | 58.6 | 65.6 | 73.2 |
| 1922 | 57.9 | 71.3 | 58.5 | 1931 | 65.6 | 38.0 | 70.8 |
| 1923 | 63.0 | 88.3 | 67.0 | 1932 | 81.4 | 18.3 | 61.0 |
| 1924 | 63.7 | 69.0 | 70.8 | 1933 | 65.0 | 28.7 | 59.0 |
| 1925 | 62.9 | 78.4 | 70.3 | 1934 | 64.6 | 31.2 | 70.0 |
| 1926 | 60.3 | 88.0 | 70.8 | 1935 | 65.4 | 38.8 | 73.0 |
| 1927 | 59.6 | 78.9 | 71.3 | 1936 | 61.1 | 59.3 | 74.0 |
| 1928 | 55.2 | 83.4 | 71.8 | 1937 | 65.6 | 71.2 | 86.0 |

**25.8** Consider the price of CD, DVD–R and DVD+/–RW discs, as reported in a *Choice* magazine (Dunn, 2008) and listed in Table 25.8. The discs were purchased in April 2005 in multi-packs. How well can the price per disc ($AUD) be explained in terms of the number of discs in the pack and the type of the media? Plot and inspect the residuals. Is this model appropriate?

**Table 25.8** Cost of computer discs along with pack size and type of media.

| Price | Pack | Media | Price | Pack | Media | Price | Pack | Media |
|---|---|---|---|---|---|---|---|---|
| 0.48 | 50 | CD | 0.57 | 50 | DVD | 3.6 | 5 | DVDRW |
| 0.6 | 25 | CD | 2.6 | 10 | DVD | 5 | 10 | DVDRW |
| 0.64 | 25 | CD | 1.59 | 10 | DVD | 2.79 | 5 | DVDRW |
| 0.5 | 50 | CD | 1.85 | 10 | DVD | 2.79 | 10 | DVDRW |
| 0.89 | 10 | CD | 1.85 | 10 | DVD | 4.37 | 5 | DVDRW |
| 0.89 | 10 | CD | 0.72 | 25 | DVD | 1.5 | 10 | DVDRW |
| 1.2 | 10 | CD | 2.28 | 10 | DVD | 2.5 | 5 | DVDRW |
| 1.3 | 10 | CD | 2.34 | 5 | DVD | 3.9 | 10 | DVDRW |
| 1.29 | 10 | CD | 2.4 | 10 | DVD | | | |
| 0.5 | 10 | CD | 1.49 | 5 | DVD | | | |

# CHAPTER TWENTY SIX

# Probability and Statistics (Non-Linear Regression)

## 26.1 INTRODUCTION

Non-linear regression is a term that is applied to models that are non-linear in the parameters. In some cases it may be possible to transform the relationship into a form that can be fitted with the multiple linear regression model. Such models are referred to as intrinsically linear, or linearisable.

If the model is not intrinsically linear, or if the transformed error structure is not thought appropriate, the principle of least squares can still be applied. The problem, though, is that a convenient explicit matrix solution for the parameter estimates is no longer available. In general numerical optimisation must instead be used to find the best parameters that fit the model.

## 26.2 NON-LINEAR REGRESSION EXAMPLE

Myers (1994) investigated the relationship between the angle of draw ($y$) and the ratio of width to depth ($x$) for 16 open-cast mining excavations in West Virginia. A scatterplot of these variables is shown in Figure 26.1.

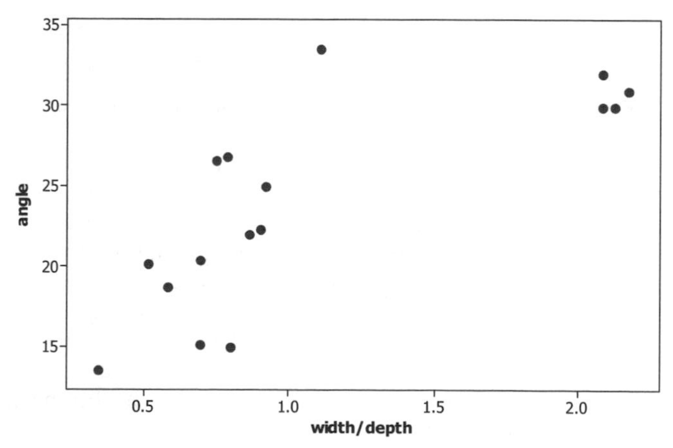


**Figure 26.1** Scatterplot of variables from mining excavations.

It is expected that the angle of draw will increase with the ratio up to a certain point but then level off. To test this, Myers fitted the non-linear model:

$$y_i = \alpha[1 - exp(-\beta x_i)] + \varepsilon_i \qquad (26.1)$$

where $\alpha$ and $\beta$ are parameters and $\varepsilon$ are the errors.

To demonstrate this model in Excel, the data are given in the first columns of the spreadsheet in Figure 26.2. The ratios of width to depth are given in column D. From the plot in Figure 26.1 reasonable initial guesses for $\alpha$ and $\beta$ are 35 and 1 respectively. The squared error for the first data point is calculated as a function of the assumed parameter values which are the numbers in fixed addresses \$E\$2 and \$F\$2 (the \$ symbols make a cell reference that does not move when a formula is copied to another cell): `=(C2-$E$2*(1-EXP(-$F$2*D2)))^2`. This formula is copied to the other cells in column G for the remaining data and the sum is calculated in G18.

| | A | B | C | D | E | F | G |
|---|---|---|---|---|---|---|---|
| 1 | width | depth | angle | width/depth | alpha | beta | error^2 |
| 2 | 610 | 550 | 33.6 | 1.1090909 | 35 | 1 | 102.9221 |
| 3 | 450 | 500 | 22.3 | 0.9 | | | 2.340711 |
| 4 | 450 | 520 | 22 | 0.8653846 | | | 2.996837 |
| 5 | 430 | 740 | 18.7 | 0.5810811 | | | 10.72739 |
| 6 | 410 | 800 | 20.2 | 0.5125 | | | 38.00559 |
| 7 | 500 | 230 | 31 | 2.173913 | | | 0.000376 |
| 8 | 500 | 235 | 30 | 2.1276596 | | | 0.690476 |
| 9 | 500 | 240 | 32 | 2.0833333 | | | 1.844182 |
| 10 | 450 | 600 | 26.6 | 0.75 | | | 66.14291 |
| 11 | 450 | 650 | 15.1 | 0.6923077 | | | 5.689669 |
| 12 | 480 | 230 | 30 | 2.0869565 | | | 0.432641 |
| 13 | 475 | 1400 | 13.5 | 0.3392857 | | | 11.76327 |
| 14 | 485 | 615 | 26.8 | 0.7886179 | | | 59.39072 |
| 15 | 474 | 515 | 25 | 0.9203883 | | | 15.54528 |
| 16 | 485 | 700 | 20.4 | 0.6928571 | | | 8.439469 |
| 17 | 600 | 750 | 15 | 0.8 | | | 18.26268 |
| 18 | | | | | | | 345.1943 |
| 19 | | | | | | | |

**Figure 26.2** Excel spreadsheet demonstrating non-linear regression.

The optimisation step will be demonstrated with the Solver Addin, which is available under the submenu Tools – Solver (it can be enabled under Tools – Addins, but if it is unavailable the user's own optimisation algorithm can be used instead). The optimisation can be performed by completing the form as shown in Figure 26.3 and clicking Solve. The least squares estimates appear in \$E\$2 and \$F\$2 and they are 32.46 and 1.511 respectively. The sum of squared residuals in G18 is 204.6. The estimated standard deviation of the errors is calculated as the square root of the quotient when 204.6 is divided by the number of data, 16, less the number of parameters estimated from the data, 2. The estimated standard deviation of the errors equals 3.82.

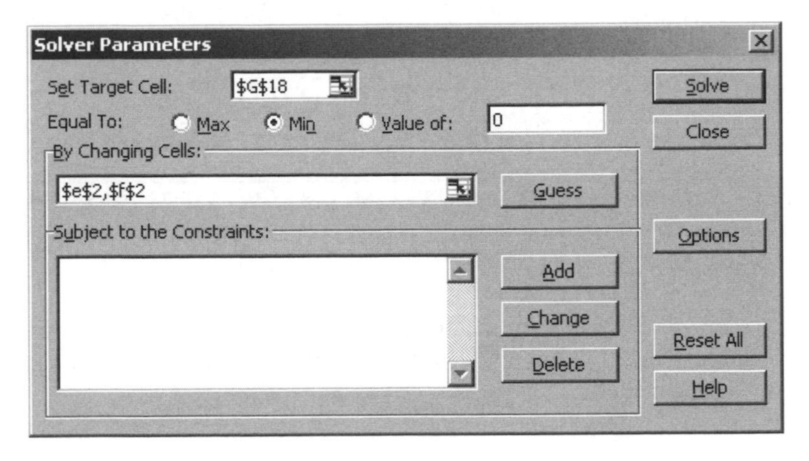


**Figure 26.3** Solver optimisation for coefficients of non-linear regression.

## PROBLEMS

**26.1** The Michaelis–Menten equation is used to relate the initial speed of an enzyme reaction, $v$, to the substrate concentration, c:

$$v = \frac{\theta_1 c}{c + \theta_2}{}_i$$

Linearise this equation with the terms $y = 1/v$, $x = 1/c$ $\beta_0 = 1/\theta_1$ and $\beta_1 = \theta_2 / \theta_1$ to show that it can be expressed as the regression $y = \beta_0 + \beta_1 x$.

**26.2** Complete the non-linear regression in Figure 26.3 and plot the fitted curve along with the observed data.

**26.3** 51 Pegasi was the first sun-like star discovered with an orbiting planet. As a planet orbits a star its gravitational pull causes a wobble in the star's orbit. This wobble causes a Doppler shift in observed light from the star, termed redshift. Measuring the changes in the frequency of light observed from the star gives an estimate of the radial orbit velocity and the period of the orbit. A set of 26 radial velocities are shown in Table 26.1 relating to 51 Pegasi (Marcy et al., 1997). 51 Pegasi is a single-planet system so the radial velocities repeat as a simple sine wave. Fit a regression model $y = \beta_0 + \beta_1 \sin(\beta_2 x + \beta_3)$. Many regressions involving sinusoidal terms can be easily linearised by knowing the period of oscillation beforehand – for example, a seasonal period. Here the primary interest is in the unknown frequency of the planet's orbit, $\beta_2$. How might the signal vary if there were two or more planets orbiting the star?

**Table 26.1** Radial velocity estimates ($ms^{-1}$) at different times for the planet 51 Pegasi.

| Day | 11.64 | 11.84 | 12.64 | 12.87 | 13.62 | 13.83 | 14.64 | 14.72 | 14.82 | 14.90 | 15.63 | 15.75 | 15.87 |
|---|---|---|---|---|---|---|---|---|---|---|---|---|---|
| Vel. | –45.5 | –39 | 27.3 | 32.5 | 63.4 | 54.8 | –1.3 | –5.5 | –10.7 | –26.3 | –50.7 | –45.8 | –57.5 |

| Day | 16.61 | 16.76 | 16.85 | 17.73 | 17.84 | 18.62 | 18.76 | 18.85 | 19.62 | 19.73 | 19.85 | 20.61 | 20.74 |
|---|---|---|---|---|---|---|---|---|---|---|---|---|---|
| Vel. | 10.6 | 12.9 | 16.4 | 63.6 | 60.5 | 26.3 | 6.0 | –7.8 | –43.6 | –59.5 | –58.6 | –20.2 | –8.1 |

**26.4** Consider the 10-year census population data for the United States in Table 26.2, in millions from 1790 through 1990 (Fox, 1997). These data can be modelled using the logistic model:

$$y_i = \frac{\beta_0}{1 + exp(\beta_1 + \beta_2 x_i)} + \varepsilon_i$$

where $y_i$ is the population size at time $x_i$; $\beta_0$ is an asymptote which the population tends towards; $\beta_1$ reflects the size of the population at time $x_0$ relative to its asymptotic size; and $\beta_2$ controls the rate of population growth.

**Table 26.2** 10-year census population data for the United States (in millions).

| Year | 1790 | 1800 | 1810 | 1820 | 1830 | 1840 | 1850 |
|---|---|---|---|---|---|---|---|
| Count | 3.929 | 5.308 | 7.24 | 9.638 | 12.866 | 17.069 | 23.192 |

| Year | 1860 | 1870 | 1880 | 1890 | 1900 | 1910 | 1920 |
|---|---|---|---|---|---|---|---|
| Count | 31.443 | 39.818 | 50.156 | 62.948 | 75.995 | 91.972 | 105.71 |

| Year | 1930 | 1940 | 1950 | 1960 | 1970 | 1980 | 1990 |
|---|---|---|---|---|---|---|---|
| Count | 122.78 | 131.67 | 150.70 | 179.32 | 203.30 | 226.54 | 248.71 |

a) Scale the times so that $x_0 = 0$ and formulate the sum of squared errors between the data and the model equation.
b) Minimise this function to find parameters of the model. Based on what the model parameters represent, how might a good set of initial values be selected for the optimisation?
c) Plot the model along with the observed data. Does it provide a good fit?
d) Plot the residuals of the model. Is there additional structure in the data not currently explained by the model?

CHAPTER TWENTY SEVEN

# Probability Distributions (Introduction)

## 27.1 INTRODUCTION

An important aspect of civil engineering work is the determination of quantities such as forces, flows, rainfall intensities or wind speeds that, by their nature, will have a range of values rather than being fixed. It is convenient to treat variables as either continuous, discrete or categorical. Continuous variables are measured on some underlying continuous scale even though they are recorded to a fixed number of decimal places. Examples include temperatures, pressures, tensile strengths and soil porosity. Discrete variables usually take non-negative integer values and in most applications are counts of numbers of occurrences. For example, the number of road traffic accidents per year, the number of hurricanes per year, the number of asbestos-type particles in fixed volumes of air and the number of defective items in random samples of a product. Categorical variables include colour, day of week, type of mineral and supplier in situations where there are several. In some cases categorical variables may be ordered in some sensible manner such as levels of agreement with a statement in a survey.

## 27.2 DISCRETE VARIABLES

Consider the following data in Table 27.1, which are the numbers of asbestos-type fibres found in 1 litre of air, 1.5 m above ground level in an auto repair shop. A line chart is shown in Figure 27.1. The height of the lines equals the proportion of samples at each count. The sum of the heights equals one.

**Table 27.1** Asbestos data for line chart.

| Number of particles in 1 litre of air | 0 | 1 | 2 | 3 | 4 | 5 | 6 or more | total |
|---|---|---|---|---|---|---|---|---|
| Frequency | 34 | 46 | 38 | 19 | 4 | 2 | 0 | 143 |
| Proportion | 0.238 | 0.322 | 0.266 | 0.133 | 0.028 | 0.014 | 0 | 1 |

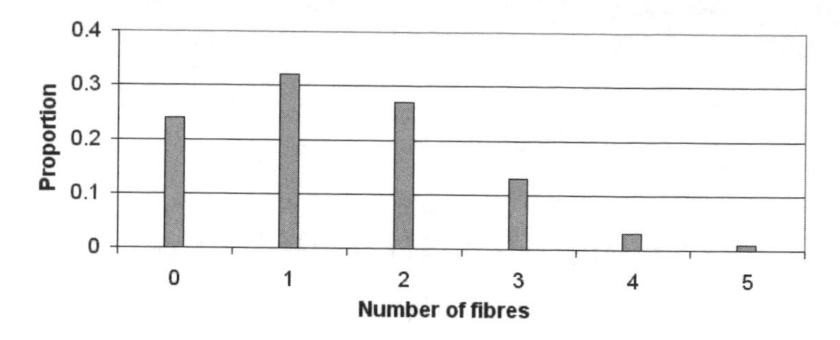


**Figure 27.1** Number of asbestos-type fibres in air samples.

## 27.3 CONTINUOUS VARIABLES

As an example of continuous data consider wind speeds collected every 30 minutes at Adelaide Airport by the Australian Bureau of Meteorology. Figure 27.2 shows data from January 1999.

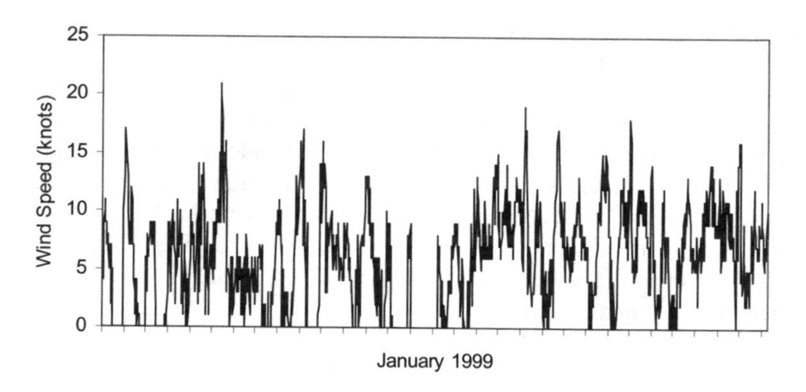


**Figure 27.2** Wind speed data for Adelaide Airport for January, 1999.

With a five-year record, the number of occurrences of wind speed that fit into a series of bins of width 2 knots can be calculated and the data can be used to generate a histogram. The areas of the rectangles of the histogram correspond to the proportion of wind speeds between 0 and 2 knots, between 2 and 4 knots, etc. (Note: Excel has a function under Tools – Data Analysis that counts the number of occurrences in a range of bins.) It follows that the vertical scale is the proportion divided by the bin-width. The group data are listed in Table 27.2 and the histogram is shown in Figure 27.3.

Table 27.2 records that wind speeds less than 2 knots occurred approximately 18% of the time and that there was no occurrence of winds greater than 24 knots during this period. When the wind speed was over 2 knots it was most likely to be between 8 and 10 knots.

**Table 27.2** Group data for wind speed histogram.

| Wind Speed (knots) | Proportion | Density |
|---|---|---|
| 0 – 2 | 0.185 | 0.093 |
| 2 – 4 | 0.072 | 0.036 |
| 4 – 6 | 0.130 | 0.065 |
| 6 – 8 | 0.154 | 0.077 |
| 8 – 10 | 0.181 | 0.091 |
| 10 – 12 | 0.120 | 0.060 |
| 12 – 14 | 0.093 | 0.047 |
| 14 – 16 | 0.042 | 0.021 |
| 16 – 18 | 0.014 | 0.007 |
| 18 – 20 | 0.007 | 0.004 |
| 20 – 22 | 0.001 | 0.001 |
| 22 – 24 | 0.001 | 0.001 |
| 24 – 26 | 0.000 | 0.000 |

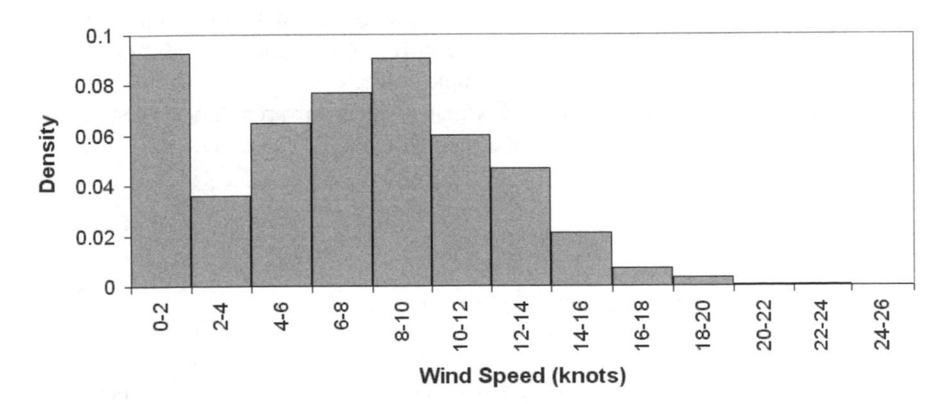


**Figure 27.3** Histogram of January wind speeds at Adelaide Airport.

One aspect of the plot is the width of the wind speed bins, which in Figure 27.3 was set at 2 knots. With additional data it would be possible to reduce this. An example, based on an extended set of Adelaide airport data, is shown in Figure 27.4 using a bin-width of 1 knot. The finer resolution that this affords will be valuable in some circumstances, although in this case it appears there is little extra to be gained from the change in bin-width.

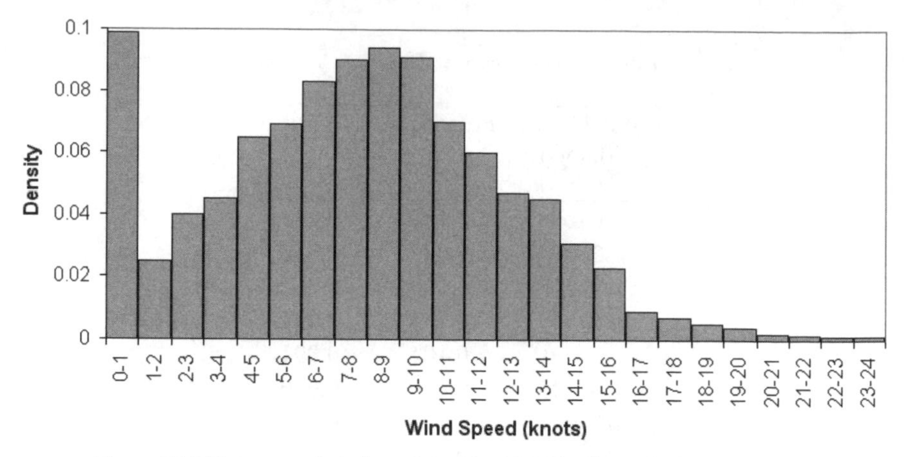


**Figure 27.4** Histogram of wind speed data for Adelaide Airport for January, 1999.

## 27.4 PROBABILITY MASS AND PROBABILITY DENSITY FUNCTIONS

Imagine Figure 27.1 being based on a very large sample. Then sample proportions will tend towards population probabilities and the figure represents a probability mass function (PMF). The set of all possible values, $x_i$, that can be taken by the variable is known as the sample space. The PMF, $p(\ )$, defines probabilities, $p(x_i)$, that the variable takes the values in the sample space.

Statistics such as the mean and variance tend towards the corresponding population values which are defined in terms of the PMF. These are:

$$\mu_x = E[x] = \sum_{i=1}^{n} x_i p_x(x_i) \tag{27.1}$$

$$\sigma_x^2 = E\left[(x-\mu_x)^2\right] = \sum_{i=1}^{n} (x_i - \mu_x)^2 p_x(x_i) \tag{27.2}$$

Applying the formulae to the data listed in Table 27.2, the population mean and variance are estimated as:

$$\hat{\mu}_x = 0(0.238) + 1(0.322) + 2(0.266)... = 1.435$$

$$\hat{\sigma}_x^2 = (0-1.435)^2 0.238 + (1-1.435)^2 0.322 + ... = 1.32$$

Imagine Figure 27.3 being based on a very large sample. Then the data are sufficiently numerous to imagine that the width of the bins is reduced to an infinitesimally small value. Thus the histogram would tend toward a smooth curve rather than juxtaposed rectangles. This curve is referred to as a probability density function (PDF). For a continuous variable, the probability that it equals a precise single value is zero and is not the height of the plot at the value. A more useful calculation is the probability that the continuous variable lies between two values $l$ and $u$. This probability is equal to the area under the PDF between $l$ and $u$. The total area under the PDF must be 1.0.

For the situation where the PDF is defined by a function $f_x(x)$ the mean and variance are defined in Equation (27.3) and Equation (27.4). The mean and variance can be estimated from group data by assuming all the data in a histogram bin are at the mid-point of that bin and proceeding as for discrete data.

$$\mu_x = E[X] = \int_{-\infty}^{\infty} x f_x(x) dx \tag{27.3}$$

$$\sigma_x^2 = E\left[(X - \mu_x)^2\right] = \int_{-\infty}^{\infty} (x - \mu_x)^2 f_x(x) dx \tag{27.4}$$

*Example*
Determine the mean and standard deviation of a PDF of the form: $f_x(x) = e^{-x}$.

*Solution*
The PDF can be plotted as shown in Figure 27.5. Note that integrating the function from 0 to infinity shows that the total area under the curve is 1.0 as required.

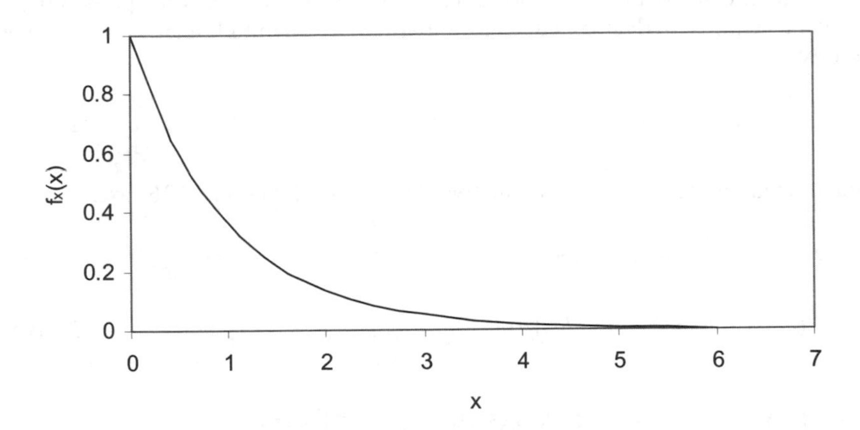


**Figure 27.5** A PDF for $f_x(x) = e^{-x}$.

$$\mu_x = \int_{-\infty}^{\infty} x f_x(x) dx = \int_{0}^{\infty} x e^{-x} dx = e^{-x}(-x-1) = 0 - (-1) = 1.0$$

$$\sigma_x^2 = \int_{-\infty}^{\infty} (x - \mu_x)^2 f_x(x) dx = 1.0$$

where the intermediate stages to obtain the variance have not been shown.

## 27.5 CUMULATIVE PROBABILITY DISTRIBUTIONS

A useful way of representing the probability distribution is as a cumulative distribution where a point on the distribution represents the probability that the value will be less than or equal to a given value. For a given probability distribution $f_x(x)$ the cumulative probability distribution is defined for a continuous function as:

$$F_X(x) = \int_{-\infty}^{x} f_X(x)dx \tag{27.5}$$

and, based on elementary calculus the relationship between the distribution and the cumulative distribution can be neatly summarised as:

$$f_X(x) = \frac{dF_X(x)}{dx} \tag{27.6}$$

For a PMF the expression is:

$$F_X(x) = \sum_{x_i \le x} p_X(x_i) \tag{27.7}$$

A significant advantage of the cumulative distribution is that the probability of an event being less than or equal to a particular magnitude can be determined directly:

$$P(X \le x) = F_X(x) \tag{27.8}$$

Fundamental properties of the cumulative distribution function (CDF) are:

$$F_X(-\infty) = 0 \quad \text{and} \quad F_X(+\infty) = 1.0 \tag{27.9}$$

$$P(a < x \le b) = F_X(b) - F_X(a) \tag{27.10}$$

## 27.6 COMMON DISTRIBUTIONS IN ENGINEERING

Due to common underlying processes, a large variety of engineering phenomena can be modelled using a small number of probability distributions. Some of these distributions are discussed in subsequent chapters. The Gaussian, normal or bell-shaped distribution, for example, is well known and has been used to describe a range of properties from the natural variation in physical properties of materials such as steel or concrete to errors in repeated measurements. However, the normal distribution is not always appropriate and it is important to be aware of alternatives. Before investigating these alternatives in detail one simple but fundamental distribution will be examined: the uniform distribution.

## 27.7 UNIFORM DISTRIBUTION

The discrete uniform distribution can be used to describe a number of phenomena such as the roll of a fair die (the singular of dice) and the tossing of a fair coin. The continuous uniform distribution commonly arises as the basis of generating pseudo-random data of selected distributions (Chapter 35).

When rolling a fair die, there is equal chance of throwing a 1, 2, 3, 4, 5 or 6 and the probability distribution for this is shown in Figure 27.6.

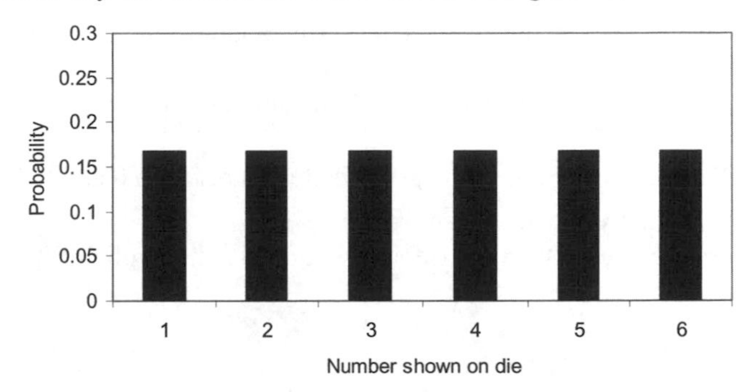


**Figure 27.6** PMF for uniform probability distribution (the roll of a die).

Applying Equations (27.1) and (27.2) it is possible to determine that the mean and variance of the distribution for the roll of a die are:

$$\mu_x = 1(1/6) + 2(1/6) + ...6(1/6) = 3.5$$

$$\sigma_x^2 = (1-3.5)^2(1/6) + (2-3.5)^2(1/6) + ... = 2.917$$

For phenomena that have a continuous uniform variation the PDF is shown in Figure 27.7. When uniform random number generators are used in Excel or Fortran this is the distribution that the numbers conform to. There is (or at least there should be) an equal chance, 0.1, of getting a number between 0 and 0.1 as there is between 0.6 and 0.7 and 0.8 and 0.9. Note that the total area under the distribution is 1.0.

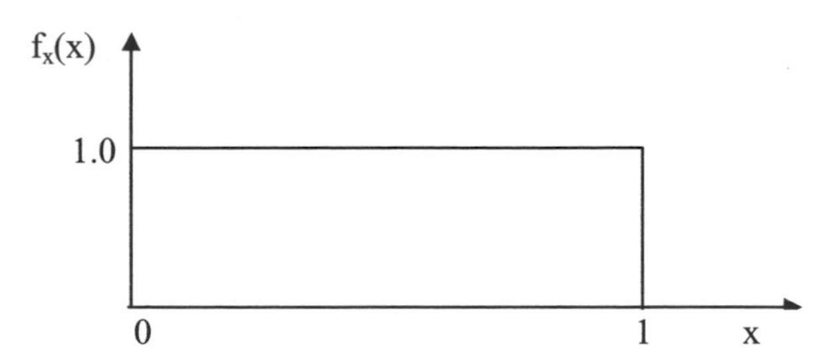


**Figure 27.7** A PDF for uniform probability distribution.

## PROBLEMS

**27.1** The random number generator available in many computing applications such as Excel or Fortran should conform to a uniform distribution as shown in Figure 27.7. Based on this:

(a) Given that the function can be written $f_x(x) = 1$, determine the mean and variance of the distribution.
(b) Calculate the probability of a uniform random number being less than 0.15.
(c) Determine the CDF. Use it to verify the answer to part (b).

**27.2** For a PDF of the form $f_x(x) = a\sin(x)$ where $x$ is between 0 and $\pi/2$:

(a) Calculate the value $a$.
(b) Determine the CDF and use it to calculate the probability of $x$ being between $\pi/6$ and $\pi/4$.

CHAPTER TWENTY EIGHT

# Probability Distributions (Bernoulli, Binomial, Geometric)

## 28.1 INTRODUCTION

When considering material properties of engineering elements, for example, it is reasonable to assume that there will be some variation that could be reflected conveniently by way of a probability distribution. There are, however, aspects of behaviour that are fundamentally different and yet may still lead to some form of probability distribution.

A variable is binary if there are only two possible outcomes. For example, the launch of a telecommunications satellite could be designated as a success or a failure, the flip of a coin can be a head or a tail. It is also often convenient to simplify a complicated measurement such as the quality of a water sample by stating that it either satisfies all specified quality criteria or that it does not. Another example might be if one were interested only in situations exceeding a certain threshold where either the threshold will be exceeded or it will not, such as a wind turbine being deactivated on wind exceeding Beaufort Force 7 (wind speeds between 28 and 33 knots). This sort of characteristic applies equally well to floods, earthquakes or failures: they either occur or they do not. These situations are described as a Bernoulli trial. The assumptions for processes that comply with a Bernoulli sequence are:

1. there are only two possible outcomes: occurrence or non-occurrence;
2. the probability of occurrence is constant; and
3. each event is independent.

Before the theory is outlined, the process will be illustrated with an example.

*Example*

The chance of at least one storm occurring during each of the three summer months is assumed to be 10%. Determine the chances of 0, 1, 2 or 3 months in which storms occur over summer.

*Solution*

Over the summer months there are a finite number of possibilities that may have occurred in terms of storms. These are listed in Table 28.1.

**Table 28.1** A full list of options regarding storm months over a 3-month period.

| | |
|---|---|
| SSS | – all three months have storms in them |
| SSN | – storms in the first two months followed by no storms in the last month |
| SNN | – storms in the first month followed by months with no storms |
| SNS | – etc. |
| NSS | |
| NSN | |
| NNS | |
| NNN | – no storms in any of the three months |

Combining these it is evident that there can be either no storm months (NNN), 1 storm month (SNN, NNS, NSN), 2 storm months (SSN, SNS, NSS) or 3 storm months (SSS). With a known probability of 0.1 of a storm occurring and, therefore, a probability of 0.9 of no storm, it is possible to generate a table of probabilities for these four situations. This is shown in Table 28.2 and the PMF is plotted in Figure 28.1 as a line chart.

**Table 28.2** Combinations that lead to a given number of storms.

| Situation | Combinations | Probability |
|---|---|---|
| 3 storms | $0.1^3$ | 0.001 |
| 2 storms | $3\ 0.1^2\ 0.9$ | 0.027 |
| 1 storm | $3\ 0.1\ 0.9^2$ | 0.243 |
| 0 storms | $0.9^3$ | 0.729 |
| TOTAL | | 1.000 |

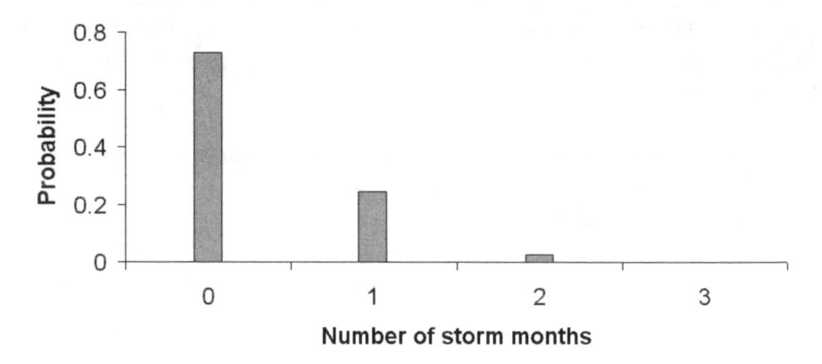


**Figure 28.1** The PMF for the number of storm months over a 3-month summer period.

## 28.2 PROPERTIES OF THE BERNOULLI DISTRIBUTION

Let $X$ represent the number of occurrences in a sequence of $n$ Bernoulli trials. Then $X$ is a discrete random variable because it assigns a discrete number to the outcome of the sequence. If the probability of an occurrence is $p$ (and the probability of non-occurrence is $1 - p$) then the chance of exactly $x$ occurrences in $n$ trials is given by the binomial PMF as follows:

$$P(X = x) = \binom{n}{x} p^x (1 - p)^{n-x} \tag{28.1}$$

where $n$ and $p$ are the parameters of the distribution and

$$\binom{n}{x} = \frac{\mathrm{n!}}{\mathrm{x!(n - x)!}} \tag{28.2}$$

is the binomial coefficient. The mean or expected value of the distribution in Figure 28.1 is $np$ and the variance is $npq$ where $q = (1 - p)$. As a check, the mean and variance of the distribution can be calculated by applying Equations. (28.1) and (28.2), respectively, as:

$$\mu_x = 0(0.729) + 1(0.243) + 2(0.027) + 3(0.001) = 0.300$$

$$\sigma_x^2 = (0 - 0.3)^2(0.729) + (1 - 0.3)^2(0.243) + (2 - 0.3)^2 0.027 + (3 - 0.3)^2(0.001) = 0.270$$

As another example, Figure 28.2 shows the number of years with floods over a 100-year period, if the chance of a flood in any particular year is 10%.

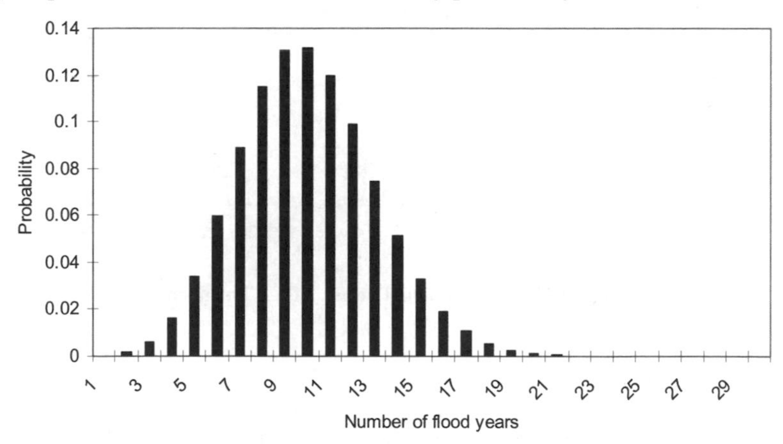


**Figure 28.2** Binomial probability distribution for number of flood years in 100-year period.

For this, the standard formula was used with $n = 100$, $p = 0.10$, and $x$ was taken from 0 to 30 to assess the likelihood of between 0 and 30 floods occurring over the 100-year period. As might be expected, with a 10% chance of a flood each year the probability of going for 100 years without a flood is vanishingly small, as is the chance of there being 30 or more floods over the 100 years. The rest of the distribution is seen to be symmetrical with a peak probability occurring for the chance of 10 floods over the 100 years.

One point of interest in Figure 28.2 is that it resembles the normal or Gaussian distribution (as discussed in Chapter 30). This is not just by chance. It can be shown that as $n$ increases the binomial distribution tends to the Gaussian distribution – a fact that is often attributed to De Moivre.

## 28.3 GEOMETRIC DISTRIBUTION

In addition to knowing the probability distribution of the number of flood years in a set period one can consider the probability distribution of time until the next flood year. Let $T$ be the number of trials until the next event in a Bernoulli sequence. If this event occurs on the $t^{\text{th}}$ trial, and it is the first, then there must be no events in the preceding $t-1$ trials. Therefore:

$$P(T = t) = pq^{t-1} \quad t = 1,2,... \tag{28.3}$$

The distribution of times until the first flood year in the example cited above is shown in Figure 28.3. There is a 10% chance of a flood year (and therefore the first) occurring in Year 1. If the first flood year is Year 2, then there must not have been a flood in Year 1 (0.9 probability) and there must have been one in Year 2 (0.1 probability), giving a total probability of 0.09 or 9%. This logic continues for all years.

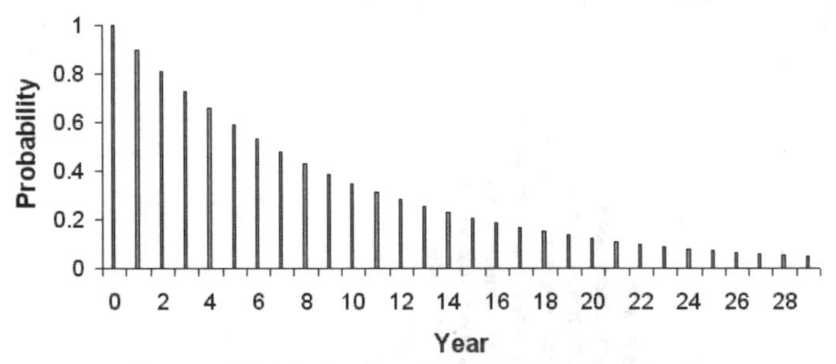


**Figure 28.3** Distribution of times until the first flood.

## 28.4 RETURN PERIOD

In a Bernoulli sequence the time until the first occurrence is a random variable. If the events are independent, then this random variable also describes the time between any two events, this time being referred to as the recurrence interval. The recurrence interval in a Bernoulli sequence has a geometric distribution with a mean that is the average recurrence interval (ARI), also less formally known as the return period. Using the definition of the mean of a distribution:

$$E[T] = \sum_{t=1}^{\infty} t.pq^{t-1} = p(1 + 2q + 3q^2 + ...) = \frac{1}{p} \quad (28.4)$$

where the sum of the infinite series can be shown to equal $1/p^2$.

Therefore, on average, the time between successive events is the reciprocal of the probability of an event occurring in any particular time period. For example, the return period for an event with a 10% chance of occurring in a year is 10 years. The chance of the 1-in-100-year event occurring in any particular year is 0.01 or 1%.

In the previous example the mean time until the first flood is the mean of the distribution shown in Figure 28.3 – this turns out to be 10 years (as expected).

## 28.5 NEGATIVE BINOMIAL DISTRIBUTION

If the time until the first occurrence is governed by the geometric distribution, then the time until subsequent occurrences is described by the negative binomial distribution. If one is interested in the $k^{th}$ occurrence in a sequence of events, then there must have been $k$–1 occurrences already. The probability of the $k^{th}$ event occurring at a particular time can be determined from:

$$P(T_k = t) = \binom{t-1}{k-1} p^k q^{t-k} \text{ for } t = k, k+1, k+2, ... \quad (28.5)$$

Note that the probability is 0 for all $t < k$ since one requires the $k^{th}$ event and this cannot occur in less than $k$ periods.

## PROBLEMS

**28.1** A researcher is conducting experiments on earthquake loading of masonry walls using a laboratory shake-table. To verify the hypothesis being investigated the researcher requires 5 trials to fail according to an intended failure mode, but due to laboratory constraints it is not possible to repeat this experiment more than 20 times. The researcher estimates a 0.4 probability of observing the intended failure mode. What is the probability of 5 satisfactory results from $n$ trials? What is the probability that 20 trials are insufficient?

**28.2** Develop a Visual Basic module to calculate the cumulative Bernoulli distribution. Apply it on a worksheet to calculate the cumulative distribution for up to 20 events in 20 trials where the probability of occurrence is 0.3. Set up a formula on the worksheet to verify the module calculations.

**28.3** Reproduce the results shown in Figure 28.2 for the number of flood years over a 100-year period. Investigate the behaviour when the chance of a flood in any particular year varies from the 10% value given.

**28.4** If the binomial distribution tends to the normal distribution as the number of trials increases, what happens to the geometric distribution, which shows the time until the first occurrence, in this case?

**28.5** If the chance of a forest fire in any particular year is 2%, what is the chance of no fires over a 20-year growing cycle?

# CHAPTER TWENTY NINE

# Probability Distributions (Poisson, Exponential, Gamma)

## 29.1 INTRODUCTION

There is an aspect of the Bernoulli process that is unsatisfactory when modelling events occurring in time. Taking floods, for example, there is surely a chance of more than a single flood in any particular month. This was dealt with (in Chapter 28) by defining a month as having no flood or having more than one flood. However, this strategy is not satisfactory for modelling the absolute number of floods. An alternative is to reduce the time increment from one month to one week or further to one day, and to assume that either no flood or precisely one flood occurs in this time increment. If this increment is sufficiently small the probability of more than one flood in the increment does become negligible.

*Example*
Using a Bernoulli process, determine the probability of 2 floods in 2 months if there are, on average, 5 floods per year and assume that at most one flood can occur in a month. (Note: to enable valid comparisons it has been assumed that there are 12 four-week months in a year.)

*Solution*
With a trial length of one month, the probability of a flood in a particular month is 5/12 = 0.417. Therefore, applying Equation (28.1) the probability of 2 floods in 2 months is:

$$P(X = x) = \binom{2}{2} 0.417^2 (1 - 0.417)^0 = 0.174$$

Reducing the trial length to 1 week, the probability of a flood in 1 week is 5/48 = 0.104 and the probability of 2 floods in any 8 weeks (assuming a 4 week month) is:

$$P(X = x) = \binom{8}{2} 0.104^2 (1 - 0.104)^6 = 0.157$$

Further reducing the trial length to 1 day, the probability of a flood in 1 day is 5/336 = 0.015 and the probability of 2 floods in 56 days is:

$$P(X = x) = \binom{56}{2} 0.015^2 (1-0.015)^{54} = 0.152$$

From the results it appears that the binomial distribution is tending towards a limiting probability as the time increment decreases. The distribution of limiting probabilities is the Poisson distribution.

**Poisson Trials: An Historical Note**

The use of the word "trial" in the Poisson process is interesting and may relate to the original research topic that Poisson worked on: the deliberations of juries at criminal and civil trials. According to Good (1986) "Poisson was scarcely responsible for introducing this [the Poisson] distribution, nor for its applications" but, given he made a number of breakthroughs that have been attributed to others, it is fair that his name be associated with this work.

## 29.2 POISSON PROCESS AND POISSON DISTRIBUTION

Many engineering problems can be modelled by occurrences at random points in space and time. For example, fatigue cracks may occur anywhere along a weld, earthquakes can occur anywhere along a fault line at any time and traffic accidents can occur at any place along a highway.

It would be possible to model the process with a Bernoulli sequence by dividing the time or space into sufficiently small intervals and assuming that an event either will occur or will not within each interval, thus constituting a trial. However, if the event can occur at any instant it may occur more than once in a given time or space interval. In such cases the occurrence may be modelled with a Poisson sequence rather than as a Bernoulli sequence. The Poisson process is based on the following assumptions:

1. an event can occur at any time or point in space;
2. events are independent; and
3. the probability of occurrence of an event in a small interval $\Delta t$ can be estimated as $\nu \Delta t$ where $\nu$ is the mean rate of occurrence of the event.

With these assumptions, it can be shown that the number of occurrences of an event in $t$ is given by the Poisson distribution:

$$P(X_t = x) = \frac{(\nu t)^x}{x!} e^{-\nu t} \qquad (29.1)$$

for any $x$, where $\nu$ is the mean occurrence rate and $t$ is the length of the trial. It is interesting to note that for a small mean occurrence rate the distribution converges with the binomial distribution. (For a small mean occurrence rate the chance of two or more events in a single trial is unlikely; hence the agreement with the Bernoulli process.) For relatively high probabilities the binomial and normal distributions converge.

**Floods and Power Failures**

In a study planning the remediation of a rural dam in Australia, Tabatabaei and Zoppou (2000) modelled the arrival of floods at the site using a Poisson distribution. This was an important aspect of the overall design because three different engineering companies had submitted tenders outlining three quite different methods of dealing with a particular structural problem and the methods had different requirements in terms of flood-free times to effect the repair.

In an unrelated study looking at power failures and their implications in Sweden, Holmgren and Molin (2006) also modelled the times of disturbances to the generation system as a Poisson process.

## 29.3 PRUSSIAN ARMY HORSES

The Poisson distribution was discovered independently in 1898 by von Bortkiewicz (Brooks, 2001) and applied to a data set that recorded the number of deaths due to kicks from horses and mules in the Prussian army over a 20-year period. The data were collected for 10 army corps giving a total of 200 trials. Over that time there were a total of 122 deaths due to kicking giving a mean death rate of 122/200 = 0.610 deaths per year per corps. In some years there were no deaths, in others one, two, three or four. The actual observed values and those estimated by applying Equation (29.1) are listed in Table 29.1 and plotted in Figure 29.1. In the output the estimated number of events is determined from:

$$e = NP(X = x) \tag{29.2}$$

where $e$ is the estimated number, $N$, is the total number of trials and $P(X{=}x)$ is the probability of the required event. The excellent agreement between the observed and estimated is evident. It can be shown that the mean or expected value of the Poisson distribution is:

$$\mu_x = vt \tag{29.3}$$

**Table 29.1** Raw and estimated number of deaths due to horse kicks in the Prussian army.

| Number of deaths | Observed | Estimated |
|---|---|---|
| 0 | 109 | 108.7 |
| 1 | 65 | 66.3 |
| 2 | 22 | 20.2 |
| 3 | 3 | 4.1 |
| 4 | 1 | 0.6 |
| 5 | 0 | 0.1 |
| 6 | 0 | 0.0 |

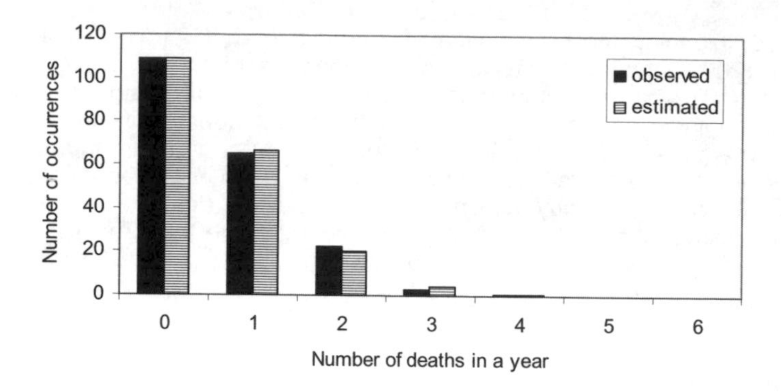


**Figure 29.1** The PMF for the number of deaths per year per corps in the Prussian army.

## 29.4 RETURN PERIODS AND PROBABILITY

A common question that arises is: what is the chance of having the 1-in-1-year event this year, or the 1-in-10-year event in the next 10 years, or more generally the 1-in-*N*-year event the next *N* years?

For a solution, it is known that the probability of the 1-in-*N*-year event occurring in any particular year is $1/N$. Therefore in the application of the Poisson process the mean rate can be calculated: $v = 1/N$. In determining the chance of an event occurring it is important to include the chance of two or more events in the same period. For this reason it is easier to write:

$$P(\text{1-in-}N\text{-year event}) = 1 - P(0 \text{ events in } N \text{ years}) \tag{29.4}$$

$$P(\text{1-in-}N\text{-year event}) = 1 - \frac{\left(\frac{1}{N}N\right)^0}{0!}e^{-1} = 1 - e^{-1} = 0.632 \tag{29.5}$$

Thus, the chance of a 1-in-100-year event occurring in the next 100 years is 63%. The chance of a 1-in-5-year event occurring in the next 5 years is 63%.

A word of warning: although the discussion of the Bernoulli and Poisson processes presented here has used events such as floods it should be noted that these may not comply with the assumption of independence between events. Once a flood has occurred it may be more likely that further flooding occurs since the storms may come from the same weather pattern and the first flood will have thoroughly wet the ground and reduced subsequent losses due to infiltration. In this case the probability of an event occurring may be affected by what happened in previous trials and is thus a conditional probability problem. If this conditional probability depends only on the preceding one trial then the resulting model is called a Markov process. This is discussed in Chapter 40.

## 29.5 EXPONENTIAL DISTRIBUTION

If the probability of an event occurring can be modelled by the Poisson process the time until the first event is modelled by the exponential distribution. The exponential distribution has the form:

$$f_{T_1}(t) = ve^{-vt} \tag{29.6}$$

The mean can be calculated:

$$\mu_t = \int tve^{-vt}\,dt = \frac{1}{v} \tag{29.7}$$

The exponential distribution is also used in a number of other applications. For example, in hydrology it can describe the correlation between two rainfall amounts with increasing distance, or in geotechnical engineering it is used to describe the size of fractures on a rock face in order to give estimates of strength.

**Radioactive Decay of Radium**

According to Rosanov (1969) the radioactive decay process whereby radium decays to radon, and in the process emits alpha particles, can be modelled as a Poisson process where a gram of radium emits alpha particles with a mean emission rate of $10^{10}$ per second.

## 29.6 GAMMA DISTRIBUTION

In the same way that the negative binomial distribution describes the time until second and subsequent occurrences for the binomial distribution, the gamma distribution describes the same events for the Poisson distribution. Put another way, the sum of *k* random variables will result in a variable that has a gamma distribution. The gamma distribution is written:

$$f_{T_k}(t) = \frac{v(vt)^{k-1}}{(k-1)!}e^{-vt} \tag{29.8}$$

where $v$ is the mean rate, $t$ the length of the trial, and one is seeking the $k^{th}$ event. The mean or expected value is $k/v$. The gamma distribution can be generalised to use real-value $k$ instead of integers:

$$f_{T_k}(t) = \frac{v(vt)^{k-1}}{\Gamma(k)!}e^{-vt} \tag{29.9}$$

where $\Gamma(\,)$ is the gamma function (set out in Appendix B).

## 29.7 WORKED EXAMPLE

If the chance of flooding in any particular year is 10% it is possible to determine the Poisson distribution of the number of floods in a 100-year period. In this case, $v = 0.1$, $t = 100$ and it is convenient to calculate the distribution for a range of $x$ values from 0 to 20. The distribution is shown in Figure 29.2.

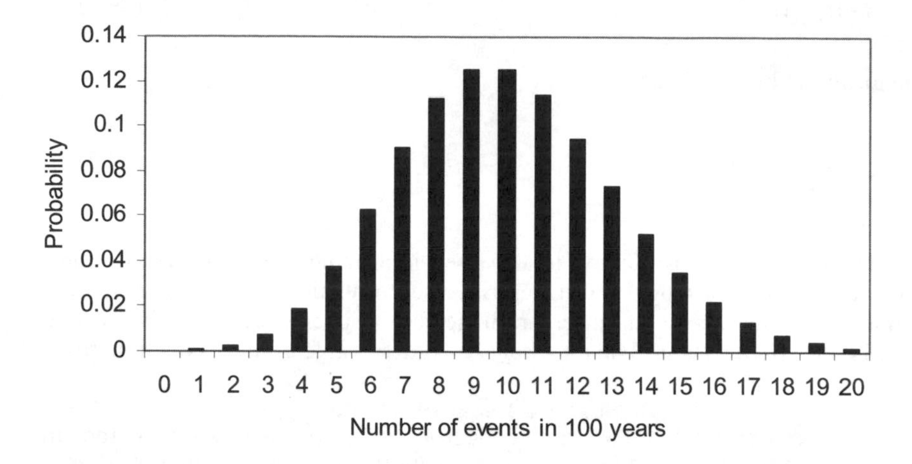


**Figure 29.2** The PMF for the number of floods expected over a 100-year period.

Applying Equation (29.3), the mean of the distribution is 0.1(100) = 10 events. The distribution of the times until the first flood in the 100-year period is given by the exponential distribution as shown in Equation (29.4). The distribution is plotted in Figure 29.3. The mean of the distribution, and therefore the expected time until the first flood, is given by Equation (29.5) as 1/0.10 = 10 years.

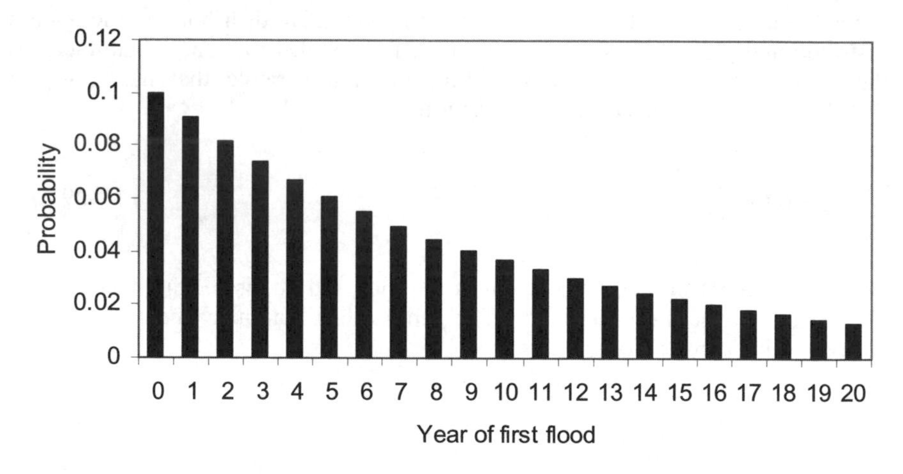


**Figure 29.3** The PMF for the time until first flood over a 100-year period.

## PROBLEMS

**29.1.** In the design of critical structures it may be required to ensure that they will only have a certain chance of being exposed to extreme conditions during their design life. For example, in the design of an offshore drilling rig it might be required to have only a 1% chance of being exposed to an extreme event over a 20-year design life. Engineers Australia (2004) provide a formula to calculate the relevant return period of that event. It can be written:

$$T = -\frac{N}{\ln\left[1 - \frac{R}{100}\right]}$$

where $T$ is the average recurrence interval or return period, $N$ the design life or planning horizon and $R$ the acceptable risk (in %). For example, given a drilling platform with a design life of 20 years and an acceptable risk of 1%, the appropriate return interval is 2000 years.

Show how this formula can be derived from a consideration of the chance of events occurring (or not occurring) through application of the Poisson probability formula.

**29.2.** Plot the gamma distribution for a wide variety of parameter values. What sort of shapes is it capable of producing?

**29.3.** Determine the analytic expression for the standard deviation of the exponential distribution.

**29.4** Generate random exponential numbers in Excel using the formula `= -1/v ln(rand())` for some parameter value $v$. Summate $k$ exponential random variables to produce a single gamma variable. Repeat this to obtain 1000 gamma random variables. Develop a histogram of the gamma variables. Plot the PDF of the gamma distribution and compare it with the histogram.

**29.5** In Excel use the menu Tools – Data Analysis – Random Number Generation to generate 1000 Poisson random numbers with a mean of 5. Compute the mean of these numbers and compare it to the parameter value. Determine the histogram of the random numbers. Compare the histogram to the PMF of the Poisson distribution using the parameter value.

**29.6** One way to simulate a Poisson process in time is to generate exponential random numbers. These numbers describe the waiting time until the next event relative to the preceding event. Can this method be used to simulate a Poisson process in 2D or 3D spatial dimensions? (Hint: Is it possible to order the points sequentially in more than one dimension?)

**29.7** As outlined in Problem 29.5 use Excel to generate 20 Poisson random numbers with mean 30. Write a program to generate $x$ pairs of uniform random coordinates $(u,v)$ where $x$ is a Poisson random number. Plot each of the 20 groups of coordinates and inspect them. These plots represent 20 replicates of a 2D Poisson process.

**29.8** If a radioactive substance decays at a rate of 1:10,000 particles per year what is the half-life of the substance (the time at which only half of the particles remain)? Plot the distribution of number of particles with respect to year.

CHAPTER THIRTY

# Probability Distributions (Normal and Log-Normal)

## 30.1 INTRODUCTION

The normal or Gaussian distribution is one that most people will have heard of. It is widely used and suitable for a wide range of situations. For this reason it is also widely abused, and may inhibit the use of other more suitable probability distributions. Nevertheless it is discussed here together with a near relation, the log-normal distribution.

The normal distribution was derived in 1733 by Abraham DeMoivre. It is bell-shaped and is also called the Gaussian distribution in honour of Karl Gauss, who also derived its equation from a study of errors in repeated measurements of the same quantity. It is particularly useful for explaining phenomena in nature, and industry, where development is under the influence of many random factors (Nordin, 1971). The normal distribution has the formula:

$$f_X(x) = \frac{1}{\sigma\sqrt{2\pi}} \exp\left[-\frac{1}{2}\left(\frac{x-\mu}{\sigma}\right)^2\right] \quad \text{for } -\infty < x < \infty \qquad (30.1)$$

where $\mu$ and $\sigma$ are the parameters which define the distribution; $\mu$ is the mean and $\sigma$ the standard deviation. Although the distribution extends infinitely in both the positive and negative directions the majority of the normal distribution is contained within ± 3 standard deviations of the mean, as shown in Figure 30.1. The equation for the distribution can be derived from first principles: refer Rozanov (1969).

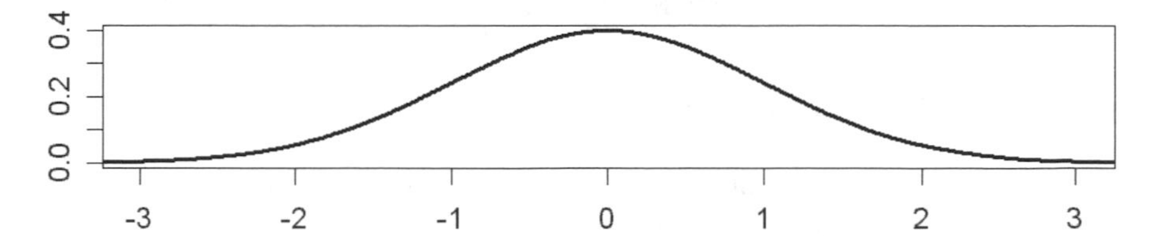


**Figure 30.1** The standard normal distribution with $\mu = 0.0$ and $\sigma = 1.0$.

The location and shape of the distribution can change depending on the mean and standard deviation respectively. The standard normal distribution is defined for the special case where the mean is 0.0 and the standard deviation is 1.0. Since the normal distribution is used so commonly there is a special notation for its cumulative distribution function; this is written $\Phi(z)$. Therefore:

$$\Phi(z) = F_Z(z) \tag{30.2}$$

and following from this, and making use of the properties of the cumulative distribution, it can be shown that:

$$P(a < x \leq b) = \Phi(b) - \Phi(a) \tag{30.3}$$

The integral of the normal distribution is not easy to calculate so the information on the cumulative distribution is usually tabulated.

*Example*
What is the probability of $x$ exceeding 1.0 if the mean is 0.0 and the standard deviation 1.0, i.e. what is $\Pr(x>1.0)$? The required area is shown in Figure 30.2.

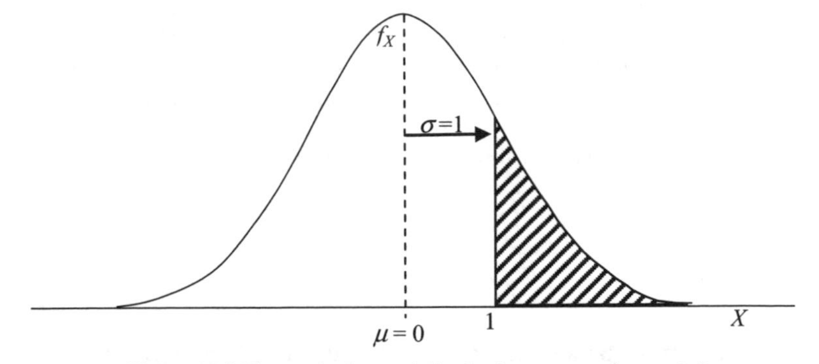


**Figure 30.2** The standard normal distribution, truncated at x = 1.0.

The probability of x exceeding 1.0 will be the area to the right of the shaded section. If the standard normal tables are consulted, the value given for a deviate of 1.0 is 0.8413. This means that 84.13% of the area lies in the segment less than 1.0 so the required answer is the complement of this – that is, 15.87%.

## 30.2 NORMAL DISTRIBUTION

It is possible to use the tabulated values for $\Phi(s)$ to calculate the probabilities of any other normal distribution where the mean is non-zero or the standard deviation is not 1.0. Suppose it is necessary to calculate:

$$P(a < x \leq b) = \frac{1}{\sigma\sqrt{2\pi}} \int_a^b exp\left[ -\frac{1}{2}\left(\frac{x-\mu}{\sigma}\right)^2 \right] dx \tag{30.4}$$

It is possible to calculate the integral directly. However, there is another way using the standard distribution. Making the substitutions:

$$z = \frac{x-\mu}{\sigma} \text{ and } dx = \sigma\, dz \tag{30.5}$$

it can be shown that this is simply the area under the standard curve between $(a - \mu)/\sigma$ and $(b - \mu)/\sigma$. That is:

$$P(a < x \leq b) = \Phi\left(\frac{b-\mu}{\sigma}\right) - \Phi\left(\frac{a-\mu}{\sigma}\right) \tag{30.6}$$

*Example*
What is P($x$ > 1.0) if the mean is –1.0 and the standard deviation 1.0? The required area is shown in Figure 30.3.

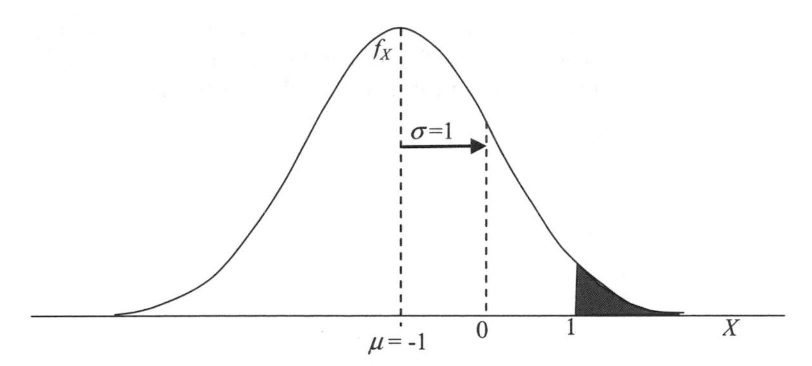


**Figure 30.3** The normal distribution, truncated at $x$ = 1.0. In this case, $\mu$ = –1.0, $\sigma$ = 1.0.

$$P(x > 1.0) = 1 - P(x < 1.0)$$

$$P(x \leq 1.0) = 1 - \Phi\left(\frac{1.0-\mu}{\sigma}\right) = 1 - \Phi\left(\frac{1.0--1.0}{1.0}\right) = 1 - \Phi(2.0)$$

Again, referring to the standard tables gives $F_z(2)$ = 0.9772, that is 97.72% of values are less than 1.0. The required probability is therefore 0.0228 or 2.28%.

## 30.3 EXCEL EXAMPLE

Excel has functions that act as look-up tables of the standard normal distribution and make it easy to work with this distribution.. The `NORMSDIST(x)` returns a cumulative probability for a given value of $x$ and `NORMSINV(pval)` gives a value of $x$ that has the probability *pval* that values from the distribution will be less than $x$. For example, the probability of a value being equal to or less than one standard deviation above the mean can be calculated as `=normsdist(1.0)` which returns the value 0.841345. The $x$ value that has 40% of values less than or equal to it can be determined from `=normsinv(0.40)` which returns the value –0.25335.

## 30.4 APPLICATION OF NORMAL DISTRIBUTION TO DIFFUSION

Diffusion of pollutants in streams (Figure 30.4) is often assumed to be described in terms of the Gaussian distribution. The equation for concentration in a 1D situation can be written:

$$\frac{\partial c}{\partial t}+u_i\frac{\partial c}{\partial x_i}=\frac{\partial}{\partial x_i}\left(D\frac{\partial c}{\partial x_i}\right) \tag{30.7}$$

where $c$ is the concentration at a point, $t$ is time, $x$ is distance along the river, $u$ is the velocity of the flow and $D$ is a diffusion coefficient. In many cases the only unknown (and the most difficult to estimate) is the diffusion coefficient $D$.

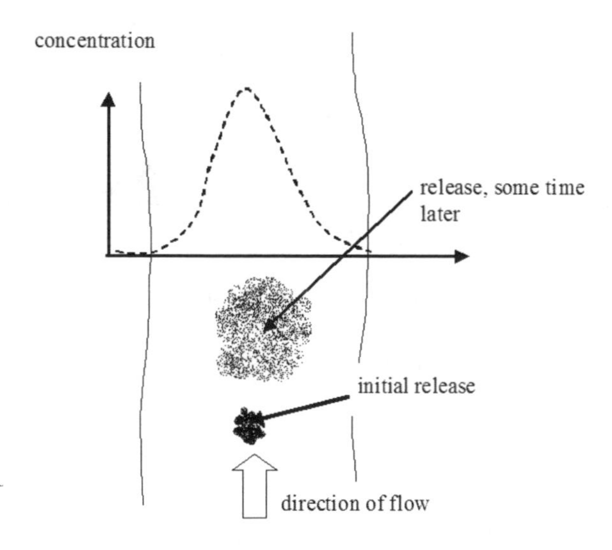


**Figure 30.4** Schematic of diffusion in river. Spread of pollutant is shown together with plot of concentrations measured across the flow. A number of sequential measurements of the spread would be necessary to estimate the diffusion coefficient.

If experiments are carried out to measure the concentration as it varies in time, the diffusion coefficient can be calculated from:

$$\sigma^2 = 2tD \qquad (30.8)$$

where $\sigma^2$ = is the variance of the distribution, $D$ is the diffusion coefficient and $t$ is the time from the start of pollutant release. To do this the concentration data along a river would be measured at a number of places and fitted to a Gaussian distribution.

## 30.5 LOG-NORMAL DISTRIBUTION

One of the major problems with the normal distribution is that it allows negative $x$ values (this may not always be appropriate). If the natural logarithms of a data set have a normal distribution, then the data have a logarithmic normal or log-normal distribution.

Since the log of a number is always positive it is useful for a number of distributions where the data will be positive. Ang and Tang (1975) list strength and fatigue life of material, rainfall intensity and time for project completion as examples of the sorts of data that might be fitted to log-normal distributions. Hardy et al. (2003) used the log-normal distribution to model the radius of tropical cyclones in a study of the wave climate on the Great Barrier Reef. The density function can be written:

$$f_x(x) = \frac{1}{\xi x\sqrt{2\pi}} \exp\left(-\frac{1}{2}\left(\frac{\ln(x)-\lambda}{\xi}\right)^2\right) \qquad (30.9)$$

where $\lambda = E(\ln x)$ and $\xi = \sqrt{\mathrm{Var}(\ln x)}$ are the mean and standard deviation of $ln(x)$ and are the parameters of the distribution. It is possible to calculate the mean ($\mu$), median ($x_m$) and standard deviation ($\sigma$) of the data $x$. This then allows the two parameters for the log-normal distribution ($\lambda,\xi$) to be determined from application of some of the following:

$$\mu = \exp(\lambda + 0.5\xi^2) \qquad (30.10)$$

$$\lambda = \ln(\mu) - \frac{1}{2}\xi^2 \qquad (30.11)$$

$$\lambda = \ln(x_m) \qquad (30.12)$$

$$\xi^2 = \ln\left(1 + \frac{\sigma^2}{\mu^2}\right) \qquad (30.13)$$

$$\text{If } \sigma/\mu \le 0.30 \text{ then } \xi \approx \sigma/\mu \qquad (30.14)$$

The normal probability tables can be used for log-normal calculations as shown in Equation (30.15).

$$P(a < x \leq b) = \Phi\left(\frac{\ln(b) - \lambda}{\xi}\right) - \Phi\left(\frac{\ln(a) - \lambda}{\xi}\right) \tag{30.15}$$

Gómez-Pujol et al. (2007) used the log-normal distribution to describe the monthly wave climate on the Spanish island of Mallorca. They were then able to show the variation of wave height throughout the year. A plot of the mean showed a seasonal variation with highest waves over the winter.

*Example*
Assuming a log-normal distribution, what is the probability of $x$ exceeding 10.0 if the mean is 5.0 and the standard deviation 1.0, i.e. $P(x>10.0)$?

*Solution*
First it is necessary to calculate $\lambda$ and $\xi$:

$$\sigma/\mu = 1.0/5.0 = 0.20 = \xi$$

$$\lambda = \ln(5.0) - \frac{1}{2}0.20^2 = 1.6094 - 0.02 = 1.5894$$

$$P(10 < x) = 1 = P(x < 10) = 1 - \Phi\left(\frac{\ln(10) - \lambda}{\xi}\right) = 1 - \Phi(3.5659) =< 0.0002$$

Note that the standard tables of the cumulative normal distribution usually go to 3.49 so it is not possible to give an exact value based on these. Using Excel gives the probability as 0.000181.

## 30.6 A WORD OF WARNING

The normal and log-normal distributions are widely understood and widely used. Part of this may be due to their ability to match some common situations, but it may also be due to the fact that they are widely understood and used. In an award acceptance speech, Bagnold (1979) warned of the inertia of convention and argued that “so strongly is the Gaussian probability idea entrenched that the implications of the real distribution and of the logarithmic system of displaying it are both still very largely ignored”.

Following a similar line of argument Sobey (1978) wrote “The log-normal distribution is the most widely used for waves, perhaps because of the relative simplicity and availability of the probability paper for this distribution, although there is no evidence to suggest that it is any more appropriate than say any of the other distributions ... [such as Gumbel, Log Pearson III or Weibull]”.

By all means use the normal and log-normal distributions, but only if they are appropriate, and not just because they are easy.

## PROBLEMS

**30.1** Use the appropriate functions in Excel to generate a set of the standard normal distribution tables; the start of one is shown in Table 30.1. For example, the probability $P(x \leq -3.42) = 0.00031$. (Hint: There will be two lines starting 0.0, one going –0.01, –0.02, ... and the other going 0.01, 0.02, ...)

**Table 30.1** The areas under the standard normal distribution.

Normal tables

| | 0.00 | 0.01 | 0.02 | 0.03 | 0.04 |
|---|---|---|---|---|---|
| -3.4 | 0.00034 | 0.00032 | 0.00031 | 0.00030 | 0.00029 |
| -3.3 | 0.00048 | 0.00047 | 0.00045 | 0.00043 | 0.00042 |
| -3.2 | 0.00069 | 0.00066 | 0.00064 | 0.00062 | 0.00060 |
| -3.1 | 0.00097 | 0.00094 | 0.00090 | 0.00087 | 0.00084 |
| -3 | 0.00135 | 0.00131 | 0.00126 | 0.00122 | 0.00118 |
| -2.9 | 0.00187 | 0.00181 | 0.00175 | 0.00169 | 0.00164 |
| -2.8 | 0.00256 | 0.00248 | 0.00240 | 0.00233 | 0.00226 |
| -2.7 | 0.00347 | 0.00336 | 0.00326 | 0.00317 | 0.00307 |
| -2.6 | 0.00466 | 0.00453 | 0.00440 | 0.00427 | 0.00415 |
| -2.5 | 0.00621 | 0.00604 | 0.00587 | 0.00570 | 0.00554 |
| -2.4 | 0.00820 | 0.00798 | 0.00776 | 0.00755 | 0.00734 |

**30.2** One hundred and forty people have been measured and found to have heights that conform to a normal distribution with a mean of 1.5 metres and a standard deviation of 0.15 metres. How many would be expected to be between 1.3 and 1.7 metres?

**30.3** One hundred and forty people have been measured and found to have heights that conform to a log-normal distribution with a mean of 1.5 metres and a standard deviation of 0.15 metres. How many would be expected to be between 1.3 and 1.7 metres? If the answer is different from that assuming a normal distribution, explain why this might be the case.

**30.4** Following on from Problem 30.3, assume that it is known that a particular person is over 1.3 metres tall; what is the probability of them being between 1.3 and 1.7 metres tall?

**30.5** In a recent research paper (Scheibehenne et al., 2007) the authors were keen to remove “outliers” from the data lest they spoil the analysis. According to the authors: “All decision times longer than two standard deviations from the mean of each participant were considered outliers and excluded from further analyses. On average 4% of all decision time measures were identified as outliers.” Discuss.

**30.6** In Excel, plot the spread of concentration in time of a material that is released into the ocean where the diffusion coefficient is believed to be 100 $m^2/s$. How long would it take for it to be spread over a width of 1 km?

**30.7** Data have been collected across a river 10 and 20 seconds after a slug of dye was released at the centre-line. The concentrations are given in Table 30.2. Assume that the diffusion process in the cross-river direction can be represented by a Gaussian distribution and that the standard deviation, s, can be determined from Equation (30.8). From the data at $t = 10$ seconds and $t = 20$ seconds, determine the diffusion coefficient.

**Table 30.2** Concentration data collected across a river.

| Offset (m) | Concentration t = 10 sec. | Concentration t = 20 sec. |
|---|---|---|
| –10 | 0.0139 | 0.0244 |
| –9 | 0.0045 | 0.0230 |
| –8 | 0.0218 | 0.0022 |
| –7 | 0.0009 | 0.0251 |
| –6 | 0.0021 | 0.0030 |
| –5 | 0.0240 | 0.0097 |
| –4 | 0.0191 | 0.0468 |
| –3 | 0.0349 | 0.0774 |
| –2 | 0.1063 | 0.1372 |
| –1 | 0.2363 | 0.1979 |
| 0 | 0.2898 | 0.2111 |
| 1 | 0.2377 | 0.1908 |
| 2 | 0.1088 | 0.1423 |
| 3 | 0.0350 | 0.0795 |
| 4 | 0.0151 | 0.0357 |
| 5 | 0.0139 | 0.0201 |
| 6 | 0.0039 | 0.0159 |
| 7 | 0.0111 | 0.0251 |
| 8 | 0.0087 | 0.0203 |
| 9 | 0.0047 | 0.0218 |
| 10 | 0.0135 | 0.0087 |

CHAPTER THIRTY ONE

# Probability Distributions (Extreme Values)

## 31.1 INTRODUCTION

Civil engineering structures need to withstand extreme or rare events such as floods, hurricanes, earthquakes and huge waves. In typical applications there are limited data to estimate the magnitudes of these rare events, and yet a design value is required. One way of generating these values is to fit a distribution to selected maximum values (e.g. largest annual values) and extrapolate in the tail of this distribution. However, such extrapolation will be sensitive to the form of the distribution. It has been found that there is theoretical justification and empirical support for assuming that annual maxima have a Gumbel distribution.

**Extreme Value Theory: E. J. Gumbel**

Emil Gumbel (1891–1966) was a statistician and a political activist. He was Professor of Mathematical Statistics at the University of Heidelberg until he moved to Paris in 1932, and later the USA. Gumbel was one of the mathematicians who developed extreme value theory. Others include E. L. Dodd, Maurice Frechet, L. H. C. Tippet and R. A. Fisher. Gumbel was interested in hydrological applications.

## 31.2 GENESIS OF AN EXTREME VALUE DISTRIBUTION

To illustrate the behaviour of selected maximum values within a set of data a simulation was carried out. A random sample of 1 million numbers from a normal distribution, with a mean and standard deviation of one, was generated. The maximum values from consecutive subgroups of 50 were extracted to represent extreme values. Figure 31.1(a) is a histogram of the normal random numbers and Figure 31.1(b) is a histogram of the extreme values. Notice that the original distribution, referred to as the parent distribution, is symmetric and bell-shaped as might be expected, whereas the distribution of extreme values is positively skewed (long tail to the right).

This procedure of generating 1 million data was repeated with an exponential distribution, having mean and standard deviation of one, as the parent distribution. Figure 31.2(a) is a histogram of the exponential random numbers and Figure 31.2(b) is a histogram of the extreme values. The exponential distribution is bounded below at zero and, despite having the same mean and standard deviation as the normal distribution, it has a much longer tail. The histogram of extreme values is similar in shape to that obtained from the normal distribution, but the mean and standard deviation are clearly higher. Unlike with the parent exponential distribution, the distribution of extreme values has no clear lower bound.

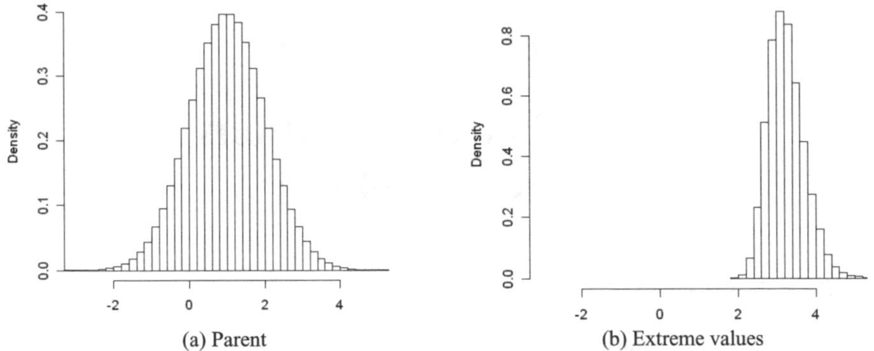


(a) Parent (b) Extreme values

**Figure 31.1** Normal parent distribution and distribution of extremes.

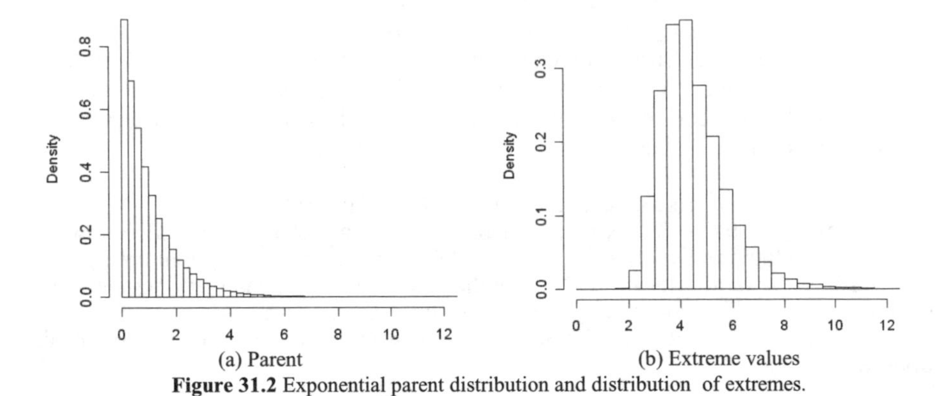


(a) Parent (b) Extreme values

**Figure 31.2** Exponential parent distribution and distribution of extremes.

The point to be noticed at this stage is that, despite the strikingly different parent distributions involved, the extreme value distributions are quite similar. This point underlies the workings of extreme value distributions and will be expanded upon in the next section.

## 31.3 THE GUMBEL DISTRIBUTION

In the previous section the maximum values from random samples of size 50 were extracted and seen to have similar-shaped histograms. The parent distributions are quite different but both have upper tails that decay at least as fast as a negative exponential. There is a mathematical proof that the distribution of the maximum in random samples of size $m$ from a distribution with an unbounded upper tail, decaying at least as fast as an exponential, tends to the Gumbel distribution as $m$ tends to infinity. The result is a reasonable approximation for moderate values of $m$.

**Asymptotic Results in Statistics**

The mathematical proofs of many results in statistics depend on the sample size tending to infinity. The best-known example is the central limit theorem, which states that in most circumstances the sum of a large number of random variables is normally distributed. The value of these results is that they provide useful approximations for moderate-sized samples.

The Gumbel distribution is commonly used to model the distribution of annual maximum flows in rivers. For rivers that do not run dry, annual maximum flows are the maximum of 365 daily average flows. The assumption that daily flows are from a distribution with an unbounded upper tail which decays exponentially is a plausible modelling assumption. The daily flows are certainly not independent, and the positive correlation between flows on consecutive days has the effect of reducing the sample size. An intuitive explanation for this is that if the flow was sampled at weekly intervals, the flows would be less correlated, but there would only be 52 values to work with. Nevertheless, the Gumbel distribution often provides a good empirical fit to yearly maxima and is widely used by engineers when they design flood defence structures or propose flood protection schemes.

The Gumbel distribution for a random variable $X$ has the cumulative distribution function (CDF):

$$F(x) = \exp\left(-\exp\left(-\frac{x-\zeta}{\theta}\right)\right) \qquad -\infty < x < \infty \qquad (31.1)$$

The probability density function (PDF) follows from differentiation (see Problem 31.1 at end of chapter). The parameters $\zeta$ and $\theta$ are the mode and a scale factor respectively. The Gumbel distribution has a skewness of 1.14. The mean and standard deviation of the distribution are related to the parameters by:

$$\mu_x = \zeta + 0.57722\theta \quad \text{and} \quad \sigma_x = 1.28255\theta \qquad (31.2)$$

The Gumbel distribution is a special case of the generalised extreme value (GEV) distribution. The GEV has an additional shape parameter that takes on the value zero for the Gumbel distribution. When its value is positive the Frechet distribution is obtained and for negative values the Weibull distribution is obtained.

## 31.4 MODELLING CYCLONES

On average about 80 tropical cyclones occur each year across the Pacific and Indian oceans. Figure 31.3 shows a distribution of the storm tracks of all cyclones in the period from 1945 to 2006 (US Navy, 2007). The cyclone properties, such as their position and sustained wind speed, are typically estimated at 6-hour intervals. The usual lifetime of a cyclone is in the order of several days.

Figure 31.4(a) shows a histogram of the sustained wind speed from 4852 tropical cyclones. For each year on record, the maximum sustained wind speed from all the tropical cyclones in that year was extracted. The histogram of the annual maxima is shown in Figure 31.4(b) together with a superimposed Gumbel distribution. The parameters of the Gumbel distribution are estimated by replacing the $\mu_x$ and $\sigma_x$ in Equation (31.2) by their sample estimates $m$ and $s$ (see Problem 31.2 at end of chapter).

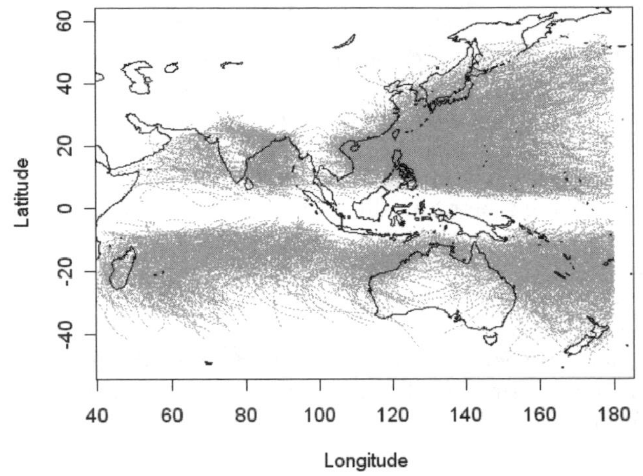


**Figure 31.3** Cyclone tracks across Indian and Western Pacific Oceans.

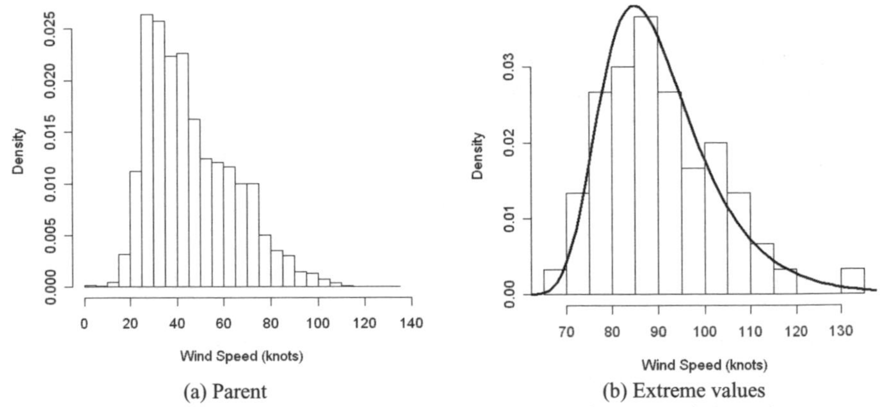


(a) Parent (b) Extreme values

**Figure 31.4** Parent and extreme value distributions of cyclone wind speed.

## 31.5 PARETO DISTRIBUTION

There are a range of natural phenomena where the relative frequency of events seems to be related to the size of the event. Take earthquakes as an example: the large quakes occur very infrequently, whereas minor earthquakes are observed to occur much more often. The situation is similar with floods, rainfall, even the size distribution of material ejected from a mining blast: a few large ones and lots of small ones.

**Historical Note: Pareto**

Professor Vilfredo Pareto observed in a book published in 1897 that the number of people in a population whose income exceeded $x$ could be modelled by the expression: $N = Cx^{-\alpha}$ for a real $C$ and a positive $\alpha$. This was popularised in the 80:20 law (80% of a country's wealth was owned by 20% of its population). Further work by others showed that the situation was slightly more complex than this but Pareto's expression has remained in use and spawned a number of approaches to probability distributions concerned with income and wealth. (Arnold, 1983)

Many of these phenomena can be described in terms of a power law:

$$f(t) = at^{-b} \tag{31.3}$$

where $a$ and $b$ are constants to be determined from the data being described. For example, analysis of earthquakes has led to the Gutenberg-Richter scaling law:

$$\log N = -bM + a \tag{31.4}$$

where $N$ is the number of earthquakes of magnitude $M$ and $a$ and $b$ are constants that are dependent on the region (of the earth). Based on empirical evidence $0.8 < b < 1.2$ (Malamud and Turcotte, 2006). Equation (31.3) can be plotted, and is shown in Figure 31.5. For Equation (31.3) to be considered a proper probability distribution the total area under the curve should be 1.0 and this will not always be true for all $a$ and $b$. A modified version of the equation is given by Corotis et al. (1983) which satisfies that constraint. It can be written:

$$f(t) = (t/t_0)^{-b}(b-1)/t_0 \tag{31.5}$$

where $b$ is an independent variable and must be greater than 1 for convergence, and $t_0$ is the minimum value recorded in the data. For a given data set, an estimate for $b$ can be determined from:

$$(b-1)^{-1} = E[\ln(t)] = \ln(t_0) \tag{31.6}$$

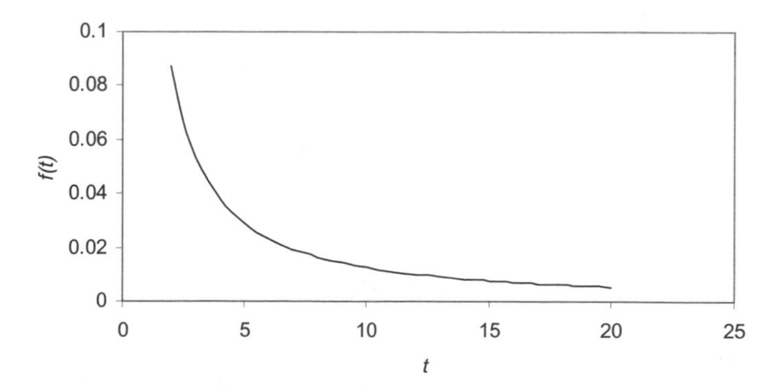


**Figure 31.5** Power law function, $f(t)=0.2t^{-1.2}$.

As an example, the data on the size of cities in Germany was taken from Gigerenzer and Goldstein (1996) and sorted into a line plot of width 100,000 residents. In all there were 83 cities with populations varying from 102,440 to 3,433,695. There were 45 cities between 100,000 and 200,000, 18 from 200,000 to 300,000 and so forth. A truncated data set (ignoring those over 1,000,000) is given in Table 31.1.

**Table 31.1** The number of German cities fitting into a range of size bins.

| Size | Number |
|---|---|
| 200000 | 45 |
| 300000 | 18 |
| 400000 | 5 |
| 500000 | 2 |
| 600000 | 7 |
| 700000 | 2 |
| 800000 | 0 |
| 900000 | 0 |
| 1000000 | 1 |

Applying Equation (31.6) gives a value for $b = 2.015$. The raw data, together with the probability distribution (suitable scaled up based on the total number of cities and the width of the histogram bands) is shown in Figure 31.6. It is evident that the distribution fits the data reasonably well.

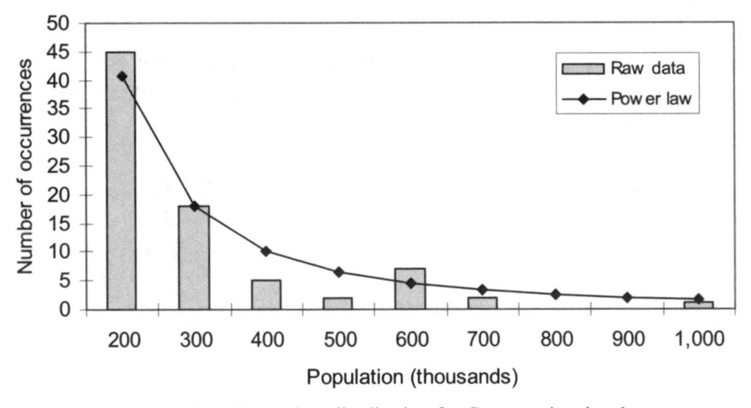


**Figure 31.6** Power law distribution for German city size data.

## 31.6 GENERALISED PARETO DISTRIBUTION

In many situations extreme values are only observed when some threshold is exceeded. Alternatively, if an entire record is available, then the peak over threshold method can be used to obtain peaks from that record. This method is depicted in Figure 31.7, where the threshold is set so that separate peaks are obtained. Note, however, that it may be necessary to discard any peaks that occur close together since they may not be independent. The peak over threshold is potentially advantageous to the method of annual maximums when there is a short data record, since it is possible to obtain more than one peak per year.

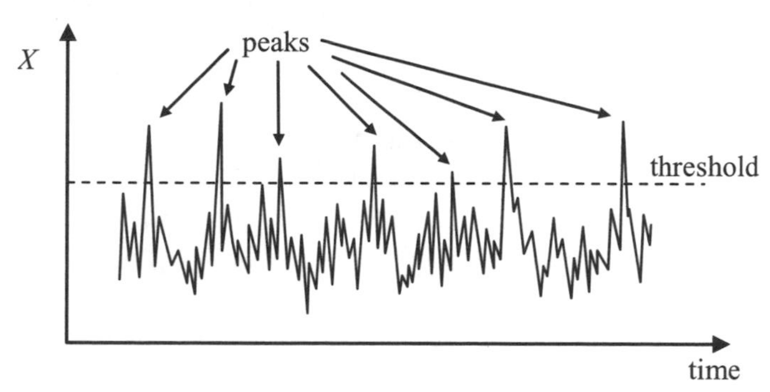


**Figure 31.7** Peak over threshold method.

The generalised Pareto distribution (GPD) has two properties which make it a reasonable choice for peak over threshold analyses. The first is that if a random variable $X$ has a GPD, then for some threshold, $L$, the conditional distribution of $X-L$ is also a GPD. The second is that it can be related to the GEV distribution. The CDF of the GPD is:

$$F_X(x)=\begin{cases} 1-(1-kx/\eta)^{\frac{1}{k}} & 0 \le x < \infty \ \ when\ k<0 \\ 1-exp(-x/\eta) & 0 \le x < \infty \ \ when\ k=0 \\ 1-(1-kx/\eta)^{\frac{1}{k}} & 0 \le x \le \eta/k \ \ when\ k>0 \end{cases} \tag{31.7}$$

where $\eta$ is the scale parameter ($\eta$>0) and $k$ is the shape parameter. If $k<-½$ the distribution has infinite variance, if $k=0$ the exponential distribution is obtained as a special case, if $k \ge ½$ the distribution has a finite upper bound and $k=1$ is the special case of a uniform distribution. For most practical applications $-½<k<½$. The $r^{th}$ moment of the distribution can be obtained as

$$E\left[(1-kX/\eta)^r\right]=1/(1+rk) \tag{31.8}$$

but the $r^{th}$ moment will not exist if $k \le -1/r$. The mean and variance can be obtained from this equation giving $\mu=\eta/(1+k)$ and $\sigma^2=\eta^2/[(1+k^2)(1+2k)]$ respectively.

Consider the peak flows from the Coquet River in England from the period 1973 to 1993 (Metcalfe, 1997). The flows have been selected as those above a threshold of 60.0 $m^3/s$ that are also separated by at least 3 days so that they can be assumed to be independent. The 25 peak flows are provided in Table 31.2.

**Table 31.2** Independent flows above 60.0 cumecs for Coquet River.

| | | | | |
|---|---|---|---|---|
| 60.2 | 66.2 | 76.4 | 86.6 | 113.2 |
| 62.5 | 66.3 | 77.4 | 89.6 | 117.1 |
| 63.1 | 66.3 | 77.5 | 91.4 | 142.5 |
| 63.8 | 71.7 | 81.3 | 101.7 | 180.7 |
| 63.9 | 72.8 | 85.2 | 106.8 | 190.9 |

The mean and variance of the peak flows are 91.0 and 1229.1 respectively. This gives parameter estimates $\hat{\eta} = 19.96$ and $\hat{k} = -0.356$. The fitted distribution is shown in Figure 31.8.

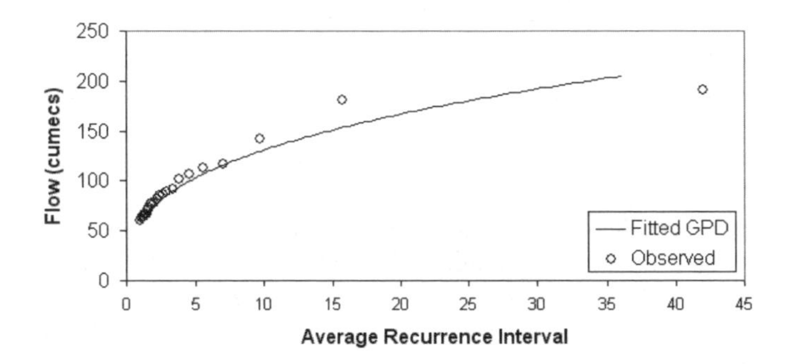


**Figure 31.8** Fitted GPD to Coquet River peak flows.

## PROBLEMS

**31.1** Differentiate $F(x)$ for $\zeta = 0$, $\theta = 1$. (Hint: The chain rule will need to be used.)

**31.2** For the 62 annual maximum cyclones, the mean wind speed is 90.8 (miles per hour) and the standard deviation is 12.4 (miles per hour). Verify that the parameters of the Gumbel distribution are $\hat{\zeta} = 85.25$ and $\hat{\theta} = 9.64$.

**31.3** For the data set described in Problem 31.2, what is the probability that the sustained wind speed of the annual maximum cyclone exceeds 100 knots?

**31.4** What wind speed will be exceeded by the annual maximum cyclone with a probability 0.01? This is referred to as the wind speed with an average recurrence interval (ARI), colloquially referred to as the return period, of 100 years. What is the probability that a wind speed with a return period of 100 years will be exceeded in the next 50 years?

## CHAPTER THIRTY TWO

# Probability Distributions (Chi-Square and Rayleigh)

### 32.1 INTRODUCTION

Some probability distributions arise when combinations of random variables are taken. For example, the sum of $k$ exponential random variables has a gamma distribution. This chapter investigates two distributions that can be formed by taking a combination of normal random variables.

### 32.2 CHI-SQUARE DISTRIBUTION

For $k$ independent normally distributed random variables, $X_i$, the sum $Z = \sum_{i=1}^{k} X_i^2$ has a chi-square distribution, denoted $Z \sim \chi_k^2$. The chi-square distribution has one parameter, $k$, which is a positive integer referred to as the degrees of freedom and the PDF is:

$$f_Z(z) = \frac{z^{(k/2)-1} e^{-z/2}}{2^{k/2}\Gamma(k/2)} \quad \text{for } z > 0 \tag{32.1}$$

The distribution has a mean $k$ and a variance of $2k$. Figure 32.1 shows the PDF of the chi-square distribution for the case $k = 3$. The chi-square distribution is used in the chi-square goodness-of-fit test and also for the distribution of the sample variance.

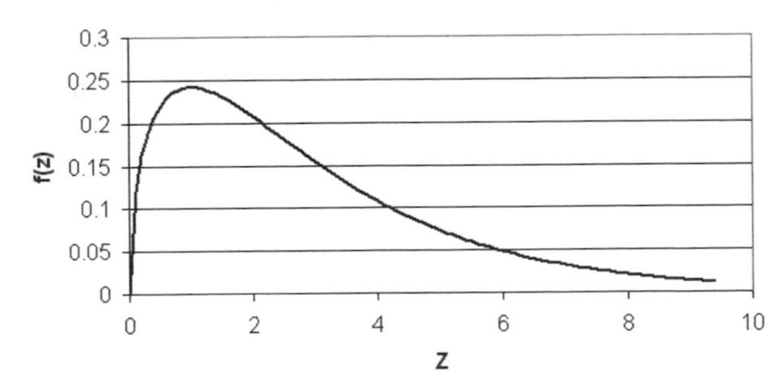


**Figure 32.1** Chi-squared distribution, $k = 3$.

## 32.3 RAYLEIGH DISTRIBUTION

Consider two independent normally distributed random variables $X \sim N(0,\sigma^2)$, $Y \sim N(0, \sigma^2)$. These random variables might represent the $x$ and $y$ components of a vector such as wind velocity. The result hypotenuse of this vector $H = \sqrt{X^2 + Y^2}$ has a Rayleigh distribution. Another example might be the real and imaginary components of a complex number that are normally distributed, as is the case for the amplitudes on a spectrum. As another example, wave heights are commonly modelled using a Rayleigh distribution. The distribution, which assumes that the energy spectrum is narrow-banded, can be written (Young, 1999) as:

$$f(H) = \frac{H}{4b^2} exp\left(-\frac{H^2}{8b^2}\right) \tag{32.2}$$

where $H$ is the wave height and $b$ the standard deviation of the distribution. The distribution, for a particular significant wave height, is shown in Figure 32.2.

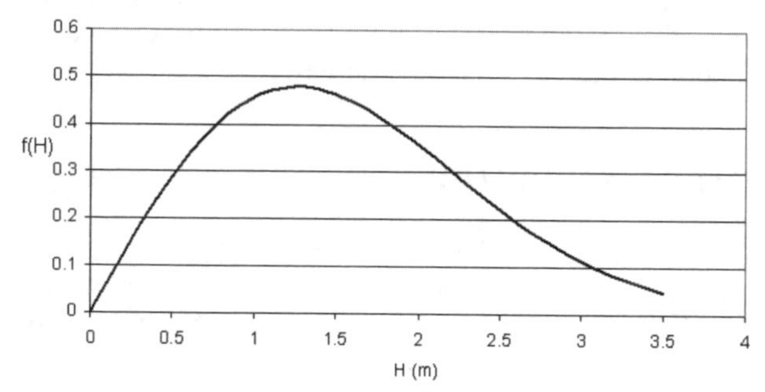


**Figure 32.2** Rayleigh distribution of wave heights, Hsig = 2.5 m.

## PROBLEMS

**32.1** Using pairs of normal random numbers in Excel `=norminv(rand(),0,1)`, compute $Z = X^2 + Y^2$ 1000 times. Make a histogram and compare it with the chi-squared PDF for $k = 2$.

**32.2** Compute a histogram of $H = \sqrt{X^2 + Y^2}$ using 1000 pairs of normal random numbers and compare it to the PDF of the Rayleigh distribution.

## CHAPTER THIRTY THREE

# Probability Distributions (Multivariate)

## 33.1 INTRODUCTION

In a population it is common to measure more than one variable, often with the aim of determining relationships between them. For example, the speed and direction of wind or the abundance, grade and location of various mineral deposits. Multivariate probability distributions allow each separate variable to be characterised along with their joint distribution that may or may not have relationships between the variables.

## 33.2 DISCRETE BIVARIATE DISTRIBUTIONS

Most of the concepts about multivariate distributions can be illustrated by the rather special case of two discrete variables. Suppose a very large number of reported water mains leaks have been classified by failure modes, *X*, including corrosion, tree root intrusion, cracks and pipe-joint failure. The pipe materials, *Y*, are concrete, cast iron and PVC. The proportions in each of the categories in Table 33.1 will be treated as probabilities.

**Table 33.1** Failure modes and material properties of pipe breakages.

| Material\Failure | Corrosion | Tree root intrusion | Crack | Joint failure | Marginal distribution for *Y* |
|---|---|---|---|---|---|
| Concrete | 0.03 | 0.18 | 0.02 | 0.06 | 0.29 |
| Cast iron | 0.25 | 0.02 | 0.08 | 0.02 | 0.37 |
| PVC | 0.00 | 0.10 | 0.14 | 0.10 | 0.34 |
| Marginal distribution for *X* | 0.28 | 0.30 | 0.24 | 0.18 | 1.00 |

The bivariate distribution is defined by the probability mass function $P_{XY}(x, y) = P(X = x \; and \; Y = y)$. For example, the probability that a randomly selected leak occurred in concrete pipe from the intrusion of tree roots is 0.18. From Table 33.1 notice that $\sum_x \sum_y P(x, y) = 1$.

The marginal distribution of *X* is simply the distribution of *X* that is obtained if all the information about *Y* is ignored. From Table 33.1 the marginal distribution of

$X$ is obtained from the bivariate distribution by adding down the columns and putting the sums in the row underneath the data. So for corrosion the marginal value is $0.03 + 0.25 + 0.00 = 0.28$. The marginal distribution of $Y$ is obtained by adding along the rows and putting the results in the right-hand column. More formally these probabilities are written:

$$P_X(x) = \sum_y P_{XY}(x, y) \qquad P_Y(y) = \sum_x P_{XY}(x, y) \tag{33.1}$$

The random variables $X$ and $Y$ are independent if, and only if, their joint probability equals the product of their marginal probabilities:

$$P_{XY}(x, y) = P_X(x)P_Y(y) \qquad \text{(independent)} \tag{33.2}$$

In the event $X$ and $Y$ are not independent, this implies that the distribution of $X$ will change with respect to $Y$ and vice versa. These are referred to as conditional distributions. In general, the conditional distribution of $X$ given $Y = y$ follows the standard rules of probability and is given by

$$P_{X|Y}(x \mid y) = \frac{P_{XY}(x, y)}{P_Y(y)} \qquad \text{(conditional)} \tag{33.3}$$

As an example consider that the engineer's interest is in leakages due to tree root intrusion. The probability distribution of material is given by the numerical entries of the "Tree root intrusion" column in Table 33.1: (0.18, 0.02, 0.1). The probabilities are scaled to add to 1, by dividing by the marginal probability 0.3 so that the conditional probabilities for concrete, iron and PVC given tree root intrusion are 0.600, 0.067 and 0.333 respectively.

## 33.3 BIVARIATE NORMAL DISTRIBUTION

The bivariate cumulative distribution function for two continuous random variables $X$ and $Y$ is $F_{XY}(x, y) = f(X \leq x \text{ and } Y \leq y)$ and the probability density function is defined as the derivative of $F_{XY}$:

$$f_{XY}(x, y) = \frac{\partial^2 F_{XY}}{\partial x \partial y} \tag{33.4}$$

Various probabilities can obtained by integrating regions of the PDF surface

$$P(x_1 \leq x < x_2, y_1 \leq y < y_2) = \int_{x_1}^{x_2} \int_{y_1}^{y_2} f_{XY}(x, y)\, dx\, dy \tag{33.5}$$

For example, setting $(x_1,y_1) = (-\infty,-\infty)$ gives the cumulative probability up to the point $(x_2,y_2)$. The marginal PDFs are obtained by integrating over one of the variables:

$$f_X(x) = \int_{y=-\infty}^{\infty} f_{XY}(x,y)\,dy \qquad f_Y(y) = \int_{x=-\infty}^{\infty} f_{XY}(x,y)\,dx \tag{33.6}$$

and similarly to Equation (33.3) the conditional distribution of $X$ given the value $Y = y$ is:

$$f_{X|Y}(x \mid y) = \frac{f_{XY}(x,y)}{f_Y(y)} \tag{33.7}$$

For the case of independence between $X$ and $Y$, $f(x/y)$ and $f(x)$ are identical (since the variable $Y$ does not provide any additional information). This observation implies that the joint density is the product of the two marginal densities:

$$f_{XY}(x,y) = f_X(x) f_Y(y) \tag{33.8}$$

To illustrate this consider the example of two independent marginal normal distributions which produce the bell-shaped curve shown in Figure 33.1(a).

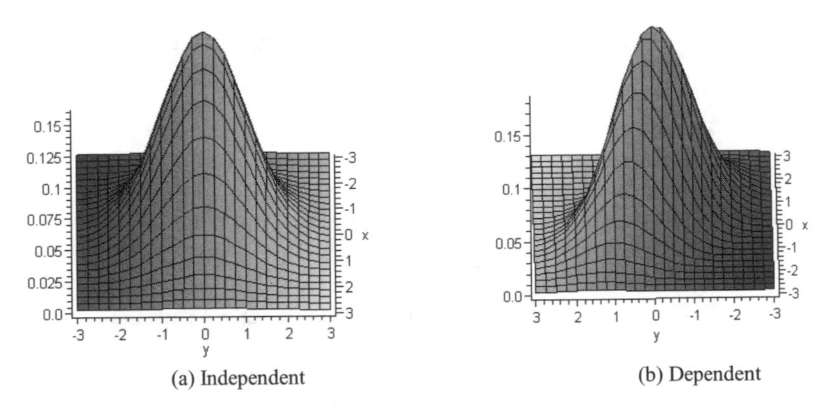


(a) Independent (b) Dependent

**Figure 33.1** Independent and dependent bivariate normal distributions.

The bivariate normal distribution (and multivariate by extension) is by far the most commonly used joint distribution. This is because it has several convenient properties for modelling the relationship between two variables. Firstly, it has a correlation parameter, $\rho$, that can be related to the sample correlation statistic from data. Secondly, all marginal and conditional distributions are themselves Gaussian. These properties are attractive enough for many users such that, in order to exploit them, any number of transforms will be applied to make their data approximately normally distributed. Figure 33.1(b) shows an example of the bivariate normal distribution with $\rho = 0.5$ where the PDF is given by:

$$f_{XY}(x, y) = \frac{1}{2\pi\sigma_x\sigma_y\sqrt{1-\rho^2}} \exp\left( -\frac{1}{2(1-\rho^2)} \left(\frac{x^2}{\sigma_x^2} + \frac{y^2}{\sigma_y^2} - \frac{2\rho xy}{\sigma_x\sigma_y}\right) \right) \quad (33.9)$$

where the location of the PDF can be changed by substituting $x' = (x - \mu_x)$ and $y' = (y - \mu_y)$. Probability contours of the PDF are plotted in Figure 33.2 for various $\rho$.

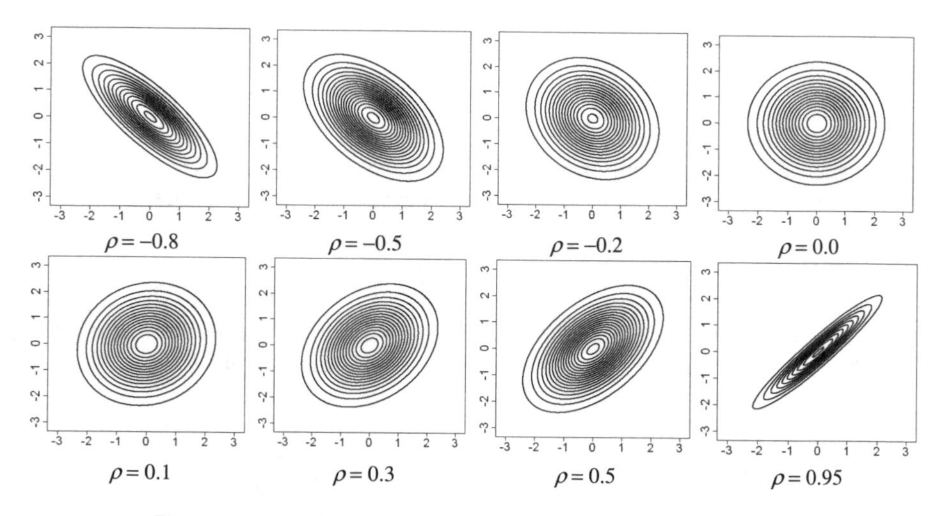


**Figure 33.2** Contours of bivariate Gaussian density function for various $\rho$.

The correlation parameter has values between 1 and −1 which respectively correspond to perfectly correlated and perfectly correlated but with opposite sign. Independent variables imply a correlation of zero, but a correlation of zero does not imply independence.

The marginal distribution of $X$ is simply the univariate Gaussian density function:

$$f_X(x) = \frac{1}{\sigma_x\sqrt{2\pi}} \exp\left( -\frac{x^2}{2\sigma_x^2} \right) \quad (33.10)$$

and similarly for the variable $Y$. The conditional distribution of $x$ given $Y = y$ is:

$$f_{X|Y}(x \mid y) = \frac{1}{\sigma_y\sqrt{2\pi(1-\rho^2)}} \exp\left( -\frac{1}{2} \frac{\left(x - \mu_x - \rho(\sigma_y/\sigma_Y)(y - \mu_Y)\right)^2}{\sigma_x^2(1-\rho^2)} \right) \quad (33.11)$$

and similarly for the variable $f_{Y|X}$. The conditional distribution $Y \mid X$ can be thought of as a slice through the joint distribution along the line $X=x$, which is then standardised to have an area of 1. An example is shown in Figure 33.3 for the values $x= 15$, $\sigma_x=3$, $\sigma_y=3$, $\mu_x=10$, $\mu_y=10$, $\rho = 0.5$.

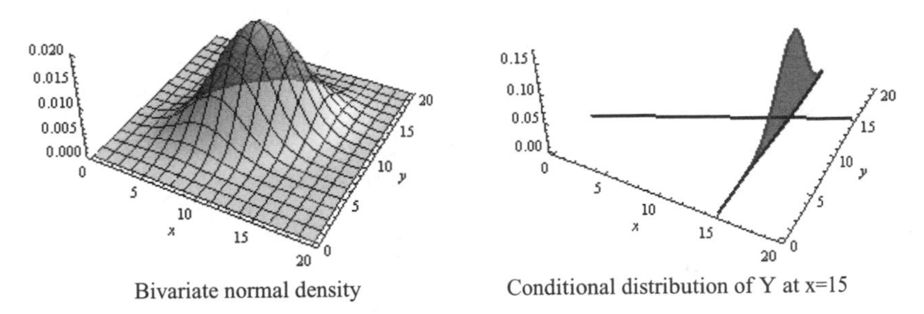


**Figure 33.3** Bivariate and conditional distributions. From Mathematica (2007).

## 33.4 MULTIVARIATE NORMAL DISTRIBUTION

The normal distribution can be easily extended into multiple dimensions, whereas many other distributions cannot. Suppose $X$ has $K$ components, and has a mean matrix $\mu$ of size $K \times 1$ and variance-covariance matrix $\Sigma$ of size $K \times K$. The $K$ elements of $\mu$ are the means of the marginal distributions of $X$, and the variances of these marginal distributions lie along the leading diagonal (top left to lower right) of $\Sigma$. The off-diagonal terms in the matrix are the covariance terms. To summarise, the elements of the mean and variance-covariance matrix are defined by:

$$\begin{aligned} \mu_k &= E[X_k] \\ \sigma_{kk}^2 &= E[(X_k - \mu_k)^2] \\ \text{cov}_{jk} &= \text{cov}_{kj} = E[(X_j - \mu_j)(X_k - \mu_k)] \end{aligned} \tag{33.12}$$

where $1 \leq j,k \leq K$ and $j \neq k$ and the positions of these elements in the matrices is:

$$\mu = \begin{pmatrix} \mu_1 \\ \vdots \\ \vdots \\ \mu_K \end{pmatrix} \qquad \Sigma = \begin{pmatrix} \sigma_{11}^2 & cov_{12} & \cdots & cov_{1K} \\ & \ddots & & cov_{2K} \\ & & \ddots & \vdots \\ cov_{K1} & & & \sigma_{KK}^2 \end{pmatrix} \tag{33.13}$$

The covariances can also be expressed as $cov_{jk}=\rho_{jk}\sigma_j\sigma_k$, where $\sigma_k = \sqrt{\sigma_{kk}^2}$ and $\rho_{jk}$ is the correlation between the variables $X_j$ and $X_k$. Any positive numbers can be specified for the variances, but the covariance terms must correspond to a correlation between $-1$ and 1. There is an additional restriction on the matrix $\Sigma$ that it must be positive definite (all eigenvalues of $\Sigma$ must be positive). As an example, this condition would disallow the possibility that as $X_1$ and $X_2$ have $\rho_{12} = 0$, $X_2$ and $X_3$ are correlated $\rho_{23} = 1$ and $X_1$ and $X_3$ are correlated $\rho_{13} = 1$, which is illogical. The PDF of $X$ is:

$$f(x) = \frac{1}{(2\pi)^{K/2} |\Sigma|^{1/2}} exp\left(-\frac{1}{2}(x-\mu)'\Sigma^{-1}(x-\mu)\right) \quad (33.14)$$

where $|\Sigma|$ is the determinant of the matrix $\Sigma$, $(x-\mu)'$ is the transpose of $(x-\mu)$ and $\Sigma^{-1}$ is the inverse of $\Sigma$.

Consider now a partition of the variables $X$ into two subgroups, $X_1$ having $m$ variables and $X_2$ having $n = K-m$. This can be represented as

$$X = \begin{pmatrix} X_1 \\ X_2 \end{pmatrix} \qquad \mu = \begin{pmatrix} \mu_1 \\ \mu_2 \end{pmatrix} \qquad \Sigma = \begin{pmatrix} \Sigma_{11} & \Sigma_{12} \\ \Sigma_{21} & \Sigma_{22} \end{pmatrix} \quad (33.15)$$

where $\mu_1$ and $\mu_2$ are the mean vectors for each partition, $\Sigma_{11}$ and $\Sigma_{22}$ are the $[m \times m]$ and $[n \times n]$ matrices, respectively, that have the covariances between all elements within each partition and $\Sigma_{12} = \Sigma_{21}'$ $[n \times m]$ are matrices that have the covariances between the $X_1$ and $X_2$ partitions. Written in this form, the marginal distributions of each partition are straightforward to determine, $X_1 \sim N(\mu_1, \Sigma_{11})$ and $X_2 \sim N(\mu_2, \Sigma_{22})$. The conditional distribution of $X_2$ given $X_1 = x_1$ is $X_{2|1} \sim N(\mu_{2|1}, \Sigma_{2|1})$ where the $[n \times 1]$ conditional means $\mu_{2|1}$ and the $[n \times n]$ conditional covariance matrix $\Sigma_{2|1}$ are given earlier. The matrix dimensions are written underneath to help verify the dimensions resulting from the matrix multiplications.

$$\begin{array}{lll} \mu_{2/1} = & \mu_2 + & \Sigma_{21}\Sigma_{11}^{-1}(x_1 - \mu_1) \\ [n \times 1] & [n \times 1] & [n \times m][m \times m][m \times 1] \\ \Sigma_{2|1} = & \Sigma_{22} - & \Sigma_{21}\,\Sigma_{11}^{-1}\,\Sigma_{12} \\ [n \times n] & [n \times n] & [n \times m][m \times m][m \times n] \end{array} \quad (33.16)$$

As an example, consider the daily readings of three air-quality values for 1 May 1973 to 30 September 1973: lower atmosphere ozone, wind speed and temperature (Chambers et al., 1983). Ozone is a pollutant in the lower atmosphere and a contributor to smog. There are 153 data points in this set and it is summarised with the three scatter lots in Figure 33.4 with ozone presented on a log axis.

From these data it is evident that there is a positive correlation between log(ozone) and temperature (intuitively, hotter weather creates more smog), there is a negative correlation between ozone and wind (wind disperses the ozone) and a negative correlation between wind and temperature. Denoting log(ozone) as O, wind as W and temperature as T, the following statistics are calculated: $m_O = 3.419$, $s_O = 0.865$, $m_W = 9.958$, $s_W = 3.523$, $m_T = 77.882$, $s_T = 9.465$ $r_{WO} = -0.538$, $r_{WT} = -0.458$ and $r_{OT} = 0.739$. Using the relationships that $cov_{jk} = \rho_{jk}\,\sigma_j\,\sigma_k$, and $\sigma_k = \sqrt{\sigma_{kk}^2}$ the trivariate normal distribution is represented as:

$$X = \begin{pmatrix} X_W \\ X_T \\ X_O \end{pmatrix} \qquad \hat{\mu} = \begin{pmatrix} 9.958 \\ 77.882 \\ 3.419 \end{pmatrix} \qquad \hat{\Sigma} = \begin{pmatrix} 12.41 & -15.272 & -1.640 \\ -15.272 & 89.59 & 6.050 \\ -1.640 & 6.050 & 0.748 \end{pmatrix} \tag{33.17}$$

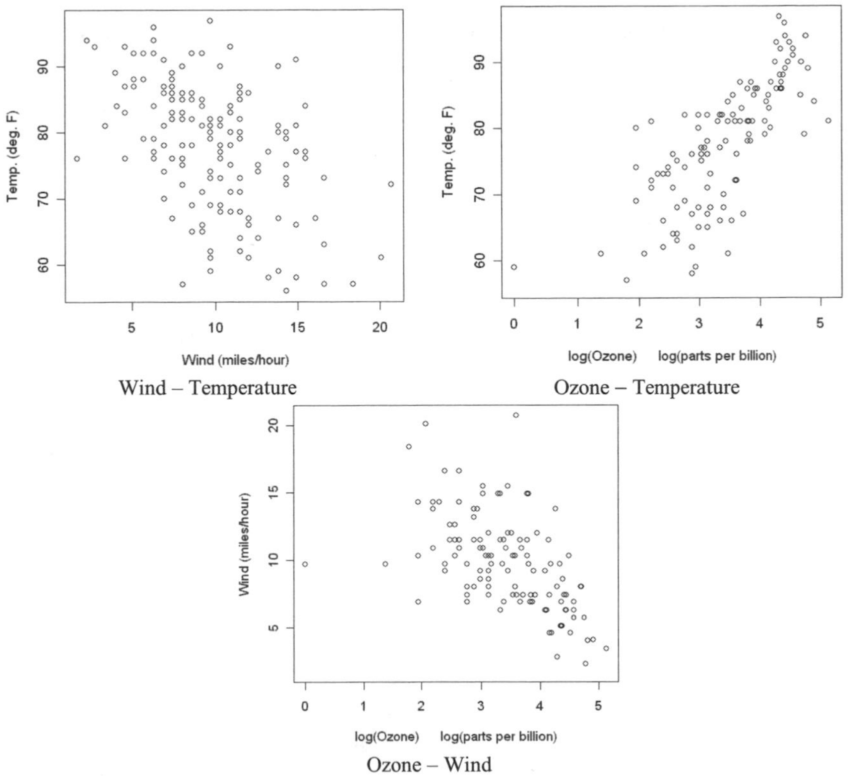

Wind – Temperature

Ozone – Temperature

Ozone – Wind

**Figure 33.4** Summary of trivariate air-quality data.

The data are plotted in Figure 33.5 along with the density contours of pairwise bivariate marginal distributions.

*Example*

One use for this model might be, given a wind measurement and a temperature measurement, to estimate the conditional distribution and use it to state the probability that the ozone will reach a certain threshold. For example, take wind = 5 miles per hour and temperature = 85° Fahrenheit: what is the probability the ozone exceeds 100 parts per billion?

*Solution*

This example requires a partition of the mean and covariance matrices between the first and second row/column:

$$X = \begin{pmatrix} X_1 \\ \hdashline X_2 \end{pmatrix} \qquad \hat{\mu} = \begin{pmatrix} 9.958 \\ 77.882 \\ \hdashline 3.419 \end{pmatrix} \qquad \hat{\Sigma} = \left(\begin{array}{cc:c} 12.41 & -15.272 & -1.640 \\ -15.272 & 89.59 & 6.050 \\ \hdashline -1.640 & 6.050 & 0.748 \end{array}\right)$$

so that $\mu_1$ is a [2 x 1]matrix, $\mu_2$ is scalar, $\Sigma_{11}$ is a [2 x 2] matrix, $\Sigma_{22}$ is a scalar and $\Sigma_{12} = \Sigma_{21}'$ are [2 x 1] matrices. Putting these matrices into Equation (33.16) gives $\mu_{2|1} = 4.132$ and $\Sigma_{2|1} = 0.302$. (Note: This scalar is a variance, not a standard deviation.) The threshold is $\log(100) = 4.605$, and from the univariate normal CDF, the probability of exceeding this threshold is 0.058.

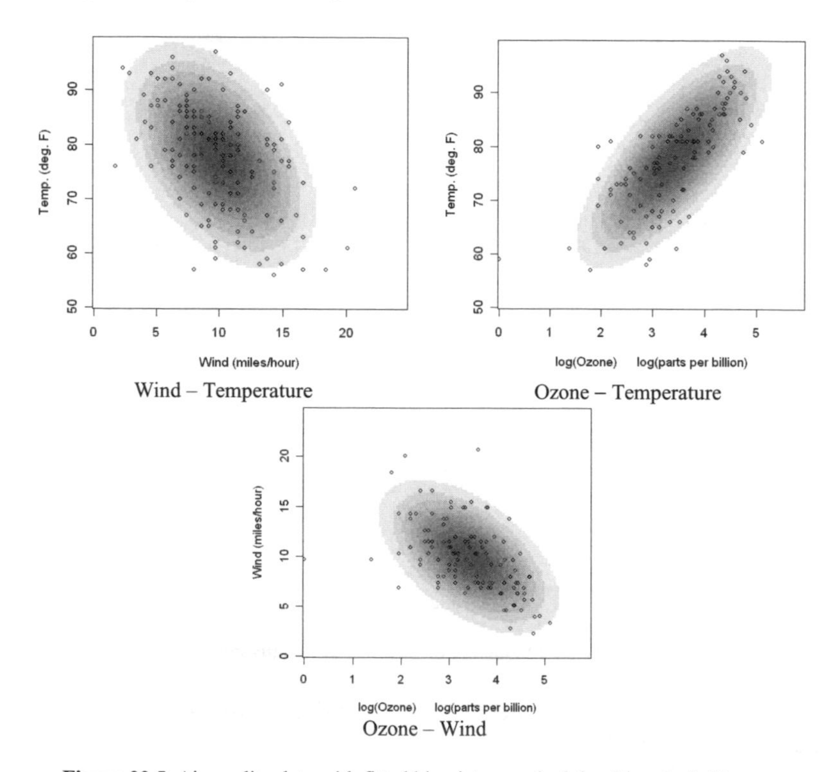


**Figure 33.5** Air-quality data with fitted bivariate marginal densities shaded in grey.

## 33.5 CONTINUOUS MULTIVARIATE DISTRIBUTIONS

A general result for multivariate distributions is that they can be factorised as a product of univariate conditional distributions:

$$f(x) = f(x_1)f(x_2 \mid x_1)f(x_3 \mid x_1, x_2)\cdots f(x_n \mid x_1, x_2, \ldots, x_{n-1}) \qquad (33.18)$$

This result can be useful in many situations. As one example, it can be used to generate a random point from a multivariate normal distribution using only a univariate random number generator:

a) generate a random normal number $x_1 \sim N(\mu_1,\sigma_1)$;
b) use Equation (30.16) to get the mean and variance of the second variable conditioned on $x_1$;
c) generate a random normal number $x_2 \sim N(\mu_{2|1},\sigma_{2|1})$;
d) repeat the process each time conditioning on all previously sampled values.

While the multivariate normal is by far the most commonly used multivariate distribution, there are others. Examples include the multivariate t-distribution, the bivariate exponential distribution, the bivariate Gumbel distribution, and, in a general setting, copula functions can be used to link any type of marginal distribution. However, comparing these distributions to the multivariate normal, there is in general less analytic convenience in representing relationships between the variables and relating them to correlations in the data.

As one example, consider the application of a bivariate Gumbel distribution to 91 maximum yearly peak flows and associated flood volume at Concordia on the Uruguay River (Adamson, 1994). The data are shown in Figure 33.6.

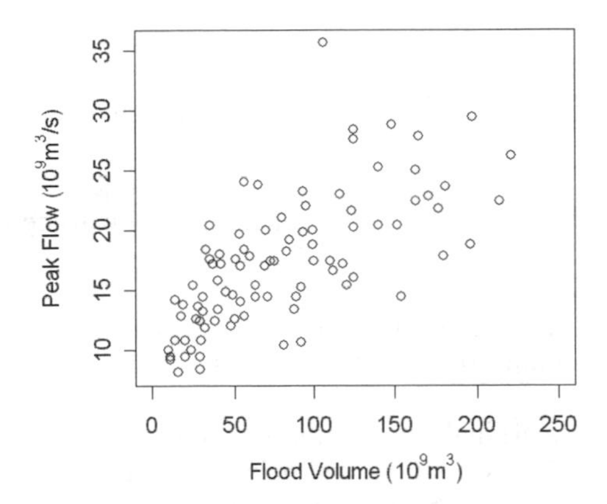


**Figure 33.6** Annual max. peak flow and associated flood volume, Uruguay River.

The data are not well fitted by the bivariate normal distribution because they are skewed in the right tail. The CDF of the bivariate Gumbel distribution is

$$F(x,y) = \exp[-(e^{-mu(x)} + e^{-mv(y)})^{1/m}] \tag{33.19}$$

where $m$ is an association parameter between the $X$ and $Y$ variables, $u(x)=(x-\zeta_x)/\theta_x$ is the standardised form of the data point $x$ with $\zeta_x$ as the location parameter and $\theta_x$ as the shape parameter of the $X$ marginal distribution. Similarly for $v(y)$ and the $Y$ variable. The method to estimate these parameters is given elsewhere but it suffices to state that $\zeta_x = 56.16$, $\zeta_y = 14.86$, $\theta_x = 42.11$ *and* $\theta_y = 4.26$. The association parameter controls the probability that large $X$ and $Y$ will occur together. It can be estimated from the proportion of points, $\hat{p}$, below the median of $x$ and below the median of $y$ using the formula:

$$\hat{m} = \ln(2)/[\ln(-\ln(\hat{p})/\ln(2))] \qquad (33.20)$$

Given that the median flood volume is $69\text{x}10^9\ \text{m}^3\text{s}^{-1}$ and the median peak discharge is $17.2\text{x}10^9\ \text{m}^3\text{s}^{-1}$, the proportion of points below both medians is 0.39, which gives $\hat{m} = 2.2626$. An example using the fitted distribution might be to estimate the probability of a flood volume greater than $100\text{x}10^9\text{m}^3$ and with a flow greater than $20\text{x}10^9\text{m}^3\text{s}^{-1}$. If $F(100,\infty)$ is the probability of flood volumes less than 100 irrespective of flow, $F(\infty,20)$ is the probability of flows less than 20 irrespective of volume and $F(100,20)$ is the probability that both volumes and flows are less than these respective values, then the probability is $1 - F(100,\infty) - F(\infty,20) + F(100,20)$, which equals 0.20. Note that the probability $F(100,20)$ is added because $F(100,\infty)$ and $F(\infty,20)$ are not mutually exclusive.

## PROBLEMS

**33.1** From Table 33.1 what is the probability of a leakage due to corrosion or joint failure?

**33.2** Given the observation of leakage due to a joint failure, what is the probability that the pipe is PVC? Given a PVC pipe has a leak, what is the probability it was due to a joint failure?

**33.3** For $\mu_x = 0$, $\mu y = 0$, $\sigma_x = 1$, $\sigma_y = 1$ calculate the bivariate normal probability density at (x,y)=1 for $\rho = -0.5, -0.2, 0.0, 0.5$ and 0.8.

**33.4** Use a numerical integration technique to integrate the standardised bivariate normal PDF up to the point (0,0) for $\rho = -0.5, -0.2, 0.0, 0.5$ and 0.8.

**33.5** Write down the marginal distribution of temperature and ozone. Now write down the conditional distribution of ozone given that the temperature is 70° Fahrenheit. What is the probability that the ozone exceeds 50 parts per billion?

**33.6** What is the probability that the ozone exceeds 50 parts per billion given that the temperature is 70° Fahrenheit and the wind speed is 5 miles per hour.

**33.7** Write a program to generate trivariate normal random numbers. Will this program work for any covariance matrix that the user inputs?

# CHAPTER THIRTY FOUR

# Monte Carlo Method (Introduction)

## 34.1 INTRODUCTION

Before attempting to explain the details of the Monte Carlo method, consider the following situation: when two dice are rolled it is possible, using simple probability theory, to determine the chance that any particular combination will come up. For example, the chance of two sixes (6&6) is 1/36. The chance of a 4 and a 5 is 2/36 or 1/18 (4&5, 5&4). The chance that the total of the two dice is 9 would be 4/36 or 1/9 (3&6, 4&5, 5&4, 6&3). Now suppose that a person was unaware of the theory. How could he or she determine the correct probabilities? One simple (but tedious) way would be to get two dice and throw them a large number of times, possibly thousands, and count the number of times two sixes come up or a 4 and a 5. Knowing the number of sought-after combinations and the total number of throws would then give an estimate of the correct value.

This is a simple illustration of the key idea behind the Monte Carlo method. A simulation is carried out involving a large number of operations whose outcome is affected in some way by chance, and a final estimate of the required probability is determined from the results.

Although there are many engineering modelling problems that can be solved using analytical probability and statistical analysis, there are many where the system is so complicated that this is not possible. Take, for example, the situation of a reservoir that is used as a water supply source, where the distribution of the withdrawal flows is known and where the distribution of inflows is also known. Given that there is full knowledge of the statistics of the inflows and outflows, it seems reasonable to assume that it should be possible to determine the reliability of the water supply in terms of the number of failures to supply (reservoir running dry) that could be expected per year or per decade. However, this problem cannot be solved using standard statistical techniques. The main obstacle to a solution is that the level of the dam at any particular time depends not only on the current inflows and outflows but also on all previous inflows and outflows and is further complicated by the possibility of the dam overflowing or running dry.

One of the ways of tackling this problem is using a Monte Carlo simulation where the water balance is programmed and a random set of inflows and outflows is applied over a long period of time. After sufficient simulations a pattern of reliability emerges. The situation with modelling reservoirs will be returned to later in the chapter.

**First Monte Carlo Simulation**

The basic philosophy of what is now called the Monte Carlo method was worked out by a mathematician, Stan Ulam, in 1946 (Hoffman, 1999). Ulam was sick and confined to bed for an extended period of time. There he passed the time playing cards, and it was while doing this that he wondered if it would be possible to determine the probability of various card combinations in the game of solitaire. He tried it from a purely mathematical perspective but found the going difficult. He then tried an approach where he just dealt many hands and counted the occurrence of various combinations (this was one of the examples cited in an original paper on the method by Metropolis and Ulam, 1949). This provided an answer, and the method was apparently named in honour of a relative who was always sneaking off to gamble at Monte Carlo. Goldstine (1990) refers to the Monte Carlo method as a "totally unexpected gem of a field". As a matter of interest Peterson (1998) gives a similar account of the name derivation but adds that John von Neumann was also involved and that at the time the method was named it was being used in the description of physics calculations (presumably hydrodynamic solutions). Others to have contributed include Everett and Richtmyer. The method was used in the development of the first atomic bomb as a means of determining the critical mass and critical density of fissionable material required to sustain an atomic reaction. According to Gleick (1987) at the time of the atomic bomb the ideas were naturally top secret and so the code name Monte Carlo was used as it gave away nothing of the method.

Monte Carlo simulations have been used in a wide variety of fields where statistical information is required but where a simulation is a convenient way of determining the required information. Examples include:

- pollutant dispersion (Webb and Mirfenderesk, 1997; Law, 2000);
- stormwater treatment facilities (Heaney and Wright, 1997);
- joint probability of high waves and high tides (Steedman, 1987; Balas and Balas, 2002);
- long-term wave climate for the Great Barrier Reef (Hardy et al., 2003);
- sensitivity analysis in wave generation (Young and Verhagen, 1996);
- modelling the spread of fires (Hargrove et al., 2000);
- sensitivity analysis of lake models (Håkanson, 2000);
- structural safety and analysis (Warner and Kabaila, 1968);
- modelling coastal cliff erosion (Hall et al., 2002);
- solution of Laplace's equation for a heated plate with specified boundary conditions (Farlow, 1993);
- risk analysis in dams (Kuo et al., 2007).

## 34.2 ILLUSTRATIVE EXAMPLE OF MONTE CARLO SIMULATION

Although there are numerous ways of determining a value for $\pi$, application of the Monte Carlo method illustrates the method quite well. It should be noted that this example is designed to illustrate the Monte Carlo method rather than determine $\pi$ efficiently. The uncertainty in the estimate after $n$ iterations is proportional to $1/\sqrt{n}$, thus the method is very inefficient if a high level of precision is required.

A circle inside a square (Figure 34.1) occupies $(\pi D^2/4)/D^2$ or $\pi/4$ of the total area. Therefore, selecting uniformly distributed random points over the total square and keeping a track of the fraction falling inside the circle should give an estimate for $\pi/4$ (assuming a true random generator is being used). As shown in Figure 34.1 (b), after 5 random numbers have been generated an admittedly crude estimate for $\pi/4 = 4/5 = 0.80$.

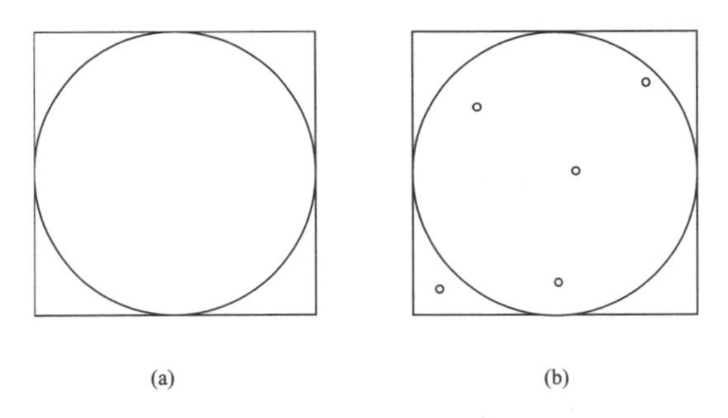


**Figure 34.1** (a) A circle transcribed inside a square occupies π/4 (78.54%) of the total area; (b) after 5 random points have been generated.

The behaviour of the estimate for π for a particular simulation is shown in Figure 34.2. Despite initial variations it is evident that the estimate tends toward a value close to 3.14. Notice that even in this simple case, a relatively large number of simulations is required to give a reliable estimate.

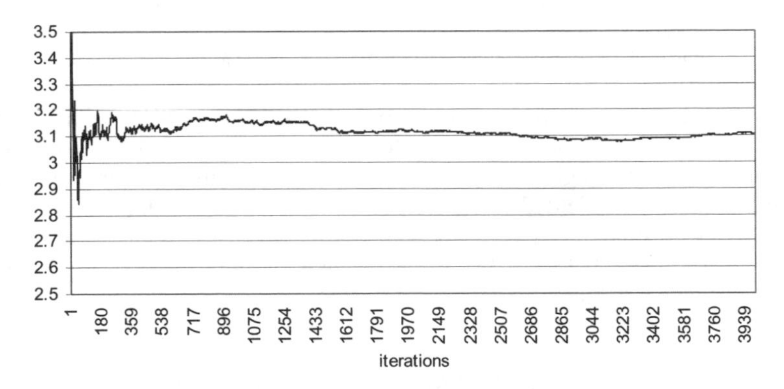


**Figure 34.2** Output of Monte Carlo method to calculate an estimate for π.

A second illustration of the Monte Carlo method is based on its application to the calculation of the integral:

$$I = \int_a^b f(x)dx \tag{34.1}$$

The procedure involves the generation of random pairs of numbers with the $x$ values between $a$ and $b$ and the $y$ values between 0 and Max[$f(x)$]. By determining whether each point is greater than or less than the function a summation is formed so that the percentage of points under the curve can be determined. The area is then this fraction multiplied by the total area of the rectangle, which is $(b-a)$*Max [$f(x)$]. As an illustration the method is applied to the problem of integrating:

$$I = \int_0^1 x^2 dx \tag{34.2}$$

where the result is shown in Figure 34.3.

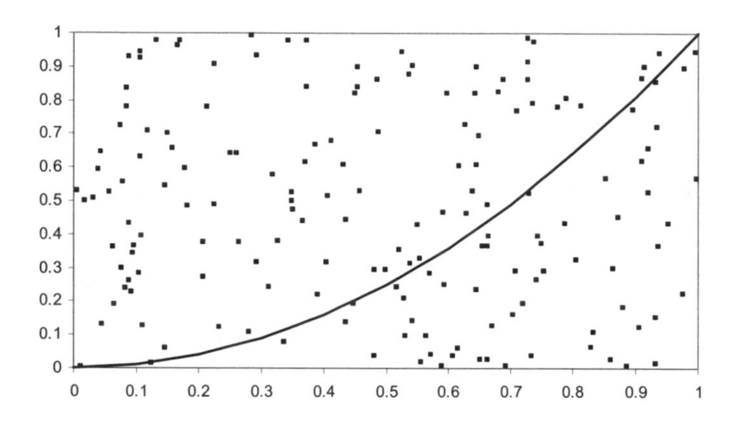


**Figure 34.3** Estimation of the integral of $f(x)=x^2$ by the Monte Carlo method.

In this case there are 168 random pairs generated, with 59 falling under the curve. The estimate of the area is therefore 59/168*1.0 = 0.351. The actual value, based on the exact integral, is 0.333. Subsequent simulations (using different sets of random numbers) yielded: 0.262, 0.315, 0.279, 0.375, 0.345, 0.357, 0.321 and so on. A set of simulations using 12,000 points gave 0.325, 0.334, 0.334, 0.333, 0.336, 0.336, 0.335, 0.330, 0.331, 0.337, which, as might be expected, are much closer to the exact value.

## 34.3 BASIC PROCEDURE FOR MONTE CARLO SIMULATION

The basic procedure for carrying out a Monte Carlo simulation involves a number of key steps:

1. develop a model of the system, its processes and statistical features;
2. generate a set of appropriate statistical inputs based on the assumed probability distributions for the various processes;
3. apply the inputs one by one, accumulating the required outputs that might include total process time, total area affected or the overall success rate of the processes being modelled.

In some ways the completely generic formula does not describe enough of the detail to make it useful, so two examples from the literature are described. In the first a risk assessment of a reservoir is described and in the second a model for the behaviour of fires is outlined. As a prelude, it is worth noting that the whole process hinges on one key factor: the provision of randomly selected input values with appropriate probability distributions. This topic is dealt with in the next chapter.

**Structural Safety Using Monte Carlo Simulation**

Structures have a zone of safety that is defined by the loads imposed on the structure and the structure's ability to resist those loads. Since loads are likely to be variable and the structural response is also variable, determining the zone of safety becomes a complex statistical problem. According to Warner and Kabaila (1968), who pioneered the application of the Monte Carlo method in structural engineering, the safety zone ($Z_i$) can be calculated as:

$$Z_i = R_i - W_i \qquad (34.3)$$

where $R_i$ is the structural response and $W_i$ is the imposed load. The subscript $i$ is used to indicate that there will be numerous load and response conditions. The response of the structure will depend on a range of material and structural properties:

$$R_i = q_i\,[X_1, X_2, X_3, \ldots X_n] \qquad (34.4)$$

where the $X_j$ account for the various material and other properties and are likely to be stochastic variables that can be described in terms of a probability distribution.

In this case the solution of $Z_i$ would involve analysis that is unlikely to be possible analytically and requires significant numerical integration if carried out using conventional statistical methods. The other alternative is to generate a range of material properties and therefore structure responses, generate a range of random loads and assess the structural safety. Warner and Kabaila (1968) did this for a simple case that had an analytical solution and were able to show good agreement. The simulation used 2,000 points and took 3 minutes to run on a computer in 1968.

Kuo et al. (2007) carried out a risk analysis of the Feitsui Reservoir in Taiwan using a variety of methods, one of which employed a Monte Carlo simulation. The dam has been in operation since 1986 and is located 30 km from Taipai City. Millions of people live downstream of the 122.5-m-high structure, which holds back a volume of over $400 \times 10^6$ m$^3$ (400 GL). Failure, where inflows exceeded the rate at which water could be discharged from the dam leading to overtopping of the structure, would be catastrophic. The reservoir routing is based on the continuity equation:

$$I - O = dS/dt \qquad (34.5)$$

which is usually written in finite difference form, making it suitable to a time-stepping solution:

$$(I_t + I_{i+1}) - (O_t + O_{t+1}) = 2/\Delta t\ (S_{t+1} - S_t) \qquad (34.6)$$

where $I$ represents the inflows, $O$ the outflows and $S$ the storage. The subscripts refer to the time step, $\Delta t$, which was taken in 1-year increments by the authors.

To start each simulation the authors selected a random storage based on possible values, then generated a range of statistically correct inflows and outflows. Solving the continuity equation allowed the storage at the next time interval to be determined. The simulation continued for 600 years of simulated time. The authors used 2000 such simulations to determine the required statistics. A flow chart of the operations that would be required is shown in Figure 34.4.

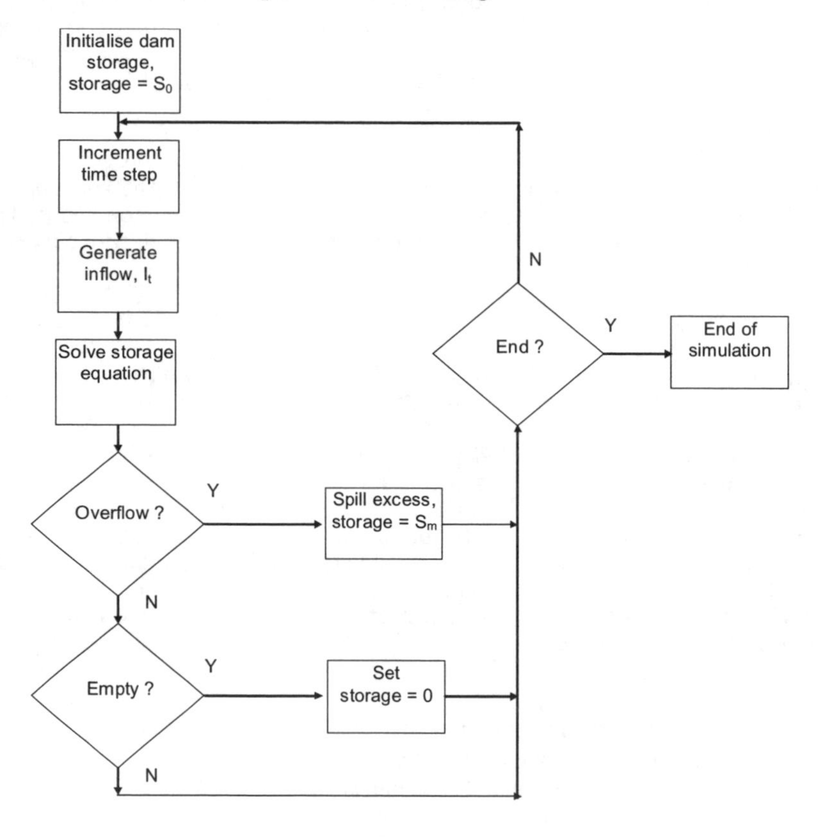


**Figure 34.4** Flow chart for reservoir simulation using the Monte Carlo method.

Hargrove et al. (2000) describe a statistical model of large fires burning in open, heterogeneous landscapes. Rather than attempting to model a particular fire, the general approach makes use of a Monte Carlo approach by assigning probabilities to key features of fire development and running the whole simulation a large number of times to assess the level of danger for key sites in the landscape. The basic procedure, programmed in Fortran, was carried out in five steps:

1. A basic unit of landscape was defined as a 50 m × 50 m square. Each unit had pre-defined values for vegetation type and moisture (and therefore the burn characteristics).
2. An area of land was defined as a rectangular grid of 500 × 500 units, giving a total area of 250,000 × 2500 m$^2$ = 62,500 ha.
3. The simulation was run over a number of time steps and at each time step and at each grid unit there was a probability of fire spreading to any of its 8 neighbours (4 immediate ones plus the four diagonal ones). At each time step there was also the chance of wind influencing the direction of spread of the fire and the chance of embers starting fires at distant grids.
4. It should be noted that the probability was not fixed but allowed to vary with a particular distribution. Therefore, each simulation would produce a different realisation.
5. The power of the Monte Carlo approach is in running the simulations a number of times and then assessing the chance of each particular grid being burnt by summing the number of simulations in which the grid is burnt and dividing by the total number of simulations.

## PROBLEMS

**34.1** Carry out a Monte Carlo simulation of the tossing of two coins and determine, using the program, the probability of the sum of the two faces totalling 7. Compare the answer with the theoretical value.

To assist in task it is worth noting that Excel has a random number generator that will give a uniformly distributed random number between 0 and 1. It is written as a function (`RAND()`) and can be used either as a simple formula in a cell `=RAND()` or as part of a longer formula `=IF(RAND()<0.166666,"1","")` which would generate a random number and if it were less than 0.166666 place a 1 in the cell, otherwise it would place a blank. Another nice feature of Excel is that every time F9 is pressed it forces a recalculation of all fields on the sheet and this causes a different random number to be generated. This allows a new simulation to be run at the press of a button. For this reason it is usually preferable to embed `RAND()` calls within cell contents rather than generating a fixed set of random numbers and then using them as required.

**34.2** Develop an Excel program to determine an estimate for the value of $\pi$ using the circle in square approach.

**34.3** Develop an Excel worksheet to determine the reliability of supply for a reservoir which has an annual inflow $I(t)$, annual demand $O(t)$, and a maximum storage of $S_{max}$. For the purposes of the simulation assume a constant annual usage of 5 units of water, inflows with a normal distribution with a mean of 5 and a standard deviation of 1 unit of water (but prevent negative inflows), and assume the level cannot fall below 0 units of water and that any overflows (where the storage exceeds the maximum) are simply lost. Assume also that the reservoir starts completely full and determine reliability over a 100-year simulation. It should be possible to organise the worksheet such that pressing F9 repeats the 100-year simulation.

**34.4** Repeat the previous question by writing a Fortran program to carry out the Monte Carlo simulation of the reservoir. In the program it should be easier to carry out multiple simulations (perhaps 1,000 100-year simulations) and to determine the overall statistics of the performance.

**34.5** Develop an Excel worksheet to plot the movement of 1000 tracking particles that are designed to illustrate the spread of dye in a wide two-dimensional flow. Each particle has a position $(x,y)$ which is updated to $(x+\Delta x, y+\Delta y)$ in the following way (Law, 2000):

$$\Delta x = u\Delta t + 2\sqrt{\alpha_x \Delta t}\,\sin(2\pi R)$$

$$\Delta y = v\Delta t + 2\sqrt{\alpha_y \Delta t}\,\cos(2\pi R)$$

where $u,v$ = uniform velocity components in $x$ and $y$, $R$ = random number (0–1) with uniform distribution, and $\Delta t$ = timestep. Note that there are two aspects that are included in the updating: a purely deterministic component to account for the mean flow ($u\Delta t$, $v\Delta t$) and a random component. The sine and cosine functions mean that the random component can be positive or negative according to the argument.

Plot the results on a scatter plot, similar to that shown in Figure 34.5.

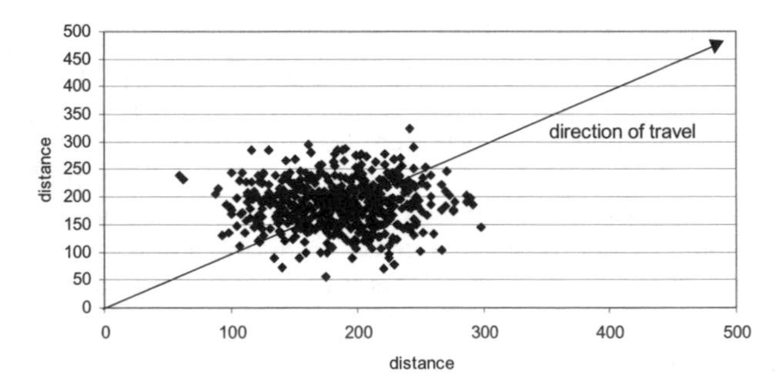


**Figure 34.5** Two-dimensional diffusion from a random walk involving 500 tracer particles.

## CHAPTER THIRTY FIVE

# Monte Carlo Method (Generation of Random Numbers)

### 35.1 INTRODUCTION

The power of the Monte Carlo method comes from its conceptual simplicity and the fact that it can be used in situations where there are no other methods of solution. Although simple, the method relies on the ability to generate random numbers that are used in the simulations. Although it may sound easy, the generation of numbers that have all the properties of a random set is extremely difficult. As R. R. Coveyou, an American mathematician, is reported to have said: "the generation of random numbers is too important to be left to chance."

In pre-computer days, people went to great lengths to generate random numbers. In 1927 Tippett listed a set of 41,600 numbers taken from the middle digits of records of area for different parishes in England. In the 1950s the RAND Corporation published a book containing random sets of 0s and 1s that had been generated mechanically (using noise from an electronic valve) and tested for randomness (Peterson, 1998; Gell-Mann, 1994). More recently, computers have been programmed to generate pseudo-random numbers and details of how these work will be given later in the chapter.

What should random numbers look like? Following an example in Gould (1991) two patterns were generated: one random, the other based on a set of quite strict rules. Both are shown in Figure 35.1.

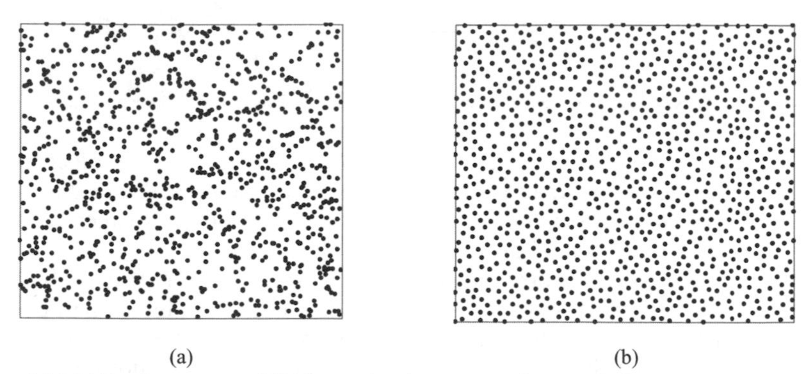

(a) (b)

**Figure 35.1** (a) Random pattern of 1000 stars showing apparent "patterns"; (b) Non-random pattern of 1000 stars generated with quite strict rules for proximity.

Despite the stars being truly random the human brain manages to find apparent patterns in Figure 35.1(a), but these are based on the randomness, rather than the lack of it. Figure 35.1(b) shows an apparently random pattern where a strict rule governing how close two stars can exist has been applied. Gould noted, if a random distribution is supposed to be uniform and without any pattern, a quite strict set of rules must be enforced to bring this about.

When generating random numbers there are a number of properties that might be expected: (i) a lack of any deterministic pattern, and (ii) a distribution of numbers that conforms to a preferred probability model. Chatfield (1984) suggests that a good way of checking for randomness is to examine the correlogram and possibly the spectral density function of a series of generated numbers. These are discussed elsewhere in the book.

## 35.2 DISTRIBUTIONS OF RANDOM NUMBERS

Most programming environments and languages come with random number generators. For example:

Matlab: `u=rand(1,n)` for *n* random numbers;
Excel worksheet: `=rand();`
Excel VBA `u=Rnd();`
Fortran 95: `call random_number(u).`

However, little information is usually given about exactly what type of random number is being generated or what type of distribution it has. In fact they are generally of a uniform distribution (Figure 35.2) between 0 and 1 or between 0 and some upper limit. Since the distribution is uniform, each number is as likely to come up as any other.

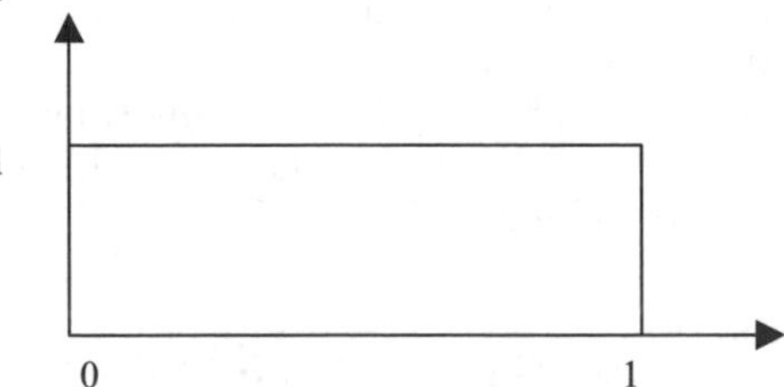


**Figure 35.2** Probability density function for a uniform distribution between 0 and 1.

There is often a need to generate numbers that have some other sort of probability distribution. Typical distributions could be the normal distribution, or a log-normal. In some cases, for example, in simulating some weather-dependent system a negative exponential distribution might be needed.

The generation of random numbers of all distributions is usually based on transforming a uniform distribution variable. This comes from the properties of the uniform distribution. Therefore the way ahead is to determine how to generate variables of a uniform distribution and then look at the transformations necessary to derive other distributions.

## 35.3 UNIFORM DISTRIBUTION

Although uniform distribution random number generators are ubiquitous it is of interest to know how they work. A general formula, for non negative integers $a$, $c$ and $m$, is:

$$x_{i+1} = ax_i + c - m.\text{Int}\left(\frac{ax_i + c}{m}\right) \tag{35.1}$$

$$u_{i+1} = \frac{x_{i+1}}{m} \tag{35.2}$$

Numbers formed using Equation (35.2) tend to repeat themselves after a while and, therefore, form a deterministic set. For this reason the period of repetition is kept as long as possible (so as to be less noticeable) and the numbers are referred to as pseudo-random numbers. For example if $a$=3, $c$=1 and $m$=5 the result is: 0.8, 0.6, 0.0, 0.2, 0.8, 0.6, 0.0, 0.2, etc. These are plotted in Figure 35.3. While it is true that the numbers may be uniformly distributed, the pattern means that they could not be considered pseudo-random.

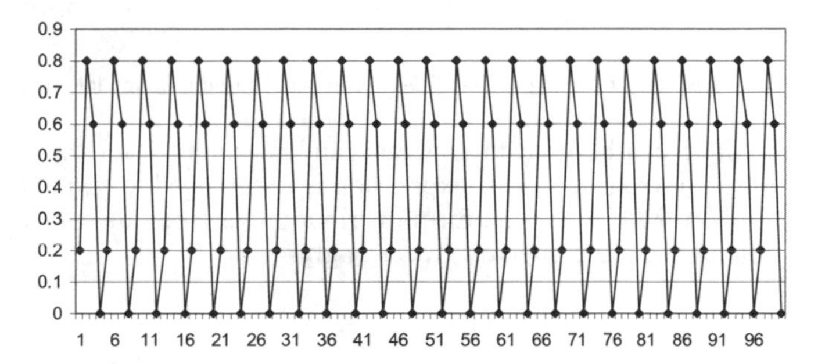


**Figure 35.3** Plot of a poor set of random numbers for $a$=3, $c$=1, $m$=5.

The period of repetition has been shown to be less than $m$, so $m$ is kept as large as possible. Taking it to the limit, good values have been found with $m=2^{35}$, $a$=129 and $c$=1. Another set has $m=2^{31}-1$, $a$=16,807 and $c$=0 (Peterson, 1998). A better set of pseudo-random numbers is shown in Figure 35.4.

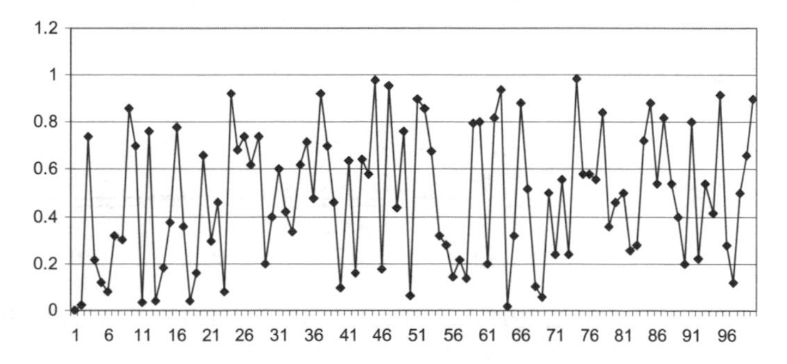


**Figure 35.4** A better set of random numbers that give a longer repetition cycle; $a$=129, $c$=1, $m$=6125.

## 35.4 EXPONENTIAL DISTRIBUTION

A common approach to generating random numbers with a particular distribution is to generate a uniformly distributed random number and then to transform it to the required distribution. One common technique for this uses the inverse cumulative distribution function. As shown in Figure 35.5, for an input value, $u$, on the interval (0,1) the corresponding quantile, $x$, is returned. If the input is uniformly randomly generated then the output will be random numbers that follow whatever distribution is described by the CDF.

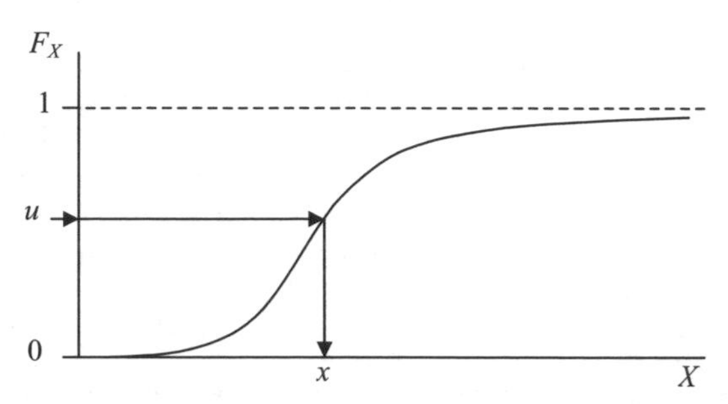


**Figure 35.5** Schematic for using inverse CDF to generate non-uniform random numbers.

This technique can be used to generate random numbers for the exponential distribution $F(x) = 1-\exp(-\lambda x)$. This expression is re-arranged to obtain the inverse form, $x = \ln(1-F(x))/\lambda$. If $u_i$ is a uniform distribution variable, then $1- u_i$ is also a random variable on the same interval. Accordingly:

$$x_i = -\frac{1}{\lambda}\ln(u_i) \tag{35.3}$$

constitutes a random exponential variable where $\lambda$ is the mean recurrence rate. Figure 35.6 shows the distribution from 1000 numbers calculated with $\lambda = 0.1$.

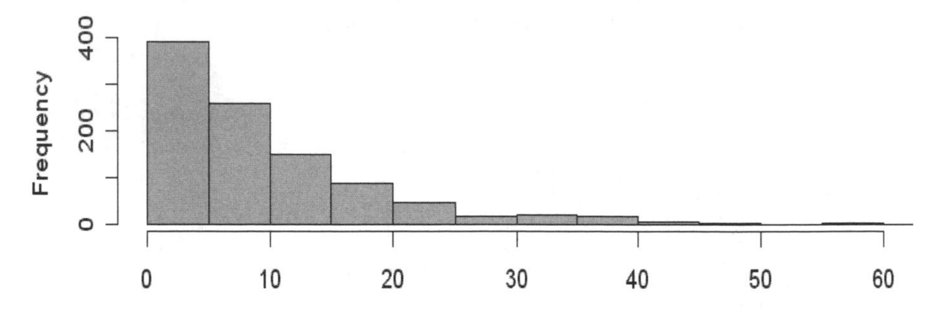


**Figure 35.6** Frequency chart for 1000 pseudo-random numbers with an exponential distribution.

## 35.5 NORMAL DISTRIBUTION

Whereas it is possible to obtain an explicit expression for the inverse CDF of the exponential, this is not possible for many distributions. One example is the normal distribution since the CDF cannot be written in a closed analytic form.

One way around this is to use numerical integration to tabulate values of the normal CDF for given values of $x$ and then use the table in reverse to look-up a randomly generated probability $u$ and return the corresponding value of $x$. Excel has an inbuilt function `norminv( )` that uses this technique. Thus, the following line could be used to generate a normal random number in Excel: `=norminv(rand(),mean,sd)`, where *mean* and *sd* are numbers representing the mean and standard deviation respectively.

Another way to use the inverse CDF with the normal distribution is to find some piecewise function that approximates the normal distribution that retains the properties of invertibility and has a high level of accuracy. To this end, numerous authors have proposed approximations with varying degrees of accuracy (Beasley and Springer, 1977; Wichura, 1988) and the approach remains highly popular because of its speed in generating random numbers.

Another approach that is specific to the normal distribution is to use a transformation to polar coordinates. The resulting equations are known as the Box-Mueller method for generating normal random numbers. If $u_1$ and $u_2$ are two independent uniform variables then:

$$x_1 = \mu + \sigma\sqrt{-2\ln(u_1)}\sin(2\pi u_2) \qquad (35.4)$$

$$x_2 = \mu + \sigma\sqrt{-2\ln(u_1)}\cos(2\pi u_2) \qquad (35.5)$$

constitute a pair of normal random variables. If only a single random number with a normal distribution is required then either $x_1$ or $x_2$ can be used. The way these numbers could be generated in Excel is:

```
=mean+sd*sqrt(-2*ln(rand()))*sin(2*pi()*rand()).
```

This method is exact, but it can be slower than methods using piecewise approximations since the *ln*( ), *sqrt*( ), *sin*( ) and *cos*( ) functions are comparatively slow to evaluate. If the scheme is implemented and 5000 numbers are generated, they can be plotted as a distribution to allow a simple visual check of the process. This is shown in Figure 35.7.

## 35.6 LOG-NORMAL DISTRIBUTION

If $Y$ is log-normally distributed then $X = ln(Y)$ is normally distributed. Therefore, one method for obtaining log-normal random numbers is to take the log of the observed data and determine the parameters $\mu$ and $\sigma$ for the normal distribution of the logged data. If $x$ is a randomly generated normal number, $X \sim N(\mu,\sigma)$, then $y = \exp(x)$ is a variable from the corresponding log-normal distribution.

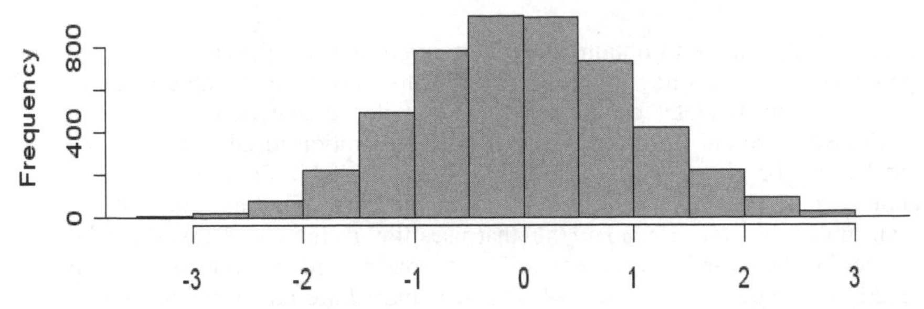


**Figure 35.7** Frequency chart of 5000 pseudo-random numbers having a normal distribution.

As an example, consider some log-normal data having a mean of 10 and a standard deviation of 2. After taking the natural logarithm of the data, the two parameters $\hat{\mu} = 2.28$ and $\hat{\sigma} = 0.2$ were estimated. 5000 numbers were generated using these parameters and then exponentiated. The final distribution is shown in Figure 35.8.

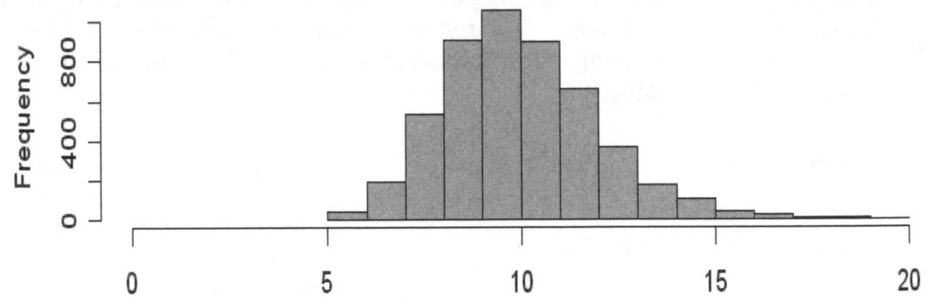


**Figure 35.8** Frequency chart of 5000 pseudo-random numbers with a log-normal distribution.

## 35.7 CENTRAL LIMIT THEOREM

The central limit theorem is often used in statistics and is perhaps a surprising result. Before stating it, the result will be illustrated with an example. Consider a set of 10,000 uniformly distributed random numbers that have been generated. Figure 35.9 shows a frequency chart of these numbers.

Now, suppose that the 10,000 are put randomly into sets of 5 and the mean of each set is calculated. The distribution of the 2000 means is shown in Figure 35.10. Despite the underlying uniform distribution the result looks very similar to the normal distribution. This is precisely what the central limit theorem states (Mandel, 1964):

> *Given a population of values with a finite variance, if independent samples are taken from this population, all of size N, then the population formed by the averages of these samples will tend to have a Gaussian (normal) distribution, regardless of what the distribution of the original population; the larger N, the greater will be this tendency.*

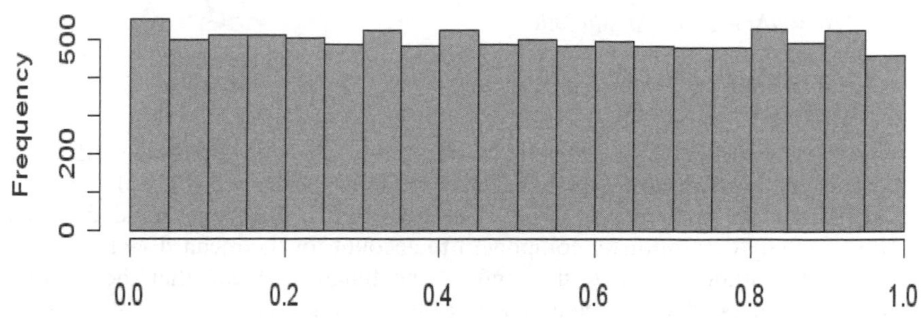


**Figure 35.9** Frequency chart of 10,000 pseudo-random numbers with a uniform distribution.

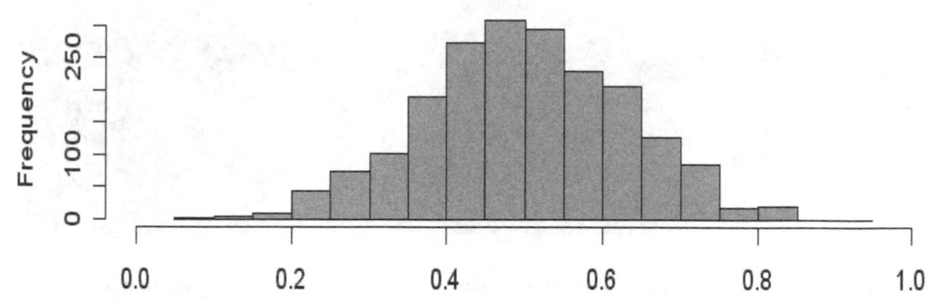


**Figure 35.10** 10,000 uniformly distributed random variables sorted into sets of 5 and the line plot of the mean of each set plotted.

## 35.8 SOLUTION OF POLLUTANT DISPERSION

The movement and spread of a pollutant in a river or stream is governed by the transport equation. For a two-dimensional flow situation this can be written:

$$\frac{\partial C}{\partial t}+u\frac{\partial C}{\partial x}+v\frac{\partial C}{\partial y}=\alpha_x\frac{\partial^2 C}{\partial x^2}+\alpha_y\frac{\partial^2 C}{\partial y^2} \tag{35.6}$$

where $C$ is the concentration, $u$ and $v$ are the mean velocities in the $x$ and $y$ directions respectively, and $\alpha_x$ and $\alpha_y$ are the diffusion coefficients in the $x$ and $y$ directions respectively. The phenomenon is best illustrated with an illustration of dye being released into the middle of a shallow open channel as shown in Figure 35.11. In addition to analytical and finite difference methods the equation can be solved using Monte Carlo simulation in what is termed "random walk". In this case a large number of individual particles are tracked through time, and the solution is built by considering the final location or distribution of each of the particles. Each particle has a position $(x,y)$ which is updated to $(x+\Delta x, y+\Delta y)$ in the following way (Law, 2000):

$$\Delta x = u\Delta t + 2\sqrt{\alpha_x \Delta t}\,\sin(2\pi R)$$
$$\Delta y = v\Delta t + 2\sqrt{\alpha_y \Delta t}\,\cos(2\pi R) \qquad (35.7)$$

where $u$, $v$ = uniform velocity components in $x$ and $y$, $\alpha_x$ and $\alpha_y$ are the diffusion coefficients in the transport equation, $R$ = random number (0–1) with uniform distribution, and $\Delta t$ = time step. There are two aspects that are included in the updating: a purely deterministic component to account for the mean flow ($u\Delta t$, $v\Delta t$) and a random component. The sine and cosine functions mean that the random component can be positive or negative according to the argument.

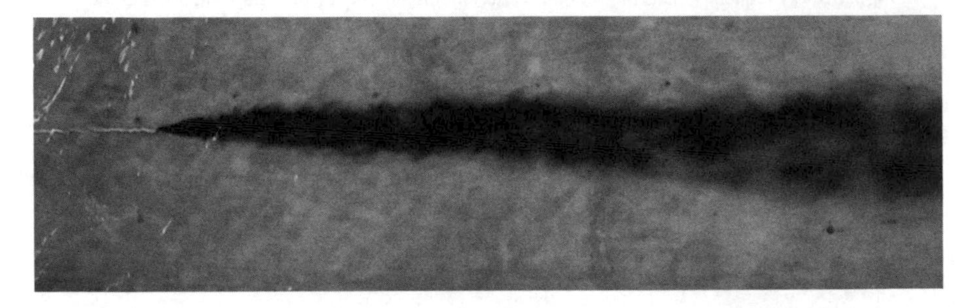

**Figure 35.11** Dye discharging into an open channel flow. Note the increasing width of dye.

Results of a typical simulation with diffusion occurring in two dimensions are shown in Figure 35.12. In the simulation 500 particles all started at (0,0) at time $t$=0. Then, by applying the rules to update the position of each point the convection (movement with the mean flow) and the diffusion (spreading) are both demonstrated as the points move with subtly different velocities.

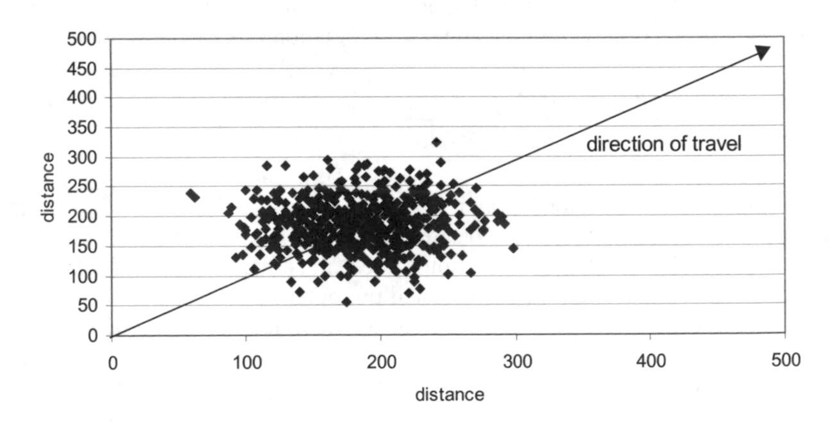


**Figure 35.12** Two-dimensional diffusion from a random walk involving 500 tracer particles.

The spread of particles can be described by plotting the density of the particle positions taken across one of the axes. This is shown in Figure 35.13.

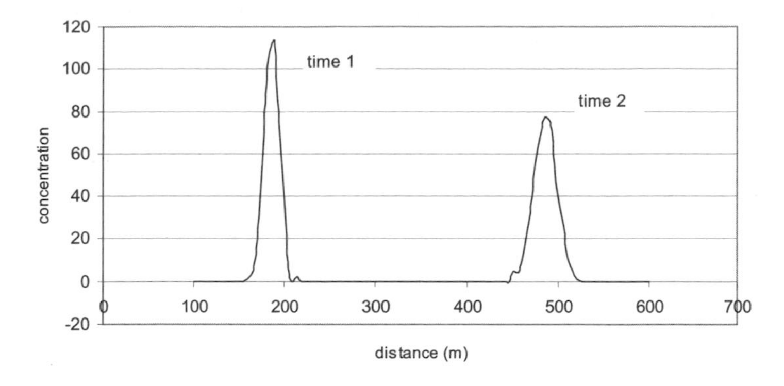


**Figure 35.13** Diffusion from solution of the two-dimensional transport equation. The spread of material as it progresses is evident from the increase in width of the distribution.

Arega and Sanders (2004) used a similar approach modelling the flow patterns in a series of tidal channels in California. The positions of the particles was updated according to:

$$\Delta x = u\Delta t + Z_1\sqrt{2\alpha_x \Delta t} \tag{35.8}$$

$$\Delta y = v\Delta t + Z_2\sqrt{2\alpha_y \Delta t} \tag{35.9}$$

where $Z_1$ and $Z_2$ were random numbers conforming to a normal distribution with zero mean and unit standard deviation.

There are a number of advantages with the random walk, the main one being the lack of numerical diffusion (which can cause inaccuracies with other numerical methods) and the fact that mass is conserved.

## PROBLEMS

**35.1** Use Excel to generate 200 random numbers. Plot them and verify that there are no easily apparent patterns visible.

Hint: Use the `=rand( )` function and verify that a new set of numbers comes with each press of F9 (which forces a sheet formula recalculation).

**35.2** Develop a line plot of the 200 random numbers generated in Problem 35.1. Use 10 bins to cover the range of values between 0 and 1. Verify that the distribution is close to uniform.

**35.3** Generate 1000 random numbers with a normal distribution where the mean is 10.0 and the standard deviation 2. Use the excel functions of `average( )` and `stdev( )` to verify that the numbers do have the correct statistical properties. Note that even with 1000 numbers there is some variation in the properties. This can be done by repeatedly pressing the F9 key, which forces a recalculation of the sheet and generates a whole new set of numbers and their statistical properties.

**35.4** Implement the generation of random numbers using the formulae given in Equations (35.1) and (35.2). Verify the fact that the pattern repeats after a length controlled by the *m* parameter.

**35.5** Generate random numbers for the Weibull distribution given by the expression:

$$F(x) = 1 - exp\left(-(\beta x)^{\alpha}\right)$$

where $\beta = 4$ and $\alpha = 0.5$. Plot them on a line plot and make observations about the basic shape and structure of the distribution.

CHAPTER THIRTY SIX

# Monte Carlo Method (Acceptance/Rejection)

## 36.1 INTRODUCTION

There are many distributions which it is difficult to obtain random numbers from. As an example consider the variable $x$ to be sampled from some function $h(x/\theta)$, with $\theta$ representing the known parameters. The function, $h(x/\theta)$, is proportional to a density function, $f(x/\theta) \propto h(x/\theta)$, but the constant of proportionality is unknown. To illustrate this, consider the function $h(x/\alpha,\beta) = x^{\alpha}(1-x)^{\beta}$. This function is proportional to the beta distribution and is plotted in Figure 36.1 for $\alpha$=1 and $\beta$=3. It is not possible to sample this function using the methodology in Chapter 35.

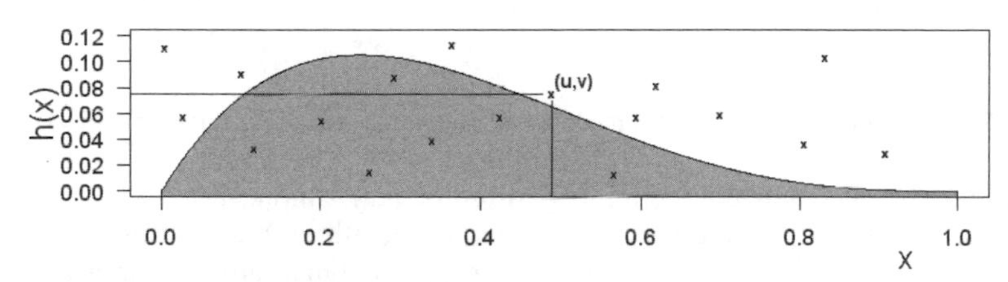


**Figure 36.1** Plot of $h(x/\alpha,\beta)$ for $\alpha$=1, $\beta$=3.

An alternative method to sample from this distribution is the method of acceptance/rejection, which is akin to Monte Carlo integration of the function $h(x/\alpha,\beta)$ (Chapter 34). This technique is very simple and is depicted graphically in Figure 36.1: a uniformly random point, $(u,v)$, is generated within the box $0 \leq u \leq 1$ and $0 \leq v \leq 0.12$. If the point lies in the grey area, $v \leq h(u)$, then $u$ is accepted as a valid random number from the beta distribution, whereas if it lands in the white area it is rejected. From Figure 36.1 it is easy to verify that in the sections where $h(x/\alpha,\beta)$ has a higher density, there will be proportionally more random numbers. The method of acceptance/rejection is therefore summarised as:

1. propose a random coordinate $u \sim U(0,x_{max})$;
2. generate a random value $v \sim U(0,h_{max})$;
3. if $v < h(u/\theta)$ then accept $u$ as a valid sample, otherwise reject it; and
4. repeat the process for the desired number of samples.

An example implementing this algorithm is shown in Figure 36.2 using Excel. The maximum value of $v$ is chosen as 0.12 as Figure 36.2 verifies that the function is always lower than this value. A histogram of the output from this example is shown in Figure 36.3.

| | A | B | C | D |
|---|---|---|---|---|
| 1 | **u** | **v** | **h(u)** | **Accepted** |
| 2 | 0.544 | 0.025 | 0.052 | 0.544 |
| 3 | 0.938 | 0.106 | 0.000 | |
| 4 | 0.148 | 0.034 | 0.092 | 0.148 |
| 5 | 0.906 | 0.107 | 0.001 | =IF(B5<C5,A5,"") |
| 6 | 0.983 | 0.099 | 0.000 | ↓ |
| 7 | 0.322 | 0.025 | =A7*(1-A7)^3 | ↓ |
| 8 | 0.877 | 0.029 | ↓ | ↓ |
| 9 | 0.094 | =RAND()*0.12 | ↓ | ↓ |
| 10 | =RAND() | ↓ | ↓ | ↓ |
| 11 | ↓ | ↓ | ↓ | ↓ |

**Figure 36.2** Excel formulae to generate 10,000 random numbers from a beta distribution $\alpha$=1, $\beta$=3.

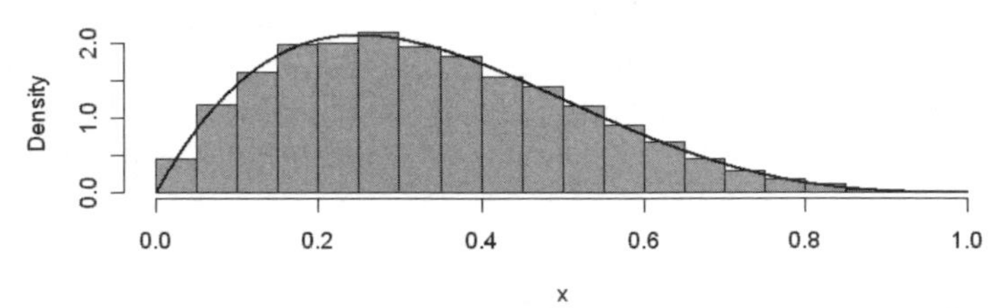


**Figure 36.3** Histogram of 10,000 random numbers from a beta distribution $\alpha$=1, $\beta$=3.

While the acceptance/rejection algorithm is very simple, it has the significant limitation that it can become very inefficient, especially in higher dimensions. The efficiency is measured as the ratio of the grey acceptance area to the area of the bounding box. There are more sophisticated variants that recommend non-rectangular bounding regions to improve efficiency, but these methods are often tailored for specific situations and are not always straightforward.

## 36.2 THE METROPOLIS ALGORITHM

In practice, there are many situations where a function is obtained that is proportional to a density function but where the constant of proportionality is unknown (or too difficult to evaluate). Typically, these functions are in multiple dimensions such that the acceptance/rejection algorithm becomes extremely inefficient. In these situations the Metropolis algorithm can be used because it is a general technique that can be applied in most situations and because it is comparatively efficient.

The Metropolis algorithm proceeds by taking a jump from a starting point to a new proposed location. The location is accepted 100% of the time if it is 'better' (has a higher function value) and if it is worse (has a lower function value) it is accepted only a portion of the time. The algorithm repeats for multiple iterations until the desired number of random data is obtained. It is because of the accept/reject rule that the algorithm will have proportionally more random numbers from the higher density regions.

As an example consider a function $h(x)$ that is proportional to the T-distribution:

$$h(x) = \left(1 + \frac{x^2}{\nu}\right)^{-\left(\frac{\nu+1}{2}\right)} \tag{36.1}$$

where $\nu$ is a degree-of-freedom parameter. To randomly generate a point $x_{i+1}$, a new point, $x_p$, is proposed by taking a random jump centred at the previous location, $x_i$. Figure 36.4 shows an example where the jump distribution is uniform, $U(x_i - ½,\ x_i + ½)$. From this it is clear that jumps are equally likely to move in either direction (as the jump distribution is symmetric), but that it is necessary to have fewer jumps that move towards the outer regions of $h(x)$. This is achieved via an acceptance/rejection rule where the acceptance ratio is formed as $\alpha = h(x_p)/h(x_i)$.

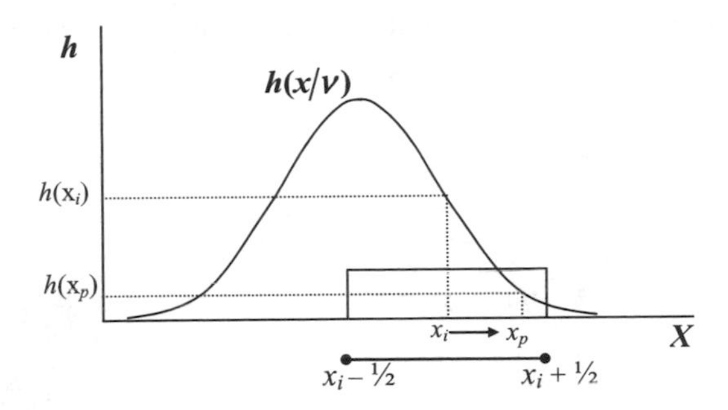


**Figure 36.4** Metropolis schematic of the T-distribution.

To summarise, the Metropolis algorithm is:

1. choose a random starting location, $x_0$ and set $i=0$;
2. sample $x_p$ from the jump distribution centred on $x_i$;
3. determine the ratio $\alpha = h(x_p)/h(x_i)$;
4. generate a random value $v \sim U(0,1)$ ;
5. if $v \le \alpha$ accept $x_p$ as a valid sample ($x_{i+1}$ becomes $x_p$) and set $i=i+1$;
6. if $v > \alpha$ reject $x_p$ so that $x_{i+1}$ becomes $x_i$ and set $i=i+1$; and
7. repeat from step 2 until finished.

This algorithm is implemented in the Matlab code shown in Figure 36.5 along with a histogram of the output after 100,000 samples. The Metropolis algorithm is a Markov chain Monte Carlo (MCMC) technique. The application of MCMC is very broad and the example here serves only as a brief introduction. Despite the simplicity of this example there are many subtleties involved when applying the technique to other problems; therefore several qualifying points need to be made:

1. The sequence of random numbers is not independent. If an independent sequence is required, then take the last value only from a sequence of, say, 100 numbers and then repeat the algorithm from a new starting position.
2. The choice of jump distribution is arbitrary with the exception in this example that it must be symmetric. The Metropolis-Hastings algorithm is a more advanced algorithm that allows for non-symmetric jump distributions.
3. If a distribution has a boundary, then the jump distribution becomes asymmetric when it gets close to the boundary (since out-of-bounds jumps are disallowed). Again the Metropolis-Hastings algorithm could be employed.
4. The parameters of the jump distribution can have a significant influence on the efficiency of the algorithm. For example, if the jump distribution being used in a particular application were U($x_i$–*0.01*, $x_i$ + *0.01*) the jumps would be very small and a much larger sample would be needed to ensure that the whole distribution has been covered adequately. On the other hand, jumps on the interval U($x_i$– *10*, $x_i$ + *10*) would cover the distribution well but would result in many rejections.
5. A related point is that the sample size must be sufficiently large to ensure that the distribution of the sample has converged to the underlying distribution. One method for checking this (other than simply specifying a huge sample size) is to use multiple samples starting at different points and to check that their summary statistics are similar.

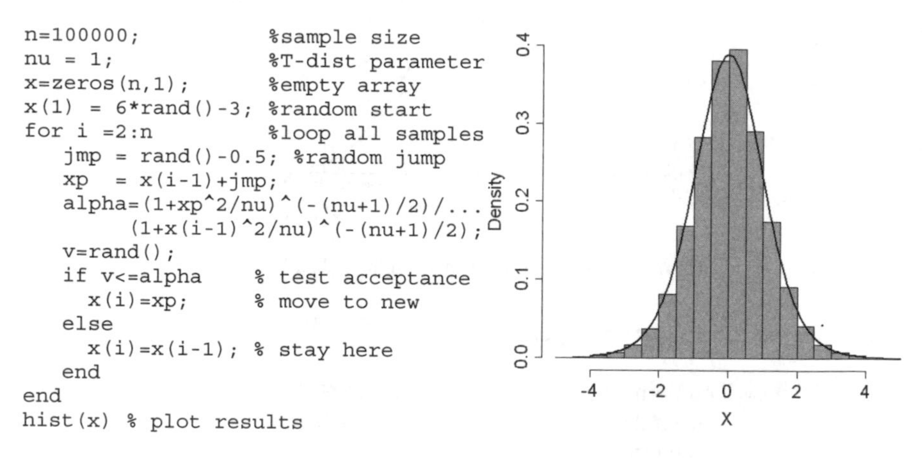


**Figure 36.5** Matlab Metropolis code and histogram output.

## PROBLEMS

**36.1** Develop a computer program to use the accept/reject algorithm to sample from the semicircle $h(x) = \sqrt{1-x^2}$ where $x \in [-1,1]$, $y \in [0,1]$. Determine the efficiency of the algorithm.

**36.2** Use the Metropolis algorithm to sample $h(x) = exp(-0.5x^2)$, which is proportional to the standard normal density N(0,1).

CHAPTER THIRTY SEVEN

# Monte Carlo Method (Metropolis Applications)

## 37.1 INTRODUCTION

Markov chain Monte Carlo (MCMC) techniques are often necessary when working with difficult multivariate probability densities. This chapter outlines some applications of the Metropolis algorithm in a multi-dimensional setting.

**Image Analysis**

Examples of image analysis include the restoration of images mapped by defects (e.g. blurred number plates from speed cameras) and matching of two similar pictures (e.g. text recognition, criminal thumbprints).

Consider a 4-megapixel (2000 x 2000 pixel) picture where each pixel has 256 possible states (i.e. 8-bit colour depth). Assuming that changing the colour of one pixel gives a new image, the number of possible images is $256^{2000 \times 2000}$. Since most of these permutations will look like speckled colour to the eye, what might meaningfully be called "pictures" are like islands in a vast sea.

MCMC formulates image analysis problems so that the probability of a pixel colour is determined with respect to a local neighbourhood. MCMC is then able to navigate the sea of speckled images incrementally and efficiently to locate a most probable matching image.

## 37.2 THE ISING MODEL

The Ising model was developed early in the 20th century as an explanation of ferromagnetism in terms of the structure of metals. The main idea is that the metal is represented as a 2D or 3D lattice of spinning electrons. When the spins are aligned there is a magnetic effect and when they are randomly oriented there is no effect. The Ising model relates the spin of an electron to those electrons in closest proximity, termed the neighbourhood.

Consider a grid of squares. The points on the corner of the squares form a lattice. Each point in the lattice ($x$) represents an electron, which can either have a down spin $-1$ or an up spin $+1$, and this value is denoted by $S_x$. Now define a binary random process by the random variable $L$ which is the set of all $S_x$ in the lattice. The neighbourhood is the set of electrons around the lattice point $x$ and is denoted

N. The neighbourhood can have an arbitrary size, with examples given in Figure 37.1.

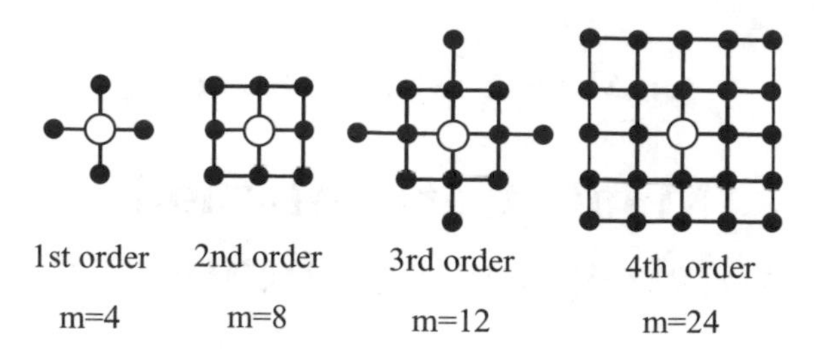


**Figure 37.1** Varying-sized neighbourhoods having *m* electrons.

The probability density function of the lattice *L* is then defined as:

$$\text{Prob}(L=l) \propto \exp\left\{\sum_{y\in \mathrm{N}} \beta\, I(s_x = s_y)\right\} \tag{37.1}$$

where $y \in \mathrm{N}$ indicates all of the electrons in the neighbourhood, $\beta$ is a control parameter and *I*( ) is an indicator function such that *I*(*A*) has the value 1 whenever A is true. So $I(s_x=s_y)$ takes the value 1 whenever the electron at *x* and the electron at *y* have the same spin and 0 if they have different spins. According to this model there is a probability contribution only when two electrons in a given neighbourhood have the same spin and the strength of this contribution is controlled by the parameter $\beta$. The main issue is that the constant of proportionality is unknown, and only the relative probability of two configurations is known. For an [$n \times n$] lattice there are $2^{n \times n}$ configurations, which makes it computationally infeasible to determine this constant for all but the smallest lattices.

This is the problem for which Nicholas Metropolis first proposed the solution to change only one electron spin at a time and accept the new configuration with a probability relative to the existing lattice. The acceptance ratio of a proposed lattice, $l_p$, based on the existing lattice $l_i$ , is given as:

$$\alpha = \frac{\text{Prob}(L=l_p)}{\text{Prob}(L=l_i)} \tag{37.2}$$

where the ratio implies that any leading constants will cancel out and need not be evaluated. If only one electron is changed at a time, then $\alpha$ need only be evaluated across one neighbourhood using Equation (37.1), since the rest of the lattice remains the same and will cancel out in the ratio. If $\alpha > 1$ the proposed lattice is more probable, and if $\alpha < 1$ the proposal is less probable than the existing lattice. Metropolis' algorithm accepts any new configuration that has equal or greater probability but if $\alpha < 1$ the proposal of a 'worse' configuration can still be accepted but only with a probability equal to α. To obtain an independent draw for a given parameter $\beta$, a random configuration of 1 and –1 values is established first and then electrons are updated one at a time. This process is repeated for a large number of iterations until the Markov chain can be assumed to have converged to the underlying distribution. The last iteration is taken as a single draw from this distribution. This algorithm is shown in Figure 37.2 using Matlab code. While the code looks lengthy, a large portion of it is dedicated to avoiding boundary issues (there are fewer neighbours at a boundary).

```
n=30;
maxit=1000000;
beta=0.8;

%Random binary pattern
%L=rand(n,n);
%L=round(L);  % 0 or 1

for i = 1:maxit
    % random coords
    x1 = floor(rand*n+1);
    x2 = floor(rand*n+1);
    % random spin
    sx = round(rand);
    %neighbourhood sum
    isum = 0; % previous
    psum = 0; % proposed
    % left neighbour
    if(x1~=1)
      if sx==L(x1-1,x2)
        psum=psum+1;
      end
      if L(x1,x2)==L(x1-1,x2)
        isum=isum+1;
      end
    end

    % right neighbour
    if(x1~=n)
      if sx==L(x1+1,x2)
        psum=psum+1;
      end
      if L(x1,x2)==L(x1+1,x2)
        isum=isum+1;
      end
    end
    % below neighbour
    if(x2~=1)
      if sx==L(x1,x2-1)
        psum=psum+1;
      end
      if L(x1,x2)==L(x1,x2-1)
        isum=isum+1;
      end
    end
    % above neighbour
    if(x2~=n)
      if sx==L(x1,x2+1)
        psum=psum+1;
      end
      if L(x1,x2)==L(x1,x2+1)
        isum=isum+1;
      end
    end

alpha=exp(beta*psum)/exp(beta*isum
);
    % accept
    if(rand<alpha)
      L(x1,x2)=sx;
    end
end
%plot
surf(L); camorbit(37.5,60)
```

**Figure 37.2** Ising model Matlab code.

The output from this model is shown in Figure 37.3. It can be seen that the $\beta$ parameter controls the patchiness of the electron spins, where increased $\beta$ generates lattices that have spins more aligned.

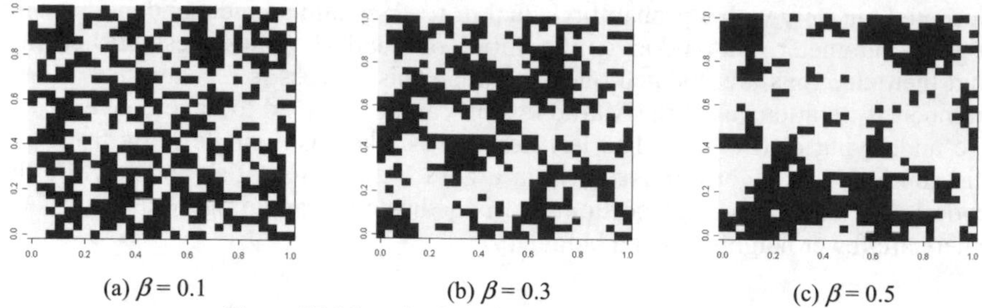

(a) $\beta = 0.1$ (b) $\beta = 0.3$ (c) $\beta = 0.5$

**Figure 37.3** Samples from the Ising model for various $\beta$.

## 37.3 THE STRAUSS MODEL

Consider the data of coordinates of Swedish pine trees in Figure 37.4. This is an example of a point pattern which has a semi-regular spacing, as there are not many trees occurring close together. There is a physical reason for this: the trees are large and they compete for nutrients, so it is less likely to find trees that are close together. This semi-regular pattern is different from a completely random pattern as highlighted in Chapter 34.

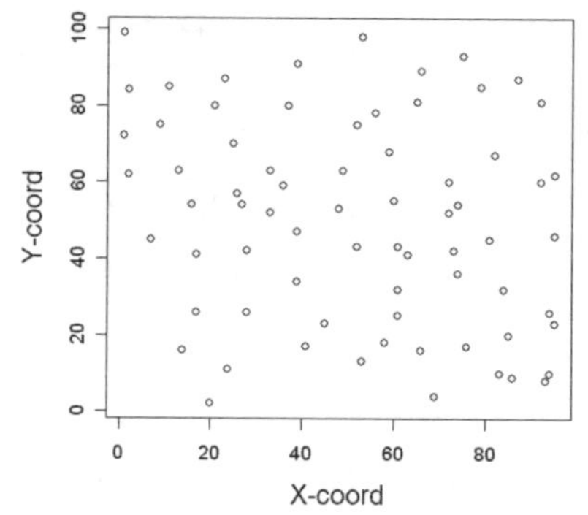


**Figure 37.4** Coordinates of Swedish pine trees.

Consider the location of $n$ coordinates inside some region where $X_k$ is the coordinate of the $k^{th}$ data point. This process can be represented using the following density function:

$$f(x \mid n) \propto \exp\left\{-\sum_{j=1}^{n}\sum_{k=j}^{n}\phi\left(\left|x_k - x_j\right|\right)\right\} \tag{37.3}$$

which involves the sum over all pairs of data points and some penalty function $\phi(\,)$ in terms of the distance between the two points $\Delta x = |x_k - x_j|$. From Equation (37.3) it can be seen that patterns that have many or larger penalties will occur with lower probability due to the negative exponential of this term. Figure 37.5 shows three different types of penalty function: (a) equal penalties at all distances (which ends up giving uniform coordinates); (b) the Strauss inhibited model; and (c) the Poisson hardcore model. The Strauss model has two parameters: $\nu$, which controls the distance at which two points are inhibited, and $\alpha$, which controls the penalty for every two points that occur within this distance. As $\alpha \to 0$ the model tends towards the uniform pattern and as $\alpha \to \infty$ the model becomes the Poisson hardcore model which disallows any two points to be closer than $\nu$ (which is not always feasible if one attempts to squeeze too many points into too small a region). For $\alpha$ on the interval $0 \leq \alpha < \infty$, the Strauss model has a finite probability that two points occur within a distance $\nu$, but larger $\alpha$ discourage the chance of this happening.

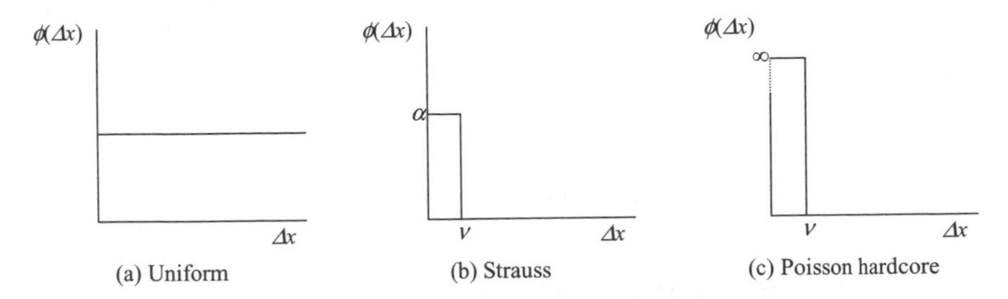


**Figure 37.5** Examples of penalty functions $\phi(\Delta x)$.

It is not straightforward to sample this model. It is recommended to attempt this task by hand: manually place 20 points within a region bounded by the coordinates (0,0) and (1,1) so that only 20% of the pairs occur within a distance 0.2 units. Numerical sampling is not straightforward either since Equation (37.3), like the Ising model, has a proportionality constant that is infeasible to evaluate. For this reason, MCMC is necessary. The code for a Metropolis algorithm is shown in Figure 37.6, where (i) an initial pattern of uniform points is generated, (ii) one point at a time is selected with a proposal of new coordinates uniformly sampled, (iii) the proposal is tested for acceptance/rejection, (iv) after a sufficiently long repetition the resulting pattern is assumed to be a sample from the specified model.

```
thresh = 0.1;
penalty = 20;
n=50;
x=rand(n,2);
maxit = 1000;
for i = 1:maxit
  % select a random point
  j = floor(rand*n+1);
  sum_1 = 0;
  for k = 1:n
   for l = k+1:n
     dist=sqrt(sum((x(k,:)-...
          x(l,:)).^2));
     if(dist<thresh)
       sum_1 = sum_1+penalty;
     end
   end
  end
  temp = x(j,:);   %new coords
  x(j,:) = rand(1,2);
  sum_2 = 0;
  for k = 1:n
   for l = k+1:n
     dist=sqrt(sum((x(k,:)- ...
          x(l,:)).^2));
     if(dist<thresh)
       sum_2 = sum_2+penalty;
     end
   end
  end
  alpha=exp(-sum_2)/exp(-sum_1);
  if(rand>alpha)  %reject
    x(j,:)=temp;
  end
end
plot(x(:,1),x(:,2),'o')
```

**Figure 37.6** Matlab implementation of the Strauss model.

Patterns having $\alpha$=0, $\alpha$=2 and $\alpha$=20 are shown in Figure 37.7, where 50 points were placed inside the unit square and a penalty was applied for patterns being less than 0.1 units apart.

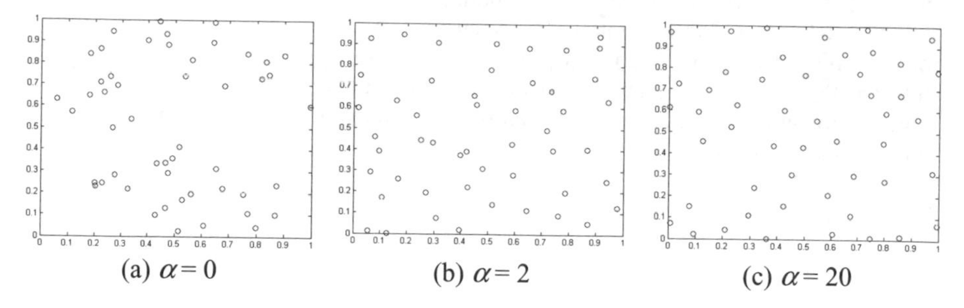


(a) $\alpha = 0$ (b) $\alpha = 2$ (c) $\alpha = 20$

**Figure 37.7** Samples from the Strauss model for various $\alpha$.

## 37.4 PARAMETER UNCERTAINTY

Another reason for the popularity of the Metropolis algorithm is because it dovetails nicely with Bayes' rule to answer questions of the parameter uncertainty and parameter correlation. Recall from that Bayes' rule can be cast in a form that relates data to parameters:

$$P(\theta|x) \propto P(x|\theta)P(\theta) \tag{37.4}$$

where $P(\theta/x)$ is the posterior distribution (parameter variability given the data), $P(x|\theta)$ is the likelihood of the data for a given set of parameters and $P(\theta)$ is the prior distribution of parameter values without knowledge of the data (say, from expert opinion or a previous experiment). This is yet another situation where the proportionality constant can be time-consuming to determine.

As an example, consider a dataset that measures the type of soil at depths 2–5 metres below the surface. An estimate of the soil type is important for designing the foundation of buildings as expansive soils can cause cracking and weak soils might require stronger foundations. The data are measured using a cone penetrometer test (CPT), which pushes a cone-tipped instrument into the ground to measure electrical resistance and friction along the cone. In this instance the data in the first two metres are ignored as this topsoil region is not of interest in this study; the soil in the range 2–5 m is assumed to be homogeneous across the site (i.e. there are no changes in soil profile). Furthermore, the tests were repeated four times, as one independent sample may be non-representative if there is high variability in the soil type. The data are summarised in Figure 37.8 using a histogram of the friction ratio from the CPT test, along with an approximate soil classification based on the friction ratio (a proper classification would include more than one variable).

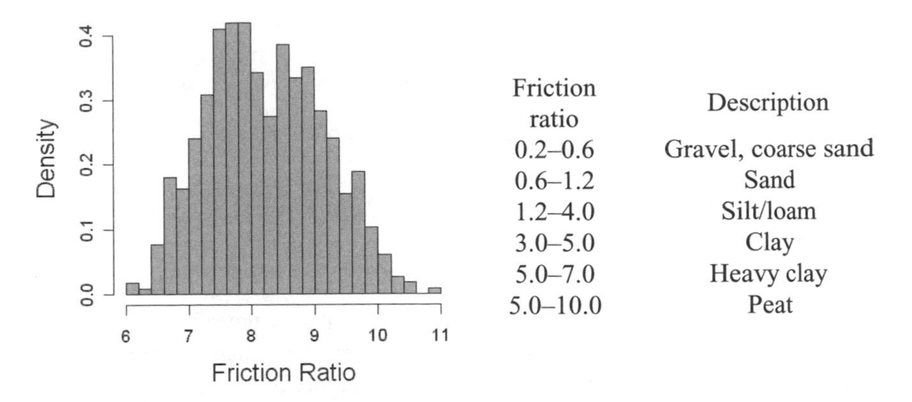


| Friction ratio | Description |
|---|---|
| 0.2–0.6 | Gravel, coarse sand |
| 0.6–1.2 | Sand |
| 1.2–4.0 | Silt/loam |
| 3.0–5.0 | Clay |
| 5.0–7.0 | Heavy clay |
| 5.0–10.0 | Peat |

**Figure 37.8** Histogram of friction ratio and soil classification.

Suppose that before any testing was conducted a geotechnical expert, based on their understanding of similar sites, estimated the following prior distribution:

$$P(\mu,\sigma) = N(\mu|7.0,1.0)N(\sigma|1.5,0.5) \tag{37.5}$$

This distribution assumes that the mean and standard deviation are normally distributed and independent of each other. The expected mean value is 7 with a 95% chance it is between 6 and 8 (1.96 standard deviations). The expected standard deviation of the data is 1.5 but the prior specifies a 95% chance it lies between 1.0 and 2.0. If the expert were more confident she could specify a narrower prior distribution (less uncertainty in her parameter estimates). There is no restriction on the type of distribution for the prior so long as it represents the prior belief about where the parameters will probably lie.

The data are assumed to follow a normal distribution (based on Figure 37.8) and the likelihood of the dataset is the product of probabilities of each separate data point, $x_i$, for a given pair of parameters, $\mu$, $\sigma$. This is given as:

$$P(x_i|\mu,\sigma) = \prod_{i=1}^{n} N(x_i|\mu,\sigma) \tag{37.6}$$

The resulting posterior distribution is then:

$$P(\mu,\sigma|x_i) \propto \prod_{i=1}^{n} N(x_i|\mu,\sigma)N(\mu|7.0,1.0)N(\sigma|1.5,0.5) \tag{37.7}$$

where the prior belief is updated by the observed measurements. If there are a large number of measurements the posterior distribution will be strongly influenced by the likelihood term; otherwise, if there are only a few, the expert's prior belief will have stronger influence. Note from Equation (37.5) that it is the parameters $\mu$ and $\sigma$ which vary randomly and the data that are 'fixed'. This can seem counterintuitive to those used to a distribution having fixed parameters, to explain some random quantity, but it is necessary to allow for uncertainty in the parameters.

The probability density in Equation (37.5) will be sampled using bivariate uniform distribution, so that a single jump will propose a new pair of parameters $\mu_p$, $\sigma_p$. The acceptance ratio needs to be calculated carefully using:

$$\alpha = \prod_{j=1}^{n} \left( \frac{N(x_j | \mu_p, \sigma_p)}{N(x_j | \mu_i, \sigma_i)} \right) \frac{N(\mu_p | 7.0, 1.0) N(\sigma_p | 1.5, 0.5)}{N(\mu_i | 7.0, 1.0) N(\sigma_i | 1.5, 0.5)} \tag{37.8}$$

This form uses a product of likelihood ratios rather than the ratio of products, where the latter would become the ratio of two very small numbers (try multiplying 100 numbers between 0 and 1 to see why). Figure 37.9 shows a Matlab implementation of the Metropolis algorithm to sample the posterior distribution.

```
% File: npdf.m
function [f] = npdf(x,m,s)
% Normal density function
f = exp(-0.5*(x- ...
m)^2/s^2)/s/sqrt(2*pi);

% File: metrosoil.m
% Assume data stored in x
ndat=length(x);
% Hyper-parameters for the prior
m=[7,1];
s=[1.5,0.5];
n=10000; %sample size
samp = zeros(n,2); %empty array
%start at mean of prior
samp(1,:)=[m(1),s(1)];
for i =1:n-1
  zi = samp(i,:);
  jmp=0.1*(rand(1,2)-0.5);
  zp = zi+jmp;
  % Prior probabilities
  prior_p=npdf(zp(1),m(1),m(2))...
           *npdf(zp(2),s(1),s(2));
  prior_i=npdf(zi(1),m(1),m(2))...
           *npdf(zi(2),s(1),s(2));
  a = prior_p/prior_i;
  % Likelihood update
  for j = 1:ndat
    a=a*npdf(x(j),zp(1),zp(2))...
          /npdf(x(j),zi(1),zi(2));
  end
  if(rand()<a) % test acceptance
    samp(i+1,:)=zp;
  else
    samp(i+1,:)= samp(i,:);
  end
end
plot(samp(:,1),samp(:,2))
```

**Figure 37.9** Metropolis sampler of posterior density in Matlab code.

Taking 10,000 samples from this distribution Figure 37.10 shows that although the Markov chain was started at the point $(\mu,\sigma)$=(7, 1.5), the majority of samples were on the intervals $\mu \approx (8.1, 8.3)$, $\sigma \approx (0.8, 1.0)$. The posterior density reflects the observed data more highly than the expert's opinion because there is a large number of data points, $n=585$. The range of parameter uncertainty is small for the same reason. The mean of the data is $m$=8.22 and the standard deviation is $s$=0.9087. Based on the likelihood, an analytic expression for the uncertainty in the mean is $\mu \sim N(m, s/\sqrt{n})$ and an approximation for the uncertainty in standard deviation is $\sigma \sim N(s, s/\sqrt{2n})$, which corresponds to the sampled region of the posterior density in Figure 37.10.

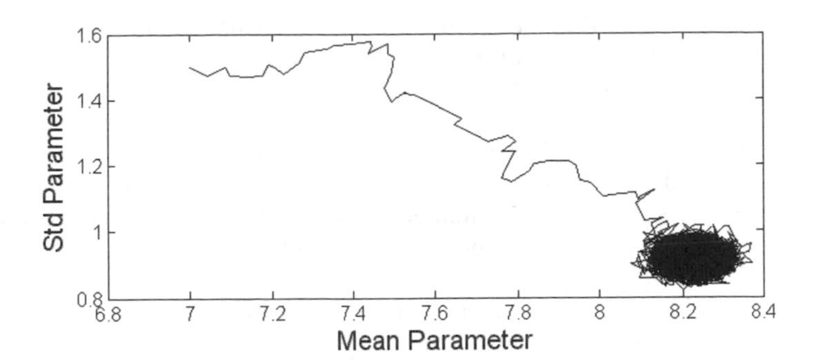


**Figure 37.10** Sample from posterior distribution.

## 37.5 CONCLUSION

The treatment of MCMC here serves only as a basic introduction to highlight the some applications of this powerful statistical and numerical tool. While the main advantage of MCMC is its ability to access otherwise difficult probability distributions, like all Monte Carlo methods, it is numerically expensive. For this reason, MCMC should only be used when other approaches are intractable. As computing power only serves to increase, and as there is no shortage of "difficult" problems, the appeal of MCMC and the reward in learning it will only continue to grow.

## PROBLEMS

**37.1** Use the Matlab code in Figure 37.2 to run the Ising model. For a grid size of [30x30] and for 1,000,000 iterations, how long does this code take to run?

**37.2** For the Ising model, initialise every cell in the grid to the value 1. For a parameter value $\beta = 0.1$, plot the surface after 100, 1,000 and 10,000 iterations. Repeat for a parameter value $\beta = 0.8$. Are 100 iterations sufficient and will the convergence depend on the parameter value?

**37.3** Using paper and pencil, manually place 20 points within a region bounded by the coordinates (0,0) and (1,1) so that only 20% of the pairs occur within a distance 0.2 units.

**37.4** Use the Matlab code in Figure 37.6 to run the Strauss model to place 50 points in the unit square for parameter values $\alpha$=2 and $\alpha$=20. Is it feasible to place 100 points within the same region for these same parameter values? What does the code do if it is not feasible?

**37.5** Change the threshold parameter in the Strauss model to 0.2. Generate patterns of 10 points using parameter values $\alpha$=2 and $\alpha$=20.

**37.6** Set the parameter $\alpha$=0 in the Strauss model and generate patterns of 50 points. What model does this correspond to? Is there a quicker method to generate equivalent patterns from this model?

**37.7** Generate 1,000 independent normal random numbers $X \sim N(7, 1.5)$. Compute the mean and standard deviation of the sample. Use the code in Figure 37.9 to generate a sample of 10,000 $(\mu,\sigma)$ parameter pairs. Compute the expected values of these pairs and compare them with the sample estimates.

**37.8** Repeat Problem 37.7, but for a sample of 100 independent normal random numbers. How does this change the output?

**37.9** Repeat Problem 37.7, but for 100 independent normal random numbers $X \sim N(8, 1)$.

CHAPTER THIRTY EIGHT

# Stochastic Modelling (Goodness of Fit and Model Calibration)

## 38.1 INTRODUCTION

The term stochastic is derived from the Greek word *stochos,* which means target, and connotes an element of chance since a person who is *stochastikos* is skilful at aiming. Today, it is used as a modelling term that simply means 'random'. Stochastic processes are distinct from deterministic processes since a future state cannot be completely determined from the previous state and there is a random component. Many processes in engineering can be considered stochastic, such as dynamic loads on a building, traffic patterns, rainfall storms, location of mineral deposits and vibrations of a machine. A stochastic model has the advantage over a purely deterministic model in that it can model random fluctuations. Probability distributions that govern these fluctuations form the basis of stochastic models and for this reason, the distributions discussed in prior chapters play an important role. However, a critical question becomes: "Which distribution to use?" This chapter addresses that question.

Before getting too involved, one basic answer is that the distributions have known properties which can be exploited. For example, the normal distribution is symmetric, the log-normal distribution is skewed, the exponential distribution has a decaying shape, and the Gumbel distribution is often suited to extreme values. Nonetheless, the more pragmatic and over-ruling criterion is to always visually assess the goodness of fit. So don't discard the Gumbel distribution as a candidate just because the data aren't extreme values! Also remember that all mathematical models are idealisations and no probability distribution is a perfect model.

## 38.2 THE METHOD OF MOMENTS

The method of moments brings together three concepts about models:

1. a statistic – a number estimated from a sample;
2. a parameter – which by taking on a specific value controls the behaviour of a model; and
3. a model property – a function that describes an aspect of the model in terms of its parameters.

For example, the exponential distribution $F(x) = 1 - \exp(-\lambda x)$, has one parameter, λ, which is the rate of decay. Some of the properties of this distribution are given in Table 38.1.

**Table 38.1** Properties of the exponential distribution.

| | | | |
|---|---|---|---|
| Mean | 1/λ | Median | ln(2) /λ |
| Std Dev. | 1/λ | Mode | 0 |
| Skewness | 2 | | |

The principle of the method of moments is to estimate population parameters by equating sample statistics with the corresponding model properties. As an example, consider the horizontal ground acceleration data from a Californian earthquake in Table 38.2.

**Table 38.2** Peak horizontal accelerations ($m/s^2$) from the 1979 Imperial Valley, California earthquake (pooled results from 14 events measured across 71 sites) (Joyner et al., 1981).

| | | | | | | | | |
|---|---|---|---|---|---|---|---|---|
| 3.52 | 0.10 | 2.74 | 0.07 | 0.05 | 1.96 | 2.13 | 0.26 | 0.10 |
| 0.14 | 0.04 | 0.71 | 1.39 | 0.03 | 1.44 | 1.12 | 0.27 | 3.82 |
| 1.92 | 0.04 | 0.12 | 0.30 | 0.84 | 1.84 | 1.47 | 0.33 | 0.30 |
| 1.32 | 1.25 | 0.06 | 0.06 | 1.76 | 2.00 | 1.45 | 0.29 | 1.27 |
| 0.61 | 4.03 | 0.03 | 0.10 | 2.01 | 3.29 | 1.10 | 0.38 | 0.11 |
| 0.53 | 0.18 | 0.18 | 0.10 | 0.72 | 0.56 | 0.42 | 0.29 | 1.18 |
| 0.14 | 4.99 | 0.47 | 0.06 | 0.44 | 0.21 | 0.56 | 1.08 | 1.67 |
| 0.18 | 4.58 | 0.11 | 0.13 | 3.67 | 1.49 | 0.29 | 0.10 | 1.37 |

Calculating the mean of this dataset, it can be related to the parameter of the exponential distribution:

$$\frac{1}{\hat{\lambda}} = \hat{\mu} = \frac{1}{72}\sum_{i=1}^{72} x_i = 1.03\ ms^{-2}, \tag{38.1}$$

where the “hat” is used to indicate that these values are estimates. If a different sample were used, then the estimate would have a different value. While the example of an exponential distribution is straightforward, issues of bias and variability of estimators arise in more complicated models. Furthermore, there may be any number of model properties to choose from that can be used to obtain parameter estimates, each giving slightly different answers. Despite this, the method of moments is commonly used because of its straightforward implementation.

## 38.3 PROBABILITY PLOTS

Having fitted a model it is important to ask how well the model fits the observed data. Probability plots are a powerful visual method that allows data to be compared to a given distribution. Consider the following ranked data sample, $X : x_1 \leq x_2 \leq \ldots \leq x_n$, the cumulative probability for the $i^{th}$ data point is well approximated by the following function:

$$\hat{F}_X(x_i) \approx \frac{i-b}{n+1-2b} \tag{38.2}$$

where various values of $b$ are proposed as offering less biased estimates, but commonly the value $b = 0$ is used. Having estimated the cumulative probability the corresponding value of $X$ is required from the chosen theoretical distribution:

$$\hat{x}_i \approx F^{-1}\left(\frac{i-b}{n+1-2b}\right) \tag{38.3}$$

where $F^{-1}$ is the inverse of the cumulative distribution function (CDF). The estimated $\hat{x}_i$ are referred to as quantiles and they are plotted against the ranked data. The effect is that if the data appear as a straight line they are well matched by the chosen distribution and any deviation from this line can be easily seen.

Consider again the Californian earthquake data. The first two columns of Table 38.3 show the index $i = 1\ldots72$ and the plotting position from Equation (38.2) where $b = 0$. The inverse CDF of the exponential distribution is easily derived as:

$$\hat{x}_i = F_X^{-1}(F_X) = -\ln(1 - F_X)/\lambda \tag{38.4}$$

and is used to obtain the quantile estimates in column three from column two.

**Table 38.3** Illustration of method to obtain a probability plot.

| $i$ | $\frac{i}{n+1}$ | $\hat{x}_i$ | $x_i$ |
|---|---|---|---|
| 1 | 0.014 | 0.014 | 0.03 |
| 2 | 0.027 | 0.028 | 0.03 |
| : | : | : | : |
| . | . | . | . |
| 70 | 0.959 | 3.274 | 4.03 |
| 71 | 0.973 | 3.69 | 4.58 |
| 72 | 0.986 | 4.401 | 4.99 |

The ranked data from Table 38.3 (column four) are plotted against the exponential quantiles and the result is shown in Figure 38.1(a). This shows that the exponential distribution is slightly underestimated at lower quantiles and overestimated at higher quantiles. An additional further step is to replace the *labels* (not the values!) on the x-axis with selected probabilities, as in column two. This can be difficult to do in some graphical packages as it often amounts to hiding the quantile value underneath a textbox that shows the equivalent probability. An example of this latter style is shown in Figure 38.1(b) and it can be seen that it is much easier to read the data value at a known probability level (e.g. the median).

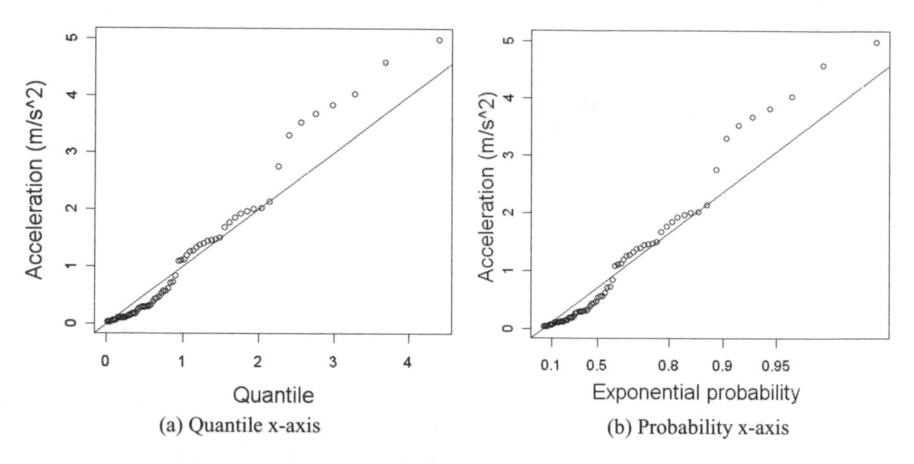


(a) Quantile x-axis

(b) Probability x-axis

**Figure 38.1** Exponential probability plots of earthquake acceleration.

The Weibull distribution could be proposed for an improved fit to the earthquake data as it has an additional parameter than the exponential, $F_X(x) = 1-\exp(-(\lambda x)^{\alpha})$. The moments of this distribution are given by the function $E[X^r]=\lambda^{-r}\,\Gamma(1+r/\alpha)$. Therefore the mean and variance are $\mu_x = \lambda^{-1}\Gamma(1+\alpha^{-1})$ and $\sigma_x^2 = \lambda^{-1}\Gamma(1+2\alpha^{-1}) - \mu_x^2$ By equating the sample mean and variance with the moments of the Weibull distribution it is not possible to obtain an explicit expression for the parameters. One method to overcome this is to form a least-squares objection function (Section 38.5) and optimise to locate the unknown parameters, giving $\hat{\alpha} = 0.70$ and $\hat{\lambda} = 0.92$ ms$^{-2}$.

An alternative, but less precise method of parameter estimation can be achieved using a probability plot. The CDF can be rearranged so that $-\ln(1-F_X) = (\lambda x)^{\alpha}$ and by taking logs again, $-\ln(-\ln(1-F_X)) = \alpha\ln(\lambda) + \alpha\ln(x)$. By using the plotting position in Equation (38.3) to estimate $F_X$ along with the ranked data $x$, after appropriately logging each term, a linear relationship is obtained so that the slope gives an estimate of $\alpha$ and the intercept, $\alpha\ln(\lambda)$, can be used to estimate $\lambda$. Figure 38.2 demonstrates this for the earthquake acceleration data which gives the parameter estimates $\hat{\alpha} = 0.8244$ and $\hat{\lambda} = 1.10$ ms$^{-2}$.

Whereas it is straightforward to compute the inverse of the exponential distribution, this is not the case for other distributions such as the normal distribution. One option for the normal distribution is to use numerical integration of the PDF or pre-tabulated values of the CDF to determine an inverse, but these methods are cumbersome. Alternative methods that perform this task much more quickly and with higher accuracy use polynomial and rational approximations to the Gaussian curve (Beasley and Springer, 1977; Wichura, 1988). Excel provides a function `norminv()` to perform the task of approximating the inverse normal distribution.

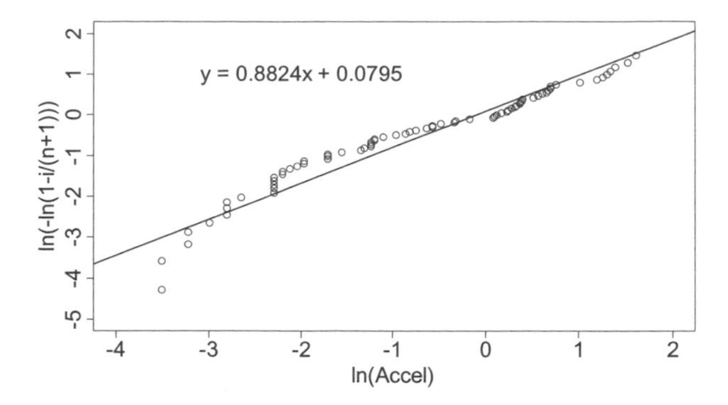


**Figure 38.2** Weibull probability plot of earthquake acceleration.

**The Michelson Morley Experiments**

In 1879 the 24-year-old Albert Michelson set up an experiment to establish the speed of light. He estimated the speed of light to be 299 850±50 km/s. A Gaussian probability plot of Michelson's 1879 estimates of the speed of light is presented.

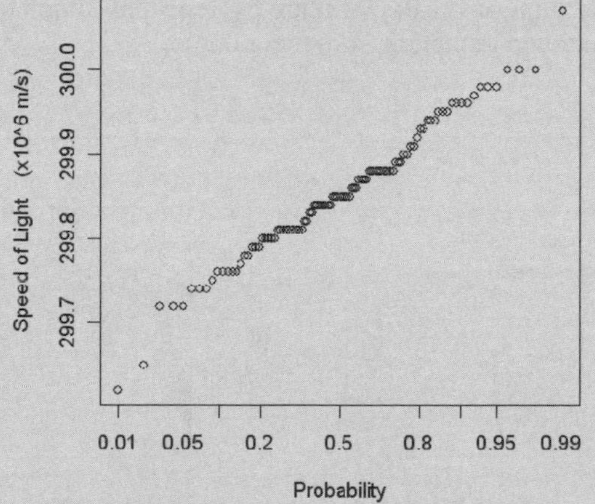


Following the success of these measurements, Michelson along with Edward Morley devised a further experiment in 1887 to detect the medium that light travelled through. Just as sound has a medium (air, water, etc.), the hypothesis of 19th century physicists was that light required a medium too. This would imply that the speed of light changed depending on whether it moved "down-wind" or "up-wind" relative to this yet unknown medium called the aether.

The Michelson-Morely experiments are perhaps remembered as the most famous "failed" experiments (for which Michelson obtained a Nobel Prize in 1907). Counter-intuitive to the expectation of the two, they reported no significant observations identifying this aether. Convinced their experiments were confounded by external factors, numerous physicists persisted for years with variations on this experiment.

Einstein later went on to establish via special relativity that there is no aether and that the speed of light in a vacuum has a constant value, c = 299 792 km/s. The variability in the experiments by Michelson, Morely and later others have since been attributed to "observation error" arising from the apparatus.

## 38.4 STATISTICAL TESTS

An alternative method for assessing the goodness of fit is to use statistical tests that provide a single number to indicate the quality of the fit. The obvious problem then becomes which test to use, so only one test, the Kolmogorov-Smirnov test, is summarised here as an example.

The KS test is based around a test statistic $D$, which is defined as the largest difference between the observed data and the fitted theoretical curve. A visual example is provided in Figure 38.3(a), but note that it is not necessary to plot the values to determine this statistic. Having obtained this statistic, a look-up table can be used to establish the probability of obtaining this $D$ value for a random sample of the same size that is truly drawn from the distribution in question. For the exponential distribution, the $D$ statistic is 0.1714 and the corresponding probability is 0.029, which is very small. This suggests that the data do not follow the exponential distribution well. Figure 38.4(b) provides a comparison for the Weibull distribution, which has a much better fit. Critical values of $D$ are tabled and available on the internet, but when parameters are estimated from the data, as they usually are, the critical values will vary and depend on the assumed distribution. It is possible to determine critical $D$ values by repeated simulation using random samples and estimating parameters from that sample.

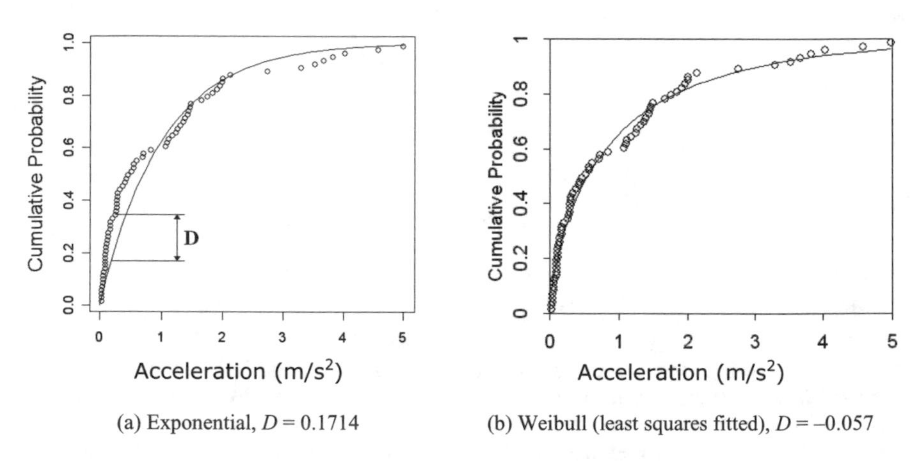


(a) Exponential, $D = 0.1714$ (b) Weibull (least squares fitted), $D = -0.057$

**Figure 38.4** KS-test $D$ statistic for earthquake data.

## 38.5 LEAST SQUARES

In many instances there may not be directly observable statistics that relate to given parameters of the model. In this scenario a least-squares approach can be used. Consider as an example the *abc* model for river runoff depicted in Figure 38.5.

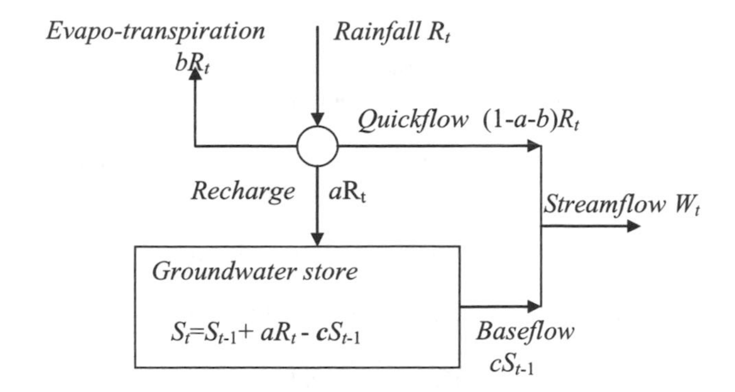


**Figure 38.5** abc model for converting rainfall to streamflow.

This model can be explained as the following sequence of events:

1. at time $t$ an amount of rainfall $R_t$ arrives on the catchment;
2. a portion $a$ of the rainfall infiltrates into the groundwater;
3. a portion $b$ is returned to the atmosphere;
4. the remaining portion $(1–a–b)$ runs into the stream;
5. the total streamflow includes a portion $c$ based on the existing groundwater store at time $t–1$; and
6. the groundwater store is updated to account for gains and losses.

This aptly named model has three parameters $a$, $b$ and $c$ and amounts to a transfer function for converting rainfall into streamflow, denoted $W_t = f(Rt \mid a,b,c,S_0)$. There is an additional nuisance parameter, $S_0$, which is the groundwater store at $t$=0, that while not of direct interest must be estimated nonetheless. A typical scenario is that one has access to observed records of rainfall, evapo-transpiration and streamflow, but not the groundwater level. This makes the task of estimating the amount of recharge to the groundwater and baseflow component of the streamflow difficult.

The solution to this problem is to formulate an optimisation problem, where the parameters $a$, $b$ and $c$ are changed until the best possible fit to the data is obtained. A simple criterion to decide the goodness of fit is the sum-of-squared differences for all timesteps $t=1..T$ between the observed streamflow $W_t$ and the modelled streamflow:

$$SS = \sum_{t=1}^{T} (\hat{W}_t - f(R_t \mid a,b,c,S_0))^2 \qquad (38.5)$$

This general procedure is named “least squares”. Although optimisation methods can become quite sophisticated, this model is simple enough for intuition to guide the user to make iterative guesses starting from an arbitrary initial guess.

## PROBLEMS

**38.1** Develop a probability plot of the earthquake data using $b = 0.2$ in Equation (38.2). Compare the plot to that obtained using $b = 0.0$.

**38.2** Develop a least-squares function for the Weibull distribution fitted to the earthquake data. Manually change the parameter values until a minimum is located.

**38.3** Is the normal distribution a good fit to the annual flows in Table 38.4?

**Table 38.4** Annual Flows in the Colorado River, 1911–1972. (McMahon and Mein, 1986)

| | | | | | |
|---|---|---|---|---|---|
| 1.66 | 10.67 | 12.2 | 14.93 | 18.01 | 22.54 |
| 3.98 | 10.73 | 12.63 | 15.21 | 18.11 | 23.04 |
| 4.85 | 10.74 | 12.99 | 15.26 | 18.13 | 24.17 |
| 7.57 | 10.79 | 13.3 | 15.77 | 18.24 | 24.18 |
| 7.64 | 10.82 | 14.04 | 15.88 | 18.64 | 24.68 |
| 8.57 | 10.88 | 14.11 | 16.03 | 19.81 | 25.22 |
| 8.68 | 11.06 | 14.3 | 16.08 | 20.89 | 25.79 |
| 9.32 | 11.39 | 14.4 | 16.19 | 21.07 | |
| 9.33 | 11.52 | 14.5 | 16.94 | 21.57 | |
| 9.57 | 11.61 | 14.71 | 17.32 | 22 | |
| 9.92 | 11.99 | 14.77 | 17.81 | 22.06 | |

**38.4** Table 38.5 gives annual maximum daily rainfall intensities (mm/hr) for Mount Bold reservoir in South Australia. How well does a Gumbel distribution fit the data?

**Table 38.5** Annual maximum rainfall intensities at Mount Bold reservoir, 1969–2004.

| | | | | | |
|---|---|---|---|---|---|
| 1.02 | 1.4 | 1.61 | 1.69 | 2.13 | 2.44 |
| 1.11 | 1.43 | 1.63 | 1.75 | 2.18 | 2.47 |
| 1.13 | 1.43 | 1.63 | 1.75 | 2.18 | 2.53 |
| 1.17 | 1.47 | 1.67 | 1.75 | 2.23 | 2.94 |
| 1.18 | 1.5 | 1.67 | 1.83 | 2.23 | 3.62 |
| 1.21 | 1.55 | 1.69 | 2.06 | 2.34 | 5.28 |

**38.5** Table 38.6 gives four months of daily Adelaide rainfall totals (mm) measured from a daily rainfall gauge. Plot the data using a log-normal probability plot.

**Table 38.6** Annual maximum rainfall intensities at Mount Bold reservoir.

| | | | | | | | | | | | |
|---|---|---|---|---|---|---|---|---|---|---|---|
| 1.4 | 2.8 | 0.4 | 2.6 | 0.6 | 2.6 | 6.4 | 9.8 | 1.4 | 3 | 15.8 | 26.6 |
| 0.2 | 3.4 | 12.8 | 3.6 | 1.4 | 4.6 | 0.2 | 6.6 | 4.8 | 1.8 | 13.6 | 0.2 |
| 0.8 | 6.4 | 12.8 | 3.6 | 0.2 | 0.2 | 12.6 | 0.2 | 0.8 | 2.4 | 6.4 | 17.6 |
| 0.4 | 12.6 | 28.6 | 2.4 | 1 | 28.6 | 4.6 | 0.2 | 0.4 | 4.8 | 0.2 | 7.4 |
| 4.2 | 0.2 | 0.6 | 0.4 | 0.2 | 8.8 | 1.8 | 0.4 | 2.8 | 1.2 | 0.2 | 29.6 |

CHAPTER THIRTY NINE

# Stochastic Modelling (Likelihood and Uncertainty)

## 39.1 INTRODUCTION

Donald Rumsfeld, the 2001–2006 US Secretary of Defence, was known to opine on the strategy of war that there are "known knowns", "known unknowns" and "unknown unknowns". It is assumed in this chapter that, while it is not possible to prepare for unknown unknowns, and the known knowns are a given, the real interest must lie in the known unknowns. Perhaps it is oxymoron to talk about "known unknowns", but it is usually possible to put a number on those unknowns, which is termed "uncertainty". In many circumstances the uncertainty can be more disconcerting than the known knowns of a process. For example, while it might be possible to plan for a 0.34-m sea-level rise on average in the next 100 years, it becomes harder when it is acknowledged there is a range of uncertainty in this estimate, 0.18–0.59 m (IPCC, 2007). While these extrapolations are based on linear trends, what is more alarming is the uncertainty in the uncertainty – that is, it remains unknown whether the uncertainty estimates should allow for long-term exponential increases instead of linear extrapolations.

Models are an approximation to some real-world processes, and it is important that their outputs are not treated as being exact. Similarly, the model parameters are not constant as they are estimated from finite observation records. This chapter investigates a method that is able to provide best estimates of a parameter along with the uncertainty in that estimate. The known parameter uncertainty can then be used to compute the uncertainty in the model output.

## 39.2 BAYES' RULE

Recall from Chapter 22 that Bayes' rule is a very powerful rule that relates the conditional probabilities between two events as given by the equation:

$$P(A|B) = \frac{P(B|A)P(A)}{P(B)} \tag{39.1}$$

where the vertical line is read as "given the event". The conditional probability P($A|B$) is the chance of an event $A$ given only the subset of events $B$. It can also be evaluated as P($A|B$) = P($A$ and $B$)/P($B$).

Many people have heard about Bayes' rule and on reaching this point often say, "So that's it: an equation?" shortly followed by, "So what is all the fuss about?" The answer is that many probabilities are interlinked and it often necessary to determine the chance of an event conditioned on the occurrence of some other event. For example, while a "fail-safe" mentality might seek to minimise the absolute chance of a catastrophic event (levee failure, building collapse), many argue that a "safe-fail", which is the probability of loss or damage given the catastrophic event, is more important.

One area where Bayes' rule comes to prominence is in relating a dataset to model parameters. This process is termed model calibration and it can be like finding a needle (good parameters) in a haystack (of all possible parameter combinations). Baye's rule casts this problem in a probability framework and interprets the various parts of Equation (39.1) using special terminology. A key difference here from the earlier discussion of Bayes' rule in Chapter 22 is that the events $A$ and $B$ in Equation (39.1) are random variables that can take on any value in a range of possible values. This implies that the probability terms are no longer single numbers between 0 and 1, but they become probability distributions that take on different values as $A$ and $B$ vary. For a set of data values, denoted $x$, and a vector of parameters, denoted $\theta$, Bayes' rule can be written in the following form:

$$P(\theta|x) = \frac{P(x|\theta)P(\theta)}{P(x)} \tag{39.2}$$

or

$$posterior = \frac{likelihood\ x\ prior}{evidence} \tag{39.3}$$

The term *prior* is a probability placed on the parameters ignorant of any data, and the term *evidence* is the probability of that set of data occurring. The prior distribution can be especially important when there are few data and it might reflect the opinions of an expert. The evidence term can be important when trying to compare two models and decide which is better. However, for now it suffices to ignore these two terms and focus on the *likelihood* and that it is proportional to the *posterior*

The likelihood is a probability of a dataset occurring given a fixed set of parameters. This is illustrated in Figure 39.1 where there are different datasets $x$, $y$ and $z$ represented as histograms and one set of parameters, $\theta_1$, which control the probability density function (solid line). It is evident that dataset $x$ is more *likely* (i.e. has a higher probability) to be explained by the parameters $\theta_1$ than $y$ or $z$. By determining this probability it is possible to develop an algorithm to objectively find the most-likely parameters. This objective criterion can be more reassuring than a subjective "it looks close enough" argument.

The posterior distribution turns things around so that the emphasis is on the probability of different parameters for a fixed data set. Figure 39.2 provides an example where models corresponding to different parameter choices, $\theta_1$, $\theta_2$ and $\theta_3$ are compared to a dataset. It is clear that given the data $x$, the model having parameters $\theta_1$ is a better (more probable) representation of the data than $\theta_2$ or $\theta_3$.

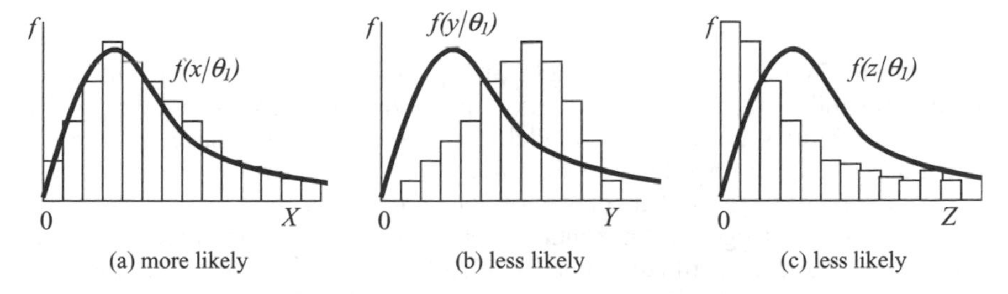


**Figure 39.1** Schematic comparison of likelihood of various datasets.

The observation of changes in probability for different parameter values is an objective basis for parameter uncertainty. Similar probabilities across a wide range of parameters suggest high uncertainty, whereas a steep change in probability from highly likely to near-zero for a small change in parameter value suggests high certainty of that particular parameter.

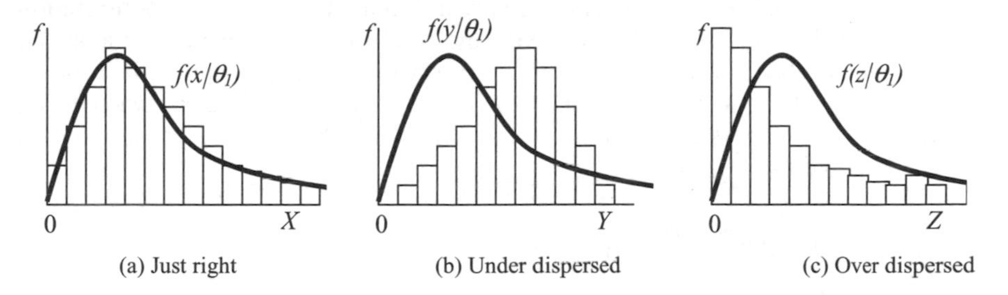


**Figure 39.2** Schematic comparison of lognormal PDF to sampled data.

Bayes' rule is therefore useful in two main areas: (i) using the likelihood to find the best parameter for data, termed maximum likelihood; and (ii) using the posterior to assess model uncertainty.

## 39.3 MAXIMUM LIKELIHOOD

A likelihood function represents the probability of data given certain parameter values. Consider a set of observed data, $\{x_1, x_2, \ldots, x_n\}$, and a PDF $f(x_i / \theta)$ specified in terms of parameters $\theta$. The probability of obtaining a particular data value $x_i$ is simply the value of the PDF evaluated at $x_i$. If each data point is independent then the joint probability of the dataset $P(x_1$ and $x_2$ and $\ldots x_n \mid \theta)$ becomes the product of their individual probabilities:

$$P(x_1, x_2, \ldots, x_n \mid \theta) = f(x_1 \mid \theta) f(x_2 \mid \theta) \ldots f(x_n \mid \theta) \tag{39.4}$$

Another way for writing this is:

$$P(x_1, x_2, ..., x_n \mid \theta) = \prod_{i=1}^{n} f(x_i \mid \theta) \qquad (39.5)$$

Consider Figure 39.1 again in light of Equation (39.4). Dataset $x$ will have a higher probability than $y$ or $z$ because it has more data points coinciding with higher probabilities given by the model. Dataset $y$ and $z$ will have more points contributing lower probabilities in Equation (39.4) and respectively fewer points near the region of the PDF that contributes the highest probabilities.

As the name suggests, "maximum likelihood" is a method where parameter values are changed until the highest likelihood is obtained. The estimated parameters that give the highest probability are said to be maximum likelihood estimates (MLEs). The estimator may be found by differentiating $P(x_1, x_2, ..., x_n \mid \theta)$ with respect to $\theta$ and setting the derivative equal to zero:

$$\frac{\partial P(x_1, x_2, ..., x_n \mid \theta)}{\partial \theta} = 0 \qquad (39.6)$$

When a likelihood function is not easily differentiable, numerical optimisation methods can be used instead of analytic methods to locate the best parameters. Because the probabilities of the individual observations are less than 1, the product of many probabilities can result in a very low number. Therefore, when using numerical methods it is common to take the log of the likelihood function to obtain the MLE:

$$\frac{\partial \log P(x_1, x_2, ..., x_n \mid \theta)}{\partial \theta} = 0 \qquad (39.7)$$

For many distributions it is possible to derive the MLEs analytically and show that they are equal to certain moments of the distribution. Examples include the exponential distribution and the normal distribution. In these situations using maximum likelihood theory offers no advantage when there is an easier alternative that offers the same result. There are, however, many situations where likelihood estimators are advantageous. One example is when a dataset contains censored data: that is, the exact value is not known, only that it exceeded some threshold. Censored data are common; for example, it may be possible to drop below the resolution of an electronic measuring instrument without knowing how far below. At the other extreme it may be possible for a rare event such as a flood to exceed the high water level of a measuring device.

*Example*

The data in Table 39.1 (Singpurwalla et al., 1975) are an excerpt from temperature and voltage-accelerated life test data for tantalum electrolytic capacitors. They are the results of the test at 62.5 V and 5° Celsius. The sample size was 174 capacitors and 18 failed within the 12,500 hours' duration of the test. One hundred and fifty six capacitors were still working at the end of the test and their lifetimes are described as censored because they exceed 12,500 h. Suppose that the data are to follow a Weibull distribution, determine the parameters of this distribution using maximum likelihood.

**Table 39.1** Hours until failure of electrolytic capacitors.

| | | | | | |
|---|---|---|---|---|---|
| 25 | 50 | 165 | 500 | 620 | 720 |
| 820 | 910 | 980 | 1270 | 1600 | 2270 |
| 2370 | 4590 | 4880 | 7560 | 8730 | 12500 |

*Solution*

The Weibull probability density function is defined as:

$$f(t) = \frac{\alpha}{\beta^{\alpha}} t^{\alpha-1} \exp(-(t/\beta)^{\alpha}) \qquad t \geq 0 \tag{39.8}$$

and the cumulative distribution function is defined as:

$$F(t) = 1 - \exp(-(t/\beta)^{\alpha}) \qquad t \geq 0 \tag{39.9}$$

where $\beta$ is a scale parameter and $\alpha$ is a shape parameter. The exponential distribution is obtained as a special case for $\alpha = 1$. The probability of the censored terms is obtained as $1-F(12500)$ since this is by definition the probability of exceeding the threshold $t$=12500. The remaining 18 failures are evaluated at $f(t)$. The likelihood function is therefore:

$$P(t_1, t_2, ..., t_{174} \mid \alpha, \beta) = \{1 - F(12500)\}^{156} \prod_{i=1}^{18} f(t_i \mid \alpha, \beta) \tag{39.10}$$

While it is possible to write this equation out in full, and differentiate it analytically, it soon becomes a cumbersome approach that is prone to mistakes. A much simpler approach is to use a numerical optimiser. Instead of using Equation (39.10), taking a log transform will ensure that the numerical values do not become too small to work with, which gives:

$$\log(P(... \mid \alpha, \beta)) = 156 \log(\{1 - F(12500)\}) + \sum_{i=1}^{18} \log(f(t_i \mid \alpha, \beta)) \tag{39.11}$$

This function has been entered into Excel, and cell C29 shown in Figure 39.3. Many versions of Excel have a built-in optimiser "Solver" that can be used to find the MLEs. If Solver is available it can be located under the menu Tools – Solver, though it may be necessary to first enable it via Tools – Add-ins – Solver. Figure 39.3 shows constraints applied to the parameters so that they remain positive. This is important; otherwise it will cause numerical issues and mislead the optimiser. Starting with an initial guess of $\alpha$=1, $\beta$=12500 and clicking Solve should verify that

Solver locates the parameters shown in Figure 39.3. If Solver is unavailable for the version of Excel being used, then it is still possible to let intuition guide the process by manually making improved guesses that increase the value in cell C29. However, for models having many more parameters a numerical optimiser is necessary since it becomes too tedious to "optimise" manually and intuition can become misleading. Matlab has an `optim` function that performs a similar task, and optimisation packages are commonly available for other programming languages.

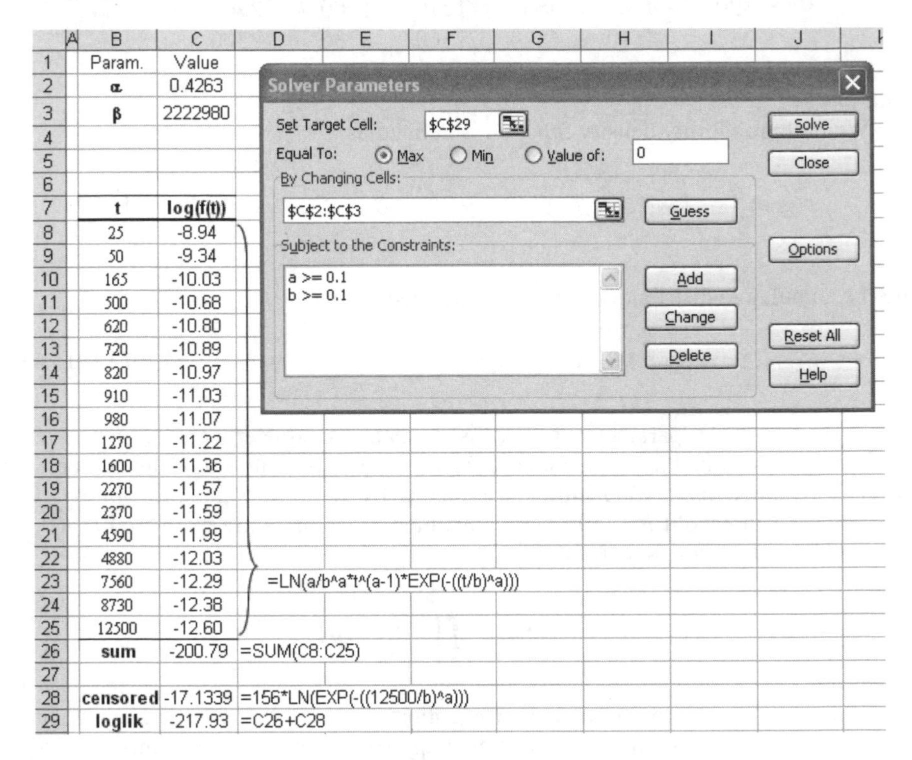


**Figure 39.3** Example of maximum likelihood optimisation using Excel Solver.

## 39.4 UNCERTAINTY

The concept of parameter uncertainty can be counter-intuitive for many. Most people newly initiated with statistics are used to thinking of a parameter as a fixed value and the distribution as specifying the random quantity of interest. By way of contrast, assessing the parameter uncertainty requires a fixed set of data but letting the parameters have a distribution of possible values. As a result, discussion about the uncertainty of a model often revolves around the variability of this parameter distribution. Higher variability implies that the parameter is not well estimated.

In the following example the mean is estimated from samples $X_1$, $X_2$ and $X_3$ that respectively have $n$=25, $n$=100 and $n$=400 random normal numbers $X$~$N(0,1)$. This process is repeated 100,000 times to build up histograms of the mean, as shown in Figure 39.4. While the population mean has a value of zero, the respective samples are not always zero since it depends on the actual numbers that were generated. While it is clear that the three histograms are centred about the zero value (i.e. they are not biased), the variability is of the most interest: as the sample size increases, the variability in the estimate decreases. In other words, the estimate becomes more certain.

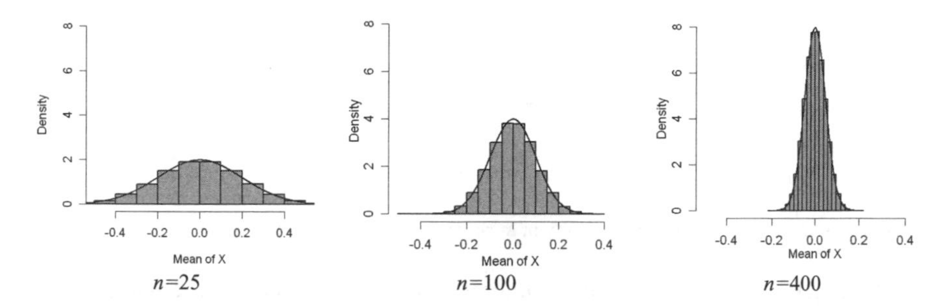


**Figure 39.4** Histograms of mean estimator for an increasing n. Exact PDF overlaid.

A well-known result for the variability of the $\hat{\mu}$ estimator for the mean of the normal distribution is:

$$\hat{\mu} \sim N\left(0, \frac{\sigma_X}{\sqrt{n}}\right) \tag{39.12}$$

where $\sigma_X$ is the standard deviation of the dataset $X$ and $n$ is the number of data points. For the examples shown in Figure 39.4 the estimate of the mean has standard deviations $\sigma_{\hat{\mu}} = 1/5$, $\sigma_{\hat{\mu}} = 1/10$ and $\sigma_{\hat{\mu}} = 1/20$ respectively and the density functions are plotted over the histograms to show the agreement. From Equation (39.12) it is possible to plan an experiment to gather enough data to limit the uncertainty in the mean to a desired tolerance (assuming the standard deviation of the data is known).

For models other than the normal distribution there is typically no neat formula for the uncertainty in various parameters. In a general setting it is the posterior distribution $f(\theta \mid x)$ that encapsulates all information about the variability of parameters given a dataset $x$. It suffices here to observe that the posterior distribution is proportional to the likelihood; thus, if it is possible to write a likelihood function, it is possible to quantify the uncertainty. The likelihood function for a normal distribution is:

$$P(x_1, x_2, \ldots, x_n \mid \mu, \sigma) = \prod_{i=1}^{n} \frac{1}{\sigma\sqrt{2\pi}} \exp\left(\frac{-1}{2\sigma^2}(x_i - \mu)^2\right) \tag{39.13}$$

and contours of this function are plotted in Figure 39.5. As there are two parameters the uncertainty distribution is bivariate. By inspecting the parameters together, one can get more information than simply looking at the uncertainty in one parameter. This is due to the possibility for parameters to be correlated. For the normal distribution it is clear from Figure 39.5 that there is, however, no association. That is, the uncertainty about the location of the $X$ does not tell anything about the uncertainty in the variability of $X$ and vice versa.

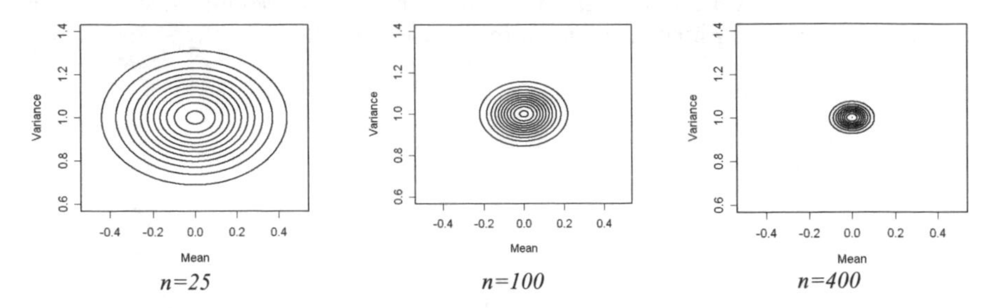


**Figure 39.5** PDF contours of uncertainty in mean and variance parameters of normal distribution.

Parameter correlations are of interest to modellers because amongst other reasons they can (i) cause numerical issues in finding good parameters, (ii) suggest possible redundant parameters if there are high correlations and (iii) explain model behaviour. As an example consider Figure 39.6, which shows a schematic of the direct shear test along with measurements.

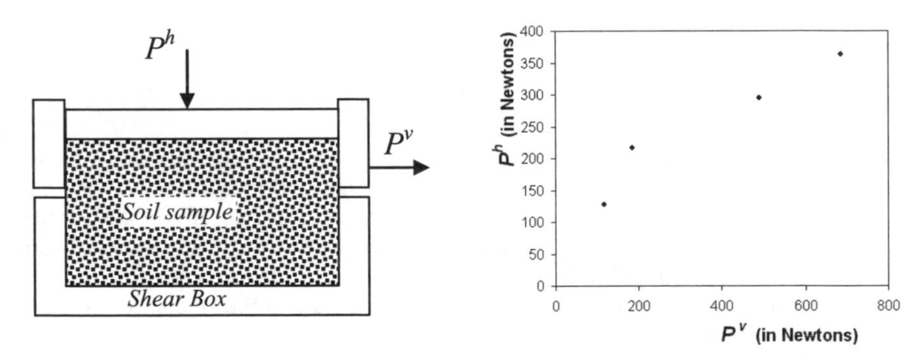


**Figure 39.6** Schematic of the direct shear test along with 4 measurements.

The following equation gives the linear relationship for the results:

$$P^h = P^v \tan(\phi) + c \tag{39.14}$$

where $P^h$ is the horizontal force on the soil (related to the shear), $P^v$ is the normal force on the soil, $c$ represents the soil cohesion and $\phi$ is the soil's internal angle of friction. This expression is a linear regression where the scatter about the line of best fit is be assumed to follow a normal distribution. In order not to overcomplicate the example with an additional parameter, the variance of this distribution is fixed at 20. The likelihood for these data is, therefore, given by:

$$P(P^h \mid \phi, c, P^v) = \prod_{i=1}^{4} N(P_i^h \mid P_i^v \tan(\phi) + c, 20) \qquad (39.15)$$

where $P^v$ is an explanatory variable so that $\phi$ and $c$ are the parameters to be estimated. Contours of this function are shown in Figure 39.7.

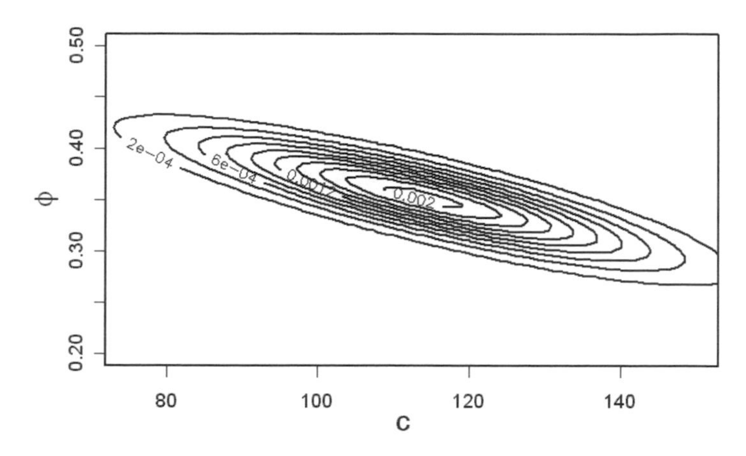


**Figure 39.7** Schematic of the direct shear test along with 4 measurements (Bowles, 1988).

The mode of the PDF in Figure 39.7 corresponds to the maximum likelihood estimates $(c, \phi) = (113.5, 0.37)$. This figure shows that not only is there uncertainty in the parameter estimates, but they are correlated. Thus, due to the parameter correlation it is possible that the soil has a lower cohesion but a steeper friction angle, say (90,0.4), or a higher cohesion but lower friction angle, say (140,0.3).

## PROBLEMS

**39.1** Analytically derive the MLE for the exponential distribution (which can be obtained from the Weibull by setting $\alpha = 1$).

**39.2** Analytically derive the MLE for the exponential distribution allowing for $n$ censored data points. Compute this estimate for the capacitor data.

**39.3** Using Excel enter the likelihood function for the Weibull and verify the result in Figure 39.3 Set $\alpha = 1$ and optimise to find the corresponding $\beta$ parameter.

**39.4** For the set $X$=(0.3, –1.1, –0.7, 1.5) evaluate the likelihood in Equation (39.13) at $(\mu,\sigma) = (0,1)$. Evaluate the likelihood at $(\mu,\sigma) = (-0.1,1)$, $(\mu,\sigma) = (0.1,1)$, $(\mu,\sigma) = (0,0.99)$ and $(\mu,\sigma) = (0,1.1)$. Verify that the likelihood values either side of the maximum are lower.

**39.5** Estimate the variance from samples $X_1$, $X_2$ and $X_3$ that respectively have $n$=25, $n$=100 and $n$=400 random normal numbers $X \sim N(0,1)$. Repeat this process 100,000 times to build up histograms of the variance. Verify the following approximation for the distribution of the variance parameter, $\sigma_X^2 \sim N(1, \sqrt{\sigma_X^2/2n})$, by computing the mean and variance of the estimates of sample variance.

## CHAPTER FORTY

# Stochastic Modelling (Markov Chains)

## 40.1 INTRODUCTION

Markov chains describe random changes in a system of discrete states over time, such that the future depends on the present but not the past. This independence of the future from the past, given the present state, is known as the Markov property after the Russian mathematician Andrei Andreevich Markov (1856–1922) who made a systematic study of such sequences.

For example, a day can be defined as wet if (say) more than 0.1 mm of rain falls, or as dry. A concrete cube, made from a sample of concrete delivered to a site, can be defined as within specification or out of specification. The contents of a reservoir can be defined to the nearest hundredth of a unit of its maximum volume. In all these cases a system is described in terms of a discrete set of states. In many applications this discrete set is an approximation to a continuous variable, as it is for the volume of water in a reservoir, but in others, such as queues, the system is naturally described in integer numbers.

Time can be modelled as discrete or continuous, and it is often convenient to distinguish the two cases by describing them as Markov chains and Markov processes respectively.

**Markov**

The Markov property is named after the Russian mathematician Andrei Andreevich Markov (1856–1922). Sergei Bernstein, who continued to develop the theories of Markov chains, said the following, "A. A. Markov's classic course on the computation of probabilities, and his original memoirs, models of accuracy and clarity of exposition, contributed to a very large extent to the transformation of the theory of probability into one of the most perfected areas of mathematics..."

(MacTutor History of Mathematics archive at the University of St Andrews, available at http://www-history.mcs.st-andrews.ac.uk/history/index.html)

## 40.2 REGULAR MARKOV CHAINS

Suppose there are $m$ states, defined so that the system will be in one and only one state at each time step. The Markov chain is defined by the transition probabilities of going from any one state $i$ to any other state $j$ in a single time step. These probabilities are written as $p_{ij}$ and are collected together in a one-step transition probability matrix M.

$$M = \begin{pmatrix} p_{11} & p_{12} & \cdots & \cdots & p_{1m} \\ p_{21} & \ddots & & & \vdots \\ \vdots & & & & \\ \vdots & & & \ddots & \vdots \\ p_{m1} & \cdots & & \cdots & p_{mm} \end{pmatrix} \tag{40.1}$$

Since the system must be in one of the states at each time step the sum of rows of the transition matrix add to 1. The state probability vector (a row array with $m$ elements rather than a physical vector) is defined at time $p_t$ as the array of probabilities of being in state $i$ at time $t$, $p_t=(p_{1,t};\ldots;p_{m,t})$.

Now consider the event of being in state $j$ at time $t+1$. It is necessary to allow for the possibility of being in any state at time $t$ and moving from that state to state $j$ in one step. The probability of being in state $j$ at time $t+1$ is equal to the sum of the products of the probabilities of being in any state $i$ from 1 up to $m$ at time $t$ with the transition probability of moving from state $i$ to state $j$:

$$p_{j,t+1} = \sum_{i=1}^{m} p_{i,t} p_{ij} \tag{40.2}$$

This can be expressed succinctly by the matrix equation:

$$p_{t+1} = p_t M \tag{40.3}$$

To complete the description of the Markov chain the initial probability vector is needed, and if it starts from a known state this will consist of a single 1 with 0s elsewhere in the vector. Then, repeated application of Equation (40.3) gives

$$p_t = p_0 M^t \tag{40.4}$$

where $p_0$ is the initial probability vector.

A state is described as *absorbing* if it is impossible to leave it and a Markov chain is said to be *irreducible* if it has no absorbing states. A state is described as *periodic*, with period $k$, if return to that given state can only occur in multiples of $k$ time steps. An irreducible Markov chain is *regular* if it has no periodic states. For a regular Markov chain, the probabilities of ending up in each of the $m$ states will tend to constant values, which are independent of the starting state, as time increases. This probability distribution is known as the stationary or equilibrium distribution. If the stationary distribution is denoted as $v$, then:

$$v = \lim_{t \to \infty} p_t . \tag{40.5}$$

As $t$ tends to infinity both $p_t$ and $p_{t+1}$ in Equation (40.3) tend toward $v$, which can therefore be found by solving $v = vM$ together with the constraint that the probabilities in $v$ add to 1. Alternatively, Equation (40.4) can be used with a very large value of $t$ and any choice of $p_0$. This method is very simple to implement numerically.

In most cases the probabilities of being in the various states after many time steps is of little practical interest. The important interpretation of $v$ is that it gives the proportion of time that the system spends in each state in the long run.

*Example*

The following data are from 21 years of daily records from a rain gauge on Emley Moor at the head of the River Dearne in Yorkshire. During the season from March until October there were: 1316 transitions from a wet day to a wet day; 691 transitions from a wet day to a dry day; 686 transitions from a dry day to a wet day; and 1749 transitions from a dry day to a dry day. Denote wet and dry days by states 1 and 2 respectively and calculate the one-day transition matrix. If it is raining today, find the probability it is raining in two days' time. What is the proportion of dry days during this season of the year?

*Solution*

Matlab is ideal for such matrix calculations (Figure 40.1). The transition probability from wet to wet is estimated by 1316/(1316+691), which is 0.656 when rounded to three decimal places. The following transition probabilities are similarly estimated and shown below in the matrix $M$. The probability it is raining in two days' time, if it is raining today, is 0.527. The long-run proportion of dry days is 0.55.

```
>>M=[0.656 0.344; 0.282 0.718]
>>p0=[1 0]
>>p2=p0*M^2  output: p2 = 0.5273 0.4727
>>v=p0*M^1000     output: v = 0.4505 0.5495
```

**Figure 40.1** Matlab code for transition matrix calculations.

**Theory of Storage: P. A. P. Moran**

This example is based on the general case analysed by Patrick Moran in his 1955 monograph *The Theory of Storage*. Moran (1917–1988) was the first professor of statistics at the Australian National University, a position that he held from 1952 until his retirement in 1983.

*Example*

A country has a rainy and a dry season each year. A dam has a capacity of 4 units of water and the state of the system at the end of the dry season can be modelled in integer units. The time step is a year. During the rainy season, 0, 1, 2, 3, or 4 units of water enter the dam with probabilities 0.4, 0.2, 0.2, 0.1, and 0.1 respectively. Overflow is regarded as lost. Provided the dam is not empty at the end of the rainy season, one unit of water is released during the dry season. If the dam starts empty, what is the probability distribution of its contents after 1, 2, and 3 years? What proportion of time will the dam be in its various possible states?

*Solution*

The first step in the solution to any Markov chain problem is to set up an appropriate state space (Figure 40.2). At the end of the dry season, the dam can contain 0, 1, 2 or 3 units of water. It cannot contain 4 units because, even if it is full at the end of the rainy season, 1 unit is released during the dry season. So the states are {0, 1, 2, 3} and the rows of the transition matrix are in that order. The transition probabilities from 0 to 0 is 0.6 because if the dam is empty at the end of this dry season it will be empty at the end of the next if there is 0 or 1 unit of rainfall. The other transition probabilities follow from similar arguments. In the long run the dam will be empty at the end of 17% of dry seasons.

```
>> M=[.6 .2 .1 .1;
      .4 .2 .2 .2;
      .0 .4 .2 .4;
      .0 .0 .4 .6]
>>p0=[1 0 0 0]
>>p1=p0*M     output: p1 = 0.600 0.200 0.100 0.100
>>p2=p1*M     output: p2 = 0.440 0.200 0.160 0.200
>>p3=p2*M     output p3 = 0.344 0.192 0.196 0.268
>>v=p3*M^100 output v = 0.174 0.174 0.261 0.391
```

**Figure 40.2** Matlab code for transition matrix calculations.

## 40.3 ABSORBING MARKOV CHAINS

Now suppose that $r$ of the $m$ states are absorbing. The states can be arranged with the absorbing states first, so that the transition matrix can be written in a partitioned form, known as the canonical form:

$$M = \begin{pmatrix} I & 0 \\ R & Q \end{pmatrix} \tag{40.6}$$

where the $r$ absorbing states account for the identity matrix in the upper left corner, and the $(m–r)$ transition rates between non-absorbing states are in the matrix $Q$. Let the matrix $N$ defined by:

$$N = (I - Q)^{-1} \tag{40.7}$$

which gives the average number of time steps spent in each non-absorbing state for each possible non-absorbing starting state. Adding the rows of $N$ will give the mean number of steps before being absorbed for each possible non-absorbing starting state. Let the matrix $B$ be defined by:

$$B = NR \tag{40.8}$$

which gives the probability of being absorbed in state $j$ for each possible non-absorbing starting state $i$.

*Example*

A geostationary satellite can be in one of four states after first launch: failed, bad deviation, slight deviation, correctly positioned. The first and last states are absorbing. If the satellite is not correctly positioned it may be possible to nudge it into this state using pulses from small rocket motors on the satellite controlled by the ground station. On one pulse, a bad deviation can be improved to a slight deviation with probability 0.7. But there are probabilities of 0.1 and 0.2 of no improvement and failure respectively. A slight deviation can be corrected with one pulse with probability 0.7. But there is a 0.2 probability of failure. What is the probability of correcting a badly positioned satellite and what is the expected number of pulses required?

*Solution*

The states are arranged in the order: failed; bad position; slight deviation; correctly positioned. The rows of the matrices in Figure 40.3 follow this ordering.

```
>> M=[1   0  0  0; 0   1  0  0;.2  0 .1 .7;.2 .7  0 .1]
>> Q=[.1 .7;0 .1]
>> N=(eye(2)-Q)^(-1) output: N = 1.111    0.864
   0           1.111
>> R=[.2  0;
      .2 .7]
>> B=N*Routput: B = 0.395  0.605
    0.222  0.778
```

**Figure 40.3** Matlab code for transition matrix calculations.

Thus if the satellite starts badly positioned the probability of correcting it is 0.60. The expected number of pulses is the sum of 1.111 and 0.864, which equals 1.98.

The theory of this section can be used to find first passage times for irreducible chains. For example, for the dam example in Section 40.2 what is the expected number of years until it is first empty given its state now? This can be found by changing 0 into an absorbing state and calculating $N$ for this redefined system (see Problem 40.3).

## 40.4 REGULAR MARKOV PROCESSES

It is often preferable to model time as a continuous variable rather than in discrete steps, and Markov processes allow this. Applications include machine maintenance, telephone and computer network traffic, and other situations that involve queues.

The principle can be explained by considering a machine which can be in one of two states: operating (1) and under repair (0). A Markov chain is set up with a time step $\delta t$ where the probability of being in a given state at time $t+\delta t$ is:

$$\begin{pmatrix} p_0(t+\delta t) & p_1(t+\delta t) \end{pmatrix} = \begin{pmatrix} p_0(t) & p_1(t) \end{pmatrix} \begin{pmatrix} p_{00} & p_{01} \\ p_{10} & p_{11} \end{pmatrix} \tag{40.9}$$

which in matrix notation becomes:

$$p(t+\delta t) = p(t)M \,. \tag{40.10}$$

Rewriting this as the change in state probabilities per time increment:

$$\frac{p(t+\delta t) - p(t)}{\delta t} = p(t)\left(\frac{M - I}{\delta t}\right) \tag{40.11}$$

and letting $\delta t \rightarrow 0$ gives:

$$\dot{p}(t) = p(t)\Lambda \tag{40.12}$$

where the matrix $\Lambda$ is known as the rate matrix. Since the rows of the transition matrix M add to 1 and the identity matrix I has 1s on the leading diagonal and 0s elsewhere, the rows of the rate matrix must add to 0.

The elements of the rate matrix are obtained from the following argument. Suppose the machine is under repair and the repair crew work at a constant rate $\theta$ repairs per time unit. Then the probability of repair in a small length of time, $\delta t$, is $\theta\delta t$, and the probability of no repair is$1-\theta\delta t$. There are two subtleties involved in this claim. Firstly, the possibility of repair and subsequent failure in time $\delta t$ has been ignored because its probability would be of order $(\delta t)^2$ and negligible – tending to 0 in the limit when dividing by $\delta t$ and letting $\delta t$ tend to 0. Secondly, the constant repair rate implies that the repair process is Markov because the probability of repair in the next $\delta t$ is independent of how long the repair has already taken. An equivalent statement is that the repair time has an exponential distribution. In terms of the transition probabilities: $p_{00} = 1-\theta\delta t$ and $p_{01} = \theta\delta t$. Now suppose that the failure rate is $\lambda$ failures per time unit. A similar argument leads to $p_{10} = \lambda\delta t$ and $p_{10} = 1-\lambda\delta t$. So, putting all this together rate matrix becomes:

$$\Lambda = \begin{pmatrix} -\theta & \theta \\ \lambda & -\lambda \end{pmatrix} \tag{40.13}$$

Equation (40.12) is two coupled linear differential equations which can be solved for the transition probabilities as a function of time. However, it is the stationary distribution that is usually of interest. This can be found by substituting $\dot{p}(t) = 0$, to obtain, in general for $m$ states:

$$v\Lambda = 0 \quad \text{where} \quad \sum_{i=1}^{m} v_i = 1 \tag{40.14}$$

For the machine repair problem this leads to the unsurprising result that in the long run the machine will be operating for a proportion of time:

$$v_1 = \frac{\theta}{\lambda + \theta} \tag{40.15}$$

This is known as the availability. If the numerator and denominator are divided by $\lambda\theta$ the availability is obtained as the ratio of mean time to failure to the sum of mean time to failure and mean time to repair. This last result is true for any distributions of repair times and times to failure.

*Example*
A contractor has two identical diggers working on a site. The time to failure is exponential with rate $\lambda$. There is one repair person. Repairs have an exponential distribution with mean $1/\theta$. Find the stationary distribution.

*Solution*
Let the state be the number of diggers that are working. If both the diggers are working the state is 2, and if one digger is working and the other is under repair the state is 1. If neither digger is working, one is under repair and the other is waiting for repair, and the state is 0. The rate matrix is:

$$\Lambda = \begin{pmatrix} -\theta & \theta & 0 \\ \lambda & -(\lambda+\theta) & \theta \\ 0 & 2\lambda & -2\lambda \end{pmatrix} \tag{40.16}$$

Notice, for example, that the probability one digger fails in time is P(digger$_1$ fails or digger$_2$ fails) = $\lambda\delta t + \lambda\delta t - (\lambda\delta t)^2$ and hence the rate from state 2 to 1 is $2\lambda$.

**Telephone exchanges: A. K. Erlang**
Agner Erlang (1878–1929) was a Danish mathematician who worked for the Copenhagen Telephone Company. He analysed the general case with *N* telephone lines, rather than diggers, and a single telephone exchange. The stationary distribution is known as the Erlang loss formula (see Problem 40.7).

## PROBLEMS

**40.1** For the Emley Moor example of Section 23.2, what is the probability it is raining in two days time, if it is dry today?

**40.2** A Markov chain is regular if some power of its transition matrix has all its entries strictly positive. Explain why this must be so. Consider a Markov chain with transition matrix as in Equation (40.17) Is it regular? If not, does Equation (40.7) have a solution, and if it does how would it be interpreted?

$$M = \begin{pmatrix} 0 & 1 & 0 \\ 0.3 & 0 & 0.7 \\ 0 & 1 & 0 \end{pmatrix} \tag{40.17}$$

**40.3** In the design of dams a typical design criterion might be the expected time from empty until the dam is full. This is an example of a first passage time. The first passage times can be calculated by making the full state into an absorbing state. Use this device to calculate the first passage time for the dam in Moran's example.

**40.4** The dwell time in a state in a Markov chain is geometric. For the Emley Moor example conduct the following simulation for 1000 days: (i) start day 0 as a dry day; (ii) sample a random uniform number; (iii) if the random number is less than the dry-to-wet probability move to the wet state, otherwise stay in the dry state; (iv) repeat from step (ii) each day using the relevant transition probability. Plot a histogram of the number of times until the state switches and verify its shape.

**40.5** Why is the dwell time in any non-absorbing state in a Markov chain geometric?

**40.6** More general distributions can be incorporated by introducing additional hidden states. Consider the Emley Moor example: introduce another wet state labelled 2, and re-label the dry state as 3. The chain moves from 3 to 1 at the end of a dry spell. It moves from state 1 to the hidden state 2 and emerges from state 2 at the end of the wet spell. The transition matrix is given in Equation (40.18). Investigate the distribution of lengths of dry spells by simulation.

$$M = \begin{pmatrix} .312 & .688 & 0 \\ 0 & .312 & .688 \\ .282 & 0 & .718 \end{pmatrix} \tag{40.18}$$

**40.7** Suppose that the contractor now has $n$ identical diggers on a site. The time to failure of a digger is exponential with rate $\lambda$. There is one repair person. Repairs have an exponential distribution with mean $1/\theta$. Show that the stationary distribution is as in Equation (40.19). This is the Erlang telegraph formula. Interpret $\lambda$ and $\theta$ where $n$ is the number of lines from a telephone exchange.

$$v_i = \frac{1}{i!}\left(\frac{\theta}{\lambda}\right)^i v_0 \qquad for\ i = 0,\ldots,n \ \ where \ \ v_0 = \left(\sum_{i=0}^{n} \frac{\theta^i}{i!\,\lambda^i}\right)^{-1} \tag{40.19}$$

CHAPTER FORTY ONE

# Stochastic Modelling (Time Series)

## 41.1 INTRODUCTION

Many of the data sets considered in earlier chapters have simply consisted of a number of values – perhaps concrete strengths based on testing a number of cylinders, or the number of floods in a particular year, or even the number of deaths per year due to kicks by horses over a 20-year period. In each case there was no need to consider the order in which the data were collected. Each data point provided a single value which could be used to determine a mean or develop a probability distribution. However, there are data where the time order is important. Take for example population size: in this situation the size is important, but equally important is the date at which the estimate was made. Other examples include share indices from world markets and climatic variations such as increasing sea levels or carbon dioxide concentration. The changes forecast from these time series, and the uncertainties inherent in these forecasts, present great challenges for engineers. Time series are analysed for three reasons: to make forecasts that facilitate decisions; to simulate realistic scenarios; and to supplement or quantify physical explanations for phenomena.

A time series is a sequence of observations of a variable made over a period of time. It is generally assumed that the time step between observations is constant. The variable can be aggregated or averaged over the time period, such as daily rainfall, or sampled from some continuous underlying electronic signal, such as water elevation in a wave tank recorded using a capacitance probe. In contrast, environmental series are usually at a much coarser time scale. The main features are likely to be a systematic pattern within years, known as seasonal variation, and a possible trend.

**Analogue to Digital**

If a signal is sampled it is essential that the sampling rate is at least twice the highest frequencies in the signal. Components at higher frequencies will not be missed, but will be mistaken for lower-frequency variation, a phenomenon known as aliasing (see Chapter 46). To avoid aliasing, higher frequencies are removed from the continuous signal by analogue circuitry known as anti-alias filters. As an example modern oscilloscopes are generally digital and can sample at gigahertz rates (1 gigahertz = 1 billion readings per second). By way of contrast, the human ear can respond to minute pressure variations in the air if they are in the audible frequency range, roughly 20 Hz–20 kHz.

## 41.2 ENVIRONMENTAL TIME SERIES APPLICATIONS

Some of the oldest and most famous time series are of environmental variables. Sunspots are slightly cooler regions of the sun that appear as dark spots and arise due to magnetic influences. Sunspots have fascinated astronomers for many centuries and a time series count of the number of sunspots going back to 1750 is shown in Figure 41.1.

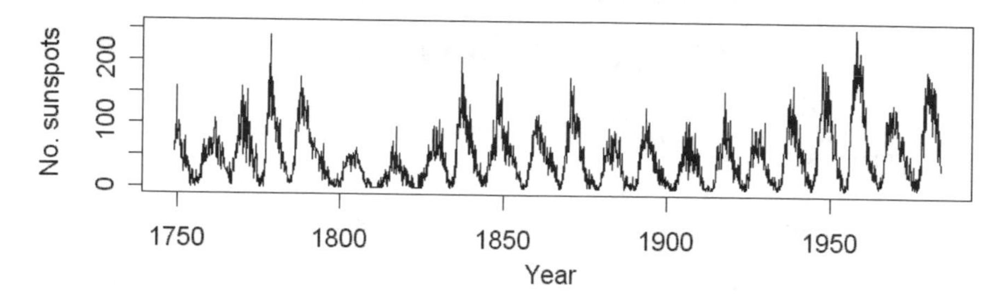


**Figure 41.1** Monthly number of sunspots. (Andrews and Herzberg, 1985)

Another example of a long environmental time series are river levels on the Nile, shown in Figure 41.2, which were used to determine taxes that would correspond to the expected harvest.

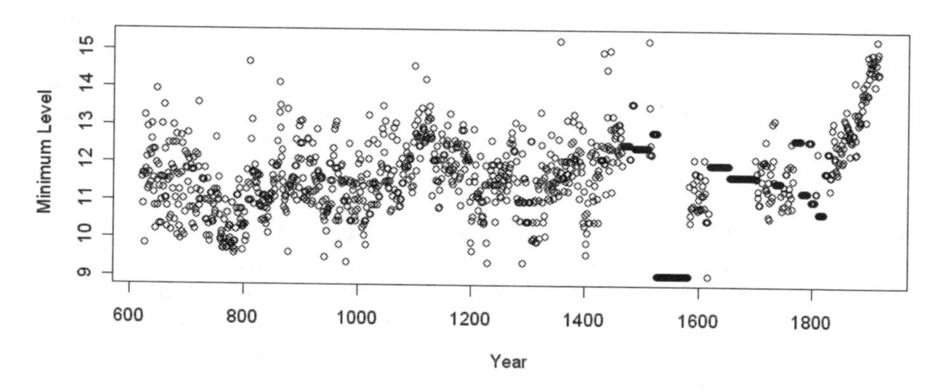


**Figure 41.2** Minimum annual levels in the Nile River. (Hipel and McLeod, 1994)

Today, monitoring environmental variables over time has great significance for the planet. For example Figure 41.3 shows changes in mean sea levels from 1992 to 2005 estimated from measurements by the TOPEX Poseidon satellite project (Leuliette et al., 2004). This figure shows increases in sea levels of approximately 3 mm per year over a 20-year period and, when combined with the increased observations of $CO_2$, is often cited as evidence of a changing climate.

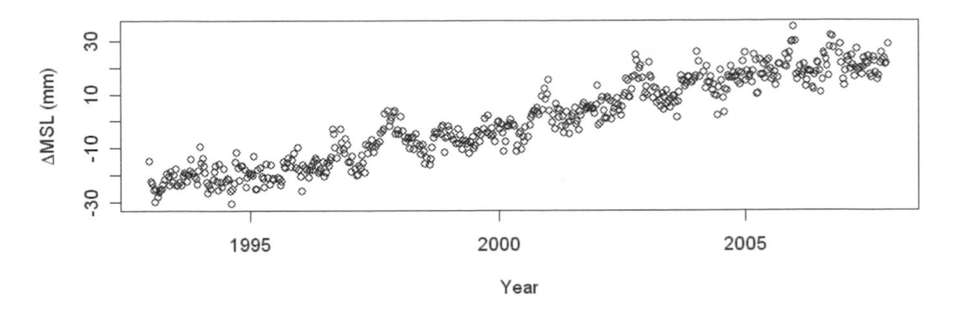


**Figure 41.3** Change in mean sea level (http://sealevel.colorado.edu/, accessed March 2008).

Another time series variable of considerable interest is the Southern Oscillation Index (SOI), shown in Figure 41.4. Although the SOI does not have an increasing trend over time, fluctuations in the SOI are strongly related to the El Niño phenomenon that is associated with the incidence of floods and droughts about the Pacific basin. El Niño events arise due to thermal gradients in the Pacific Ocean. Warmer waters off the coast of South America create moist air and higher rainfall over the Americas, with comparatively drier conditions in Australia. When the warm waters are nearer to the coast of Australia, the Americas experience drier conditions and Australia has comparatively more rain.

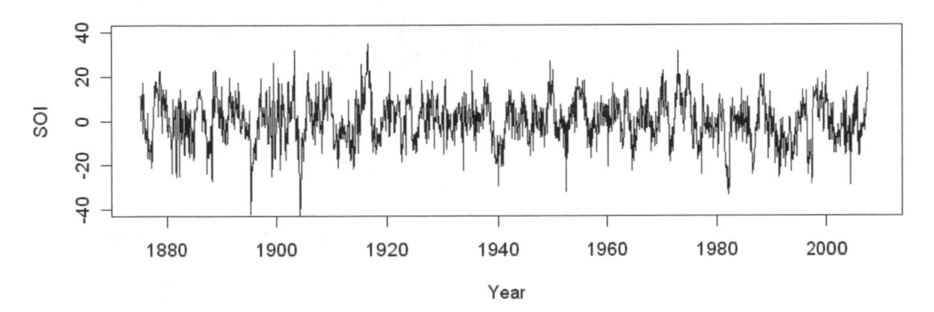


**Figure 41.4** Southern Oscillation Index. (Australian Bureau of Meteorology, 2008)

Figure 41.5 gives monthly mean atmospheric carbon dioxide concentrations for January 1959 until December 2003, measured at the Mauna Loa Observatory, which is at an altitude of 3400 m, in Hawaii. The data are available from the USA National Oceanic and Atmospheric Administration. The top frame is a time-series plot of the original data. The second frame is a moving average of the original data which smoothes out both seasonal and random variation and shows the increasing trend in $CO_2$. An alternative method to detect a linear trend is to fit a simple regression model. The third frame is the seasonal component, which shows a variation that repeats itself on an annual basis. The fourth frame is the remaining variability which is not otherwise accounted for. This is referred to as random variation. Despite deterministic trends being removed, it is possible that the random variation has remaining structure that needs to be modelled. That is, the random quantities may be correlated in time.

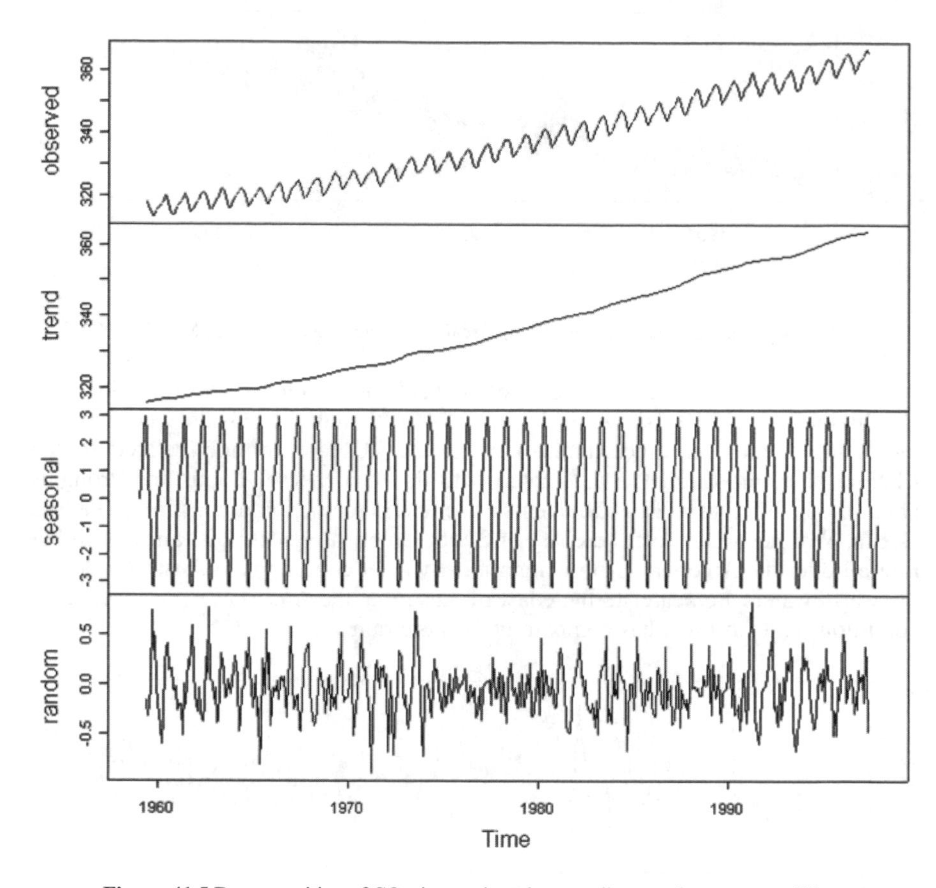


**Figure 41.5** Decomposition of $CO_2$ time series where readings are in parts per million.

## 41.3 CONCEPTS OF TIME SERIES

The general approach to analyse a time series is to estimate a trend and seasonal effects, remove these and investigate the residuals. These residuals may be correlated over time – for example, if the current value is above the mean, then if the data is positively correlated the next value is also more likely than not to be above the mean. An observed time series is considered to be a particular realisation of some underlying random process and a realisation should be thought of as one of the infinite number of time series that might have occurred. This hypothetical infinite collection is referred to as the ensemble and Figure 41.6 gives a schematic example of five possible time series within the ensemble.

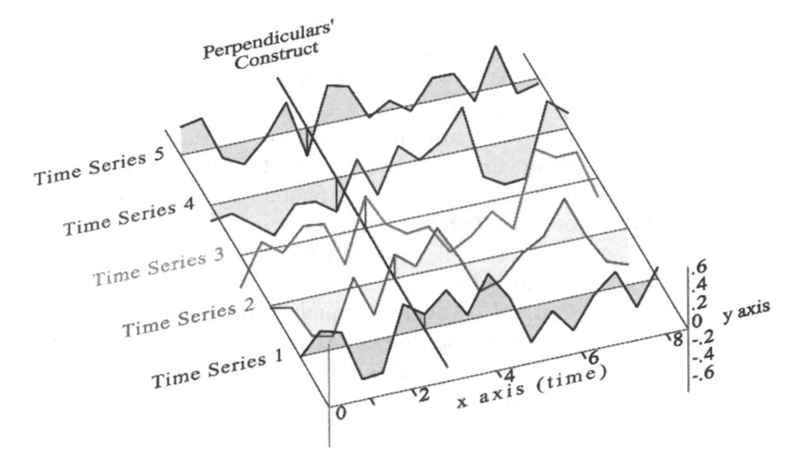


**Figure 41.6** Realisations of a random process.

The aim of time series analysis is to remove all deterministic trends and identify a suitable model for the random process. The random components can then be simulated and all of the deterministic trends are added back in to generate a replicate of the process. A random process, represented by a time series model, is said to be second-order stationary if the mean and variance do not change over time, and the correlation between variables separated by some time lag depends only on that lag and not on absolute time. A stationary stochastic process is ergodic in the mean if the time average of any realisation is the same as the theoretical average across the ensemble at a point in time. This concept is depicted for the perpendicular line in Figure 41.6.

**Ergodicity**

To determine whether a random process is ergodic many realisations would be necessary. This is not feasible for environmental or economic time series, where the historical record is the only realisation. In contrast, in a laboratory experiment to investigate turbulence it is possible to obtain many realisations of the underlying random process. The question is whether these realisations all have the same mean or have a mean that depends, for example, on the initial conditions. It is not always true that the process is ergodic.

## 41.4 MODELLING TRENDS AND SEASONAL EFFECTS

There are many ways to model trends and seasonal effects in time series. However, with some ingenuity, multiple regression is often an adequate means for doing so. Consider again the data in Figure 41.1.

To model the $CO_2$ data the first step is to fit a linear trend and a sine wave of period 1 year. If time $t$ in months is mean-adjusted to have a mean value of 0, this will be an advantage if a quadratic $t^2$ term is included. The sine wave of period 1 year is given by a linear combination of sine and cosine functions having arguments $2\pi t/12$.

Also calculate the sine and cosine functions with argument 4 $\pi t/12$ to allow the addition of a sine wave with a period of 6 months, if this seems necessary. A Matlab script for the regression is provided in Figure 41.7.

The output from this script provides T-ratios which indicate that all the coefficients are large compared with their standard deviations. Therefore, the corresponding variables are worth retaining in the model. The standard deviation of $CO_2$ is 14.97 and the standard deviation of the residuals is 1.70, so the trend and seasonal effects account for a substantial proportion of the variability.

Now the command `plot(t,R)` will plot the residuals against time $t$. The clear quadratic curve in Figure 41.8 indicates that an improved fit would be obtained by adding a $t^2$ term. However, although adding a quadratic term will improve the fit to 1959–2003, one should be very wary about extrapolating it. The next step is to repeat the regression including the quadratic term. The Matlab script is unchanged apart from building the $X$ matrix: `X=[col1 t t2  C1 S1]`.

```
%Time series regression for CO2 (Data stored in CO2)
%Create time indexes
n=length(CO2);
t=1:n; t=t';
col1=ones(n,1);
%Polynomial terms
t=t-mean(t);
t2=t.^2;
%Periodic terms
C1=cos(2*pi*t/12);
S1=sin(2*pi*t/12);
C2=cos(4*pi*t/12);
S2=sin(4*pi*t/12);
%Make predictor matrix
X=[col1 t C1 S1];
%Calculate coefficients
B=(X'*X)^(-1)*X'*CO2;
F=X*B; %Calculate trend
R=CO2-F; %Residuals
%Calculate variance Resid.
temp=size(X);np=temp(1,2);
se2=sum(R.^2)/(n-np);
%Calc. variance predictors
V=(X'*X)^(-1)*se2;
%Calc diagnostic stats
Tratio=B./sqrt(diag(V))
stdCO2 =std(CO2)
stdR=std(R)
```

*output:*
*Tratio = 1.0e+003 *(4.258 0.186 0.0098 –0.0230)*
*stdCO2 = 14.97*
*stdR=1.70*

**Figure 41.7** Matlab regression script for $CO_2$ data.

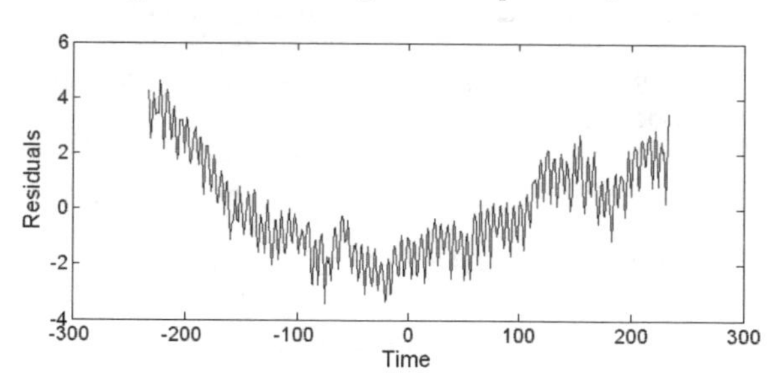


**Figure 41.8** Quadratic trend in residuals.

The next step is to estimate the correlation between the residuals at time $t$ and the residuals at time $t+k$, which is known as the autocorrelation at lag $k$. This is repeated for values of $k$ from 1 up to 20 (Figure 41.9). The correlogram is a plot of the autocorrelations against their lag and is shown in Figure 41.10.

```
n=length(R); % no. of data
R=R-mean(R); % subtract mean
csp0=sum(R.^2);
for k=1:20 %repeat for all lags
  csp(k)=0;
  %summate lag-k covariance
  for i=1:n-k % for all data
    csp(k)=csp(k)+R(i)*R(i+k);
  end
end
csp=csp/csp0; %standardise
var=csp0/(n-1); %variance
%plot results
lag=1:20;       %x-values
plot(lag,csp,'o')
xlabel('lag')
ylabel('autocorrelation')
```

**Figure 41.9** Matlab script to estimate correlogram.

The high correlations at lags 6, 12, and 18 suggest that it is necessary to add the higher-frequency sine wave. Thus, repeat the regression including the two extra explanatory variables at the period $4\pi t/12$. The Matlab script is unchanged apart from building the $X$ matrix: `X=[col1 t t2  C1 S1 C2 S2]`.

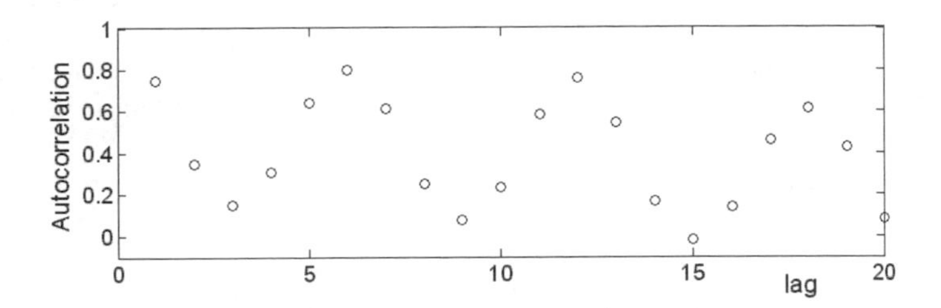


**Figure 41.10** Correlogram of $CO_2$ concentrations (regression excluding 6-month period components).

The plot of the residuals in Figure 41.11(a) now looks more like a realisation of a stationary stochastic process. The correlogram in Figure 41.11(b) suggests that the seasonality has been adequately modelled since the correlations decay toward zero without any strong seasonal component. The autocorrelations of this residual series will be modelled using an autoregressive process in the following section. Although the slow decay of these correlations is best modelled by a long-memory process, only auto-regressive models that have shorter memory will be investigated.

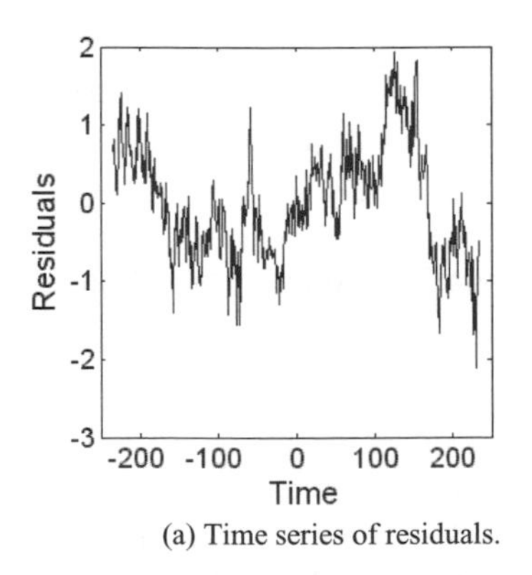


(a) Time series of residuals.

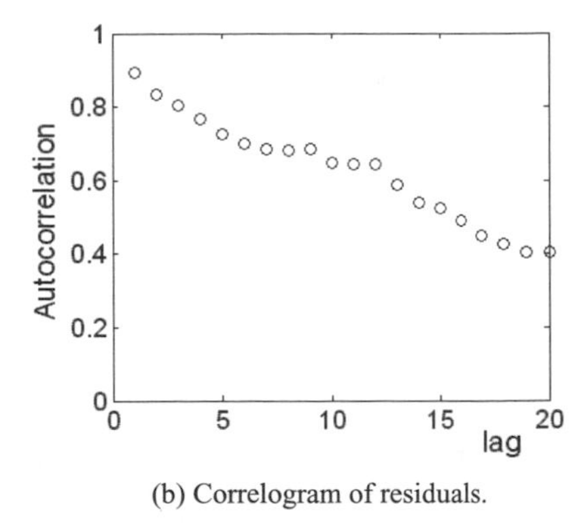


(b) Correlogram of residuals.

**Figure 41.11** Analysis of $CO_2$ residuals (regression including 6-month period components).

## 41.5 AUTOREGRESSIVE PROCESSES

A time series that is a sequence of independent random variables from some probability distribution is called discrete white noise (DWN). The theoretical autocorrelations of DWN are 0 for any non-zero lag. Sample estimates will vary somewhat about 0, especially in small samples. DWN is the basis for constructing autoregressive (AR) processes. An AR process of order 1, AR(1), is defined by:

$$x_t = \alpha\, x_{t-1} + \varepsilon_t \tag{41.1}$$

where $x_t$ is the mean-adjusted time-series variable at time $t$, $x_{t-1}$ is the preceding value, $\alpha$ is a constant between $-1$ and 1 for the process to be stationary and $\varepsilon_t$ is DWN with a mean of zero. This model has two parameters, $\alpha$ and the variance of the DWN, $\sigma_\varepsilon^2$.

The correlogram of the AR(1) model is $\alpha^k$ at lag $k$ and therefore has an exponentially decaying shape. A straightforward estimate for $\alpha$ is the lag-one autocorrelation. This is $\hat{\alpha} = 0.89$ for the $CO_2$ residuals (see lag-one from Figure 41.11(b)). The residuals after fitting the AR(1) process are given by:

$$x_t - \hat{\alpha}\, x_{t-1} = \hat{\varepsilon}_t \tag{41.2}$$

and if an AR(1) model is suitable the residuals should appear as a realisation of DWN. The variance of the residuals is an estimate of $\sigma_\varepsilon^2$. The value of $\sigma_\varepsilon^2$ for the $CO_2$ residuals is 0.105.

An AR(2) model is an extension of an AR(1) model and is defined by:

$$x_t = \alpha_1\, x_{t-1} + \alpha_2\, x_{t-2} + \varepsilon_t \tag{41.3}$$

where the terms are as previously defined. For many values of the parameters $\alpha_1$ and $\alpha_2$, the correlogram of the AR(2) process is a damped sinusoid. It is convenient to fit AR processes by using a regression routine on the triples $(x_{t-2}, x_{t-1}, x_t)$. An alternative method is to solve the Yule-Walker equations:

$$\begin{bmatrix} 1 & r_1 & r_2 \\ r_1 & 1 & r_1 \\ r_2 & r_1 & 1 \end{bmatrix} \begin{bmatrix} 1 \\ -\hat{\alpha}_1 \\ -\hat{\alpha}_2 \end{bmatrix} = \begin{bmatrix} \hat{\sigma}_\varepsilon^2 / s^2 \\ 0 \\ 0 \end{bmatrix} \tag{41.4}$$

for the parameters $\hat{\alpha}_1, \hat{\alpha}_2$ and $\hat{\sigma}_\varepsilon^2$ where $r_k$ is the correlation estimate at lag-$k$ and $s^2$ is the variance of the data being modelled.

*Example*

Consider the time series of wave tank data shown in Figure 41.12(a). They are vertical displacements (mm) from the still water level recorded at 0.1 second intervals by a probe at the centre of the tank. The fluctuations measured by the probe are due to the superposition of many waves travelling in different directions.

Modelling the time series of wave displacements can be important for designing structures such as bridges or off-shore platforms. It is common practice to use scale models of large engineering structures in wind tunnels and wave tanks, so although the scale in this example is in millimetres it may correspond to much larger waves. Figure 41.12(b) shows the correlogram of the wave data and the damped sinusoid suggests that an AR(2) model may be appropriate.

*Solution*

Using the Matlab script to calculate the correlogram, the variance of the wave data is 70955, the lag-one correlation is 0.471 and the lag-two correlation is –0.261, thus Equation (41.4) becomes:

$$\begin{bmatrix} 1 & 0.471 & -0.261 \\ 0.471 & 1 & 0.471 \\ -0.261 & 0.471 & 1 \end{bmatrix} \begin{bmatrix} 1 \\ -\alpha_1 \\ -\alpha_2 \end{bmatrix} = \begin{bmatrix} \sigma_\varepsilon^2/70955 \\ 0 \\ 0 \end{bmatrix} \tag{41.5}$$

which is 3 parameters with 3 unknowns. Solving this system gives $\hat{\alpha}_1 = 0.763$, $\hat{\alpha}_1 = -0.621$ and $\hat{\sigma}_\varepsilon^2 = 33956$. After fitting this model the correlogram of the residuals should be checked to verify that they approximate DWN.

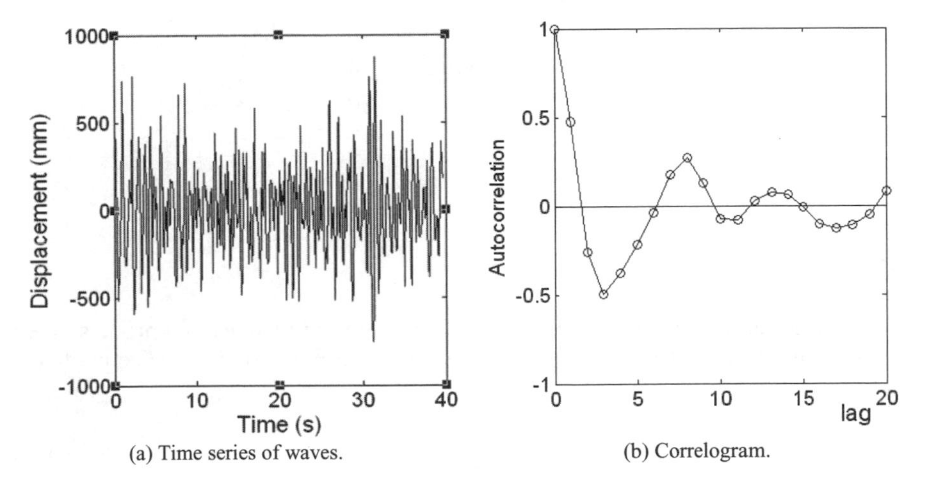


(a) Time series of waves.

(b) Correlogram.

**Figure 41.12** Analysis of wave tank data.

## 41.6 SIMULATION

To simulate a time series the starting point is to make a histogram of the residuals after fitting the AR process. A plausible probability distribution must then be chosen. It is convenient if this distribution is normal and it may be worth using a transform such as logarithm or square-root of the data to achieve normality. Initial values of zero for the process variable are used to start the procedure. Subsequent values are computed using the autoregressive equation with an additional value of

DWN. This continues for the desired length of the simulation. The trends and seasonal components are then added back in the reverse order to which they were removed. An example is shown in Figure 41.13 using the wave tank parameters in Excel. The *alpha1* and *alpha2* names in the autoregressive expression correspond to cells G4 and G5 respectively. The DWN data are assumed normal and are generated using the formula `=norminv(rand( ),0,var_e)` where *var_e* is the name corresponding to cell G6. The names for these cells can be defined under the menu item Insert – Name – Define. The `rand( )` function will generate a new uniformly distributed random number every time the F9 key is pressed. This gives a new realisation.

| | A | B | C | D | E | F | G |
|---|---|---|---|---|---|---|---|
| 1 | | | | | | | |
| 2 | | **Time step** | **DWN** | **x_t** | | | |
| 3 | | initial | - | 0 | | **Parameter** | **Value** |
| 4 | | initial | - | 0 | | $\alpha_1$ | 0.761 |
| 5 | | 1 | -364.23 | -364.23 | | $\alpha_2$ | -0.621 |
| 6 | | 2 | 530.87 | 253.69 | | $\sigma^2_\varepsilon$ | 33956 |
| 7 | | 3 | -230.58 | 188.66 | | | |
| 8 | | 4 | 148.95 | 134.99 | | | |
| 9 | | 5 | -131.60 | =alpha1*D10+alpha2*D9+C11 | | | |
| 10 | | 6 | 91.34 | ↓ | | | |
| 11 | | 7 | -67.31 | | | | |
| 12 | | 8 | 156.73 | | | | |
| 13 | | 9 | =norminv(rand(),0,sqrt(var_e)) | | | | |
| 14 | | ↓ | ↓ | | | | |
| 15 | | | | | | | |
| 16 | | | | | | | |

**Figure 41.13** Excel implementation of an AR(2) model.

## 41.7 FORECASTING

AR models can be used to make forecasts of likely future values of a process given the most recent past. Consider the AR(1) model from Equation (41.2). To predict $x_t$ at time $x_{t-1}$ use:

$$x_t = \hat{\alpha} x_{t-1} \tag{41.6}$$

where $\alpha$ and $x_{t-1}$ are both known values and the DWN has a mean value of zero. Since the DWN has the variance $\hat{\sigma}^2_\varepsilon$ the exact value of $x_t$ is not known and the forecast is given by a probability distribution. If the DWN data are normal the one-step-ahead forecast is specified by the distribution $X_t \sim N(\hat{\alpha} x_{t-1}, \hat{\sigma}^2_\varepsilon)$. Moving forward another time step, the expected value at time $t+1$ can be determined from time $t$, which is $E[x_{t+1}] = \hat{\alpha}^2 x_{t-1}$. The variability at this time step will be increased since the exact value of the process $x_t$ is itself unknown. If the DWN is Gaussian, then the variance is $(1+\hat{\alpha}^2)\sigma^2_\varepsilon$ and the forecast is $X_{t+1} \sim N(\hat{\alpha}^2 x_{t-1}, (1+\hat{\alpha}^2)\hat{\sigma}^2_\varepsilon)$. By extension, the forecast $k$ steps ahead is:

$$X_{t+k-1} \sim N(\hat{\alpha}^k x_{t-1}, \frac{1-\hat{\alpha}^{2k}}{1-\hat{\alpha}^2}\hat{\sigma}_\varepsilon^2). \quad (41.7)$$

The mean of this distribution decays exponentially toward zero for increasing $k$ and the variance increases toward $s^2$ which is the variance of the data being modelled. Thus unless the value $|\alpha|$ is exceptionally close to 1 the forecast will quickly come to reflect an ignorant knowledge of the future, with a the mean of the forecast close to zero and the variance of the forecast tending toward variance of the data itself after only a few time steps. To illustrate this a forecast of an AR(1) model having $\hat{\alpha} = 0.9$ and $\hat{\sigma}_\varepsilon = 0.3$ was performed for 16 time steps as shown in Figure 41.14. The variability is indicated by the dashed lines which approximately show the 95% region within which the forecast value is likely to lie. It is clear that the forecast tends toward zero and that the variability increases when the range of the forecast is increased. The forecast process is similar for the AR(2) model with the exception that the mean and variance of the forecast are determined by regression on the two preceding values.

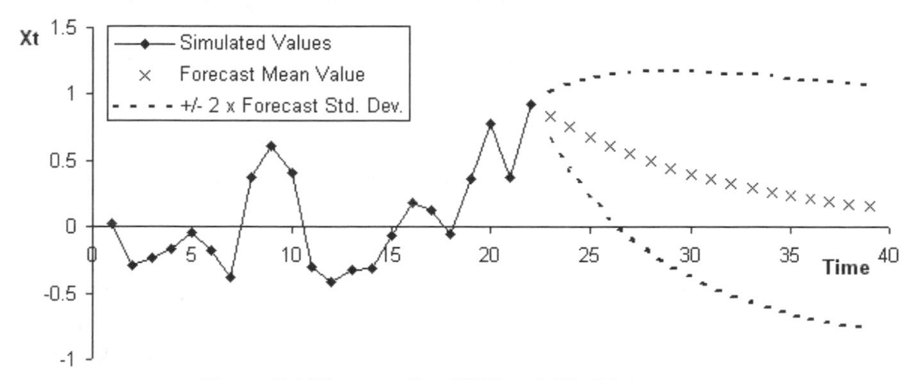


**Figure 41.14** Forecast of an AR(1) model for 16 time steps.

## 41.8 SUMMARY

Time series are unique in that events can be ordered and models can be developed to exploit this. Quite often time series are recorded with linear and seasonal trends so that it is necessary to decompose the time series into deterministic and random components, which are modelled separately before being recombined to give the simulation model. Linear regression was demonstrated to be suitable for estimating deterministic trends and auto-regressive models were shown to allow the random fluctuations of a dataset to be correlated in time. The auto-regressive model specifies that the current value is determined by the most recent history of values along with a random error. This procedure can be extended to allow future values to be forecast, but for most environmental series where the correlation is not strong the usefulness of this forecast diminishes quickly. The time series models demonstrated here are suitable for linear trends and for non-skewed stationary random fluctuations. There are, however, a variety of time series models that allow for the broad range of phenomena observed in measured time series.

## PROBLEMS

**41.1** For the SOI data in Figure 41.5 fit an AR(1) model. Build a simulation model in Excel. Plot various realisations of this model.

**41.2** Fit an AR(2) model to the SOI data. Compute the correlogram of the residuals and compare this correlogram to the that of the AR(1). Which model gives a better fit to the data and why?

**41.3** Use the Matlab script to fit a linear trend to the sea level data in Figure 41.3. Inspect the correlogram of the residuals for any periodic components. Would a periodic trend give an improved fit?

**41.4** Fit an AR(1) model to the residuals of the sea level data and build a simulation model. Do not forget to include the trend after modelling the AR process.

**41.5** Using a fitted AR(1) model to the sea level data? What is the probability that the anomaly will be above 40 mm 5 years from the end of the record? This will require the forecast distribution to be determined.

**41.6** Use the Matlab regression script to fit a regression of sine waves at 11, 12 and 13-year frequencies. Plot a histogram of the residuals. Are they normal? How would you build and AR model?

**41.7** The correlogram in Figure 41.11(b) decays rather more slowly than exponential decay and an AR(2) model gives an improved fit. Estimate parameters for this model, simulate from the model and calculate the correlogram. Compare this correlogram to the correlogram from an AR(1) model.

CHAPTER FORTY TWO

# Optimisation (Local Optimisation)

## 42.1 INTRODUCTION

Optimisation can be described as the problem of locating the best solution from a wide range of possibilities – in other words, finding the value of $x$ such that the function $f(x)$ is a maximum (or minimum). Whether it is a maximum or a minimum is governed by the problem; for example, the maximum reliability of a system or the minimum error in a model.

There are many different types of optimisation algorithms and it is best to understand a little about their different approaches. One way optimisation can be thought of is finding the root of the derivative, $f'(x) = 0$. Therefore, a root-finding method such as the Netwon-Raphson method can be applied, giving:

$$x_{i+1} = x_i + \frac{f''(x)}{f'(x)} \tag{42.1}$$

This method requires knowledge of the derivative and second derivative. Another similar method which also uses derivative information is the method of steepest ascent, which takes a step proportional to the gradient:

$$x_{i+1} = x_i + \lambda\, f'(x) \tag{42.2}$$

where $\lambda$ is some number to control the step size. Both of these algorithms are hill climbers that, given a starting point, will keep moving up a function and then stop at the point of zero derivative. The algorithms are also deterministic in that the same starting point will always give the same end-point. Often, however, this end-point turns out to be a local maximum as depicted in Figure 42.1. For this reason these optimisers are referred to as local optimisers. Short of evaluating every point in the function there is no guarantee that the best solution, the 'global optimum', has been found. Note also that it is possible that a local (or global) optimum can occur at the boundary of a search region. In this instance the derivative is not zero and it implies that a better solution could be found if the search region is extended.

What is often done in practice is to introduce a stochastic component; for example, a local optimisation could be repeated from multiple random starting points, from which the best final solution is chosen. While there is no guarantee that the global optimum is found, the usual practice is to hope that a near-global solution is found. After multiple searches it may be that the best found solutions are of a similar quality such that it is assumed a more exhaustive search will be unlikely to find an improvement in $f(x)$. While users often apply this logic as a justification that they *must* be at a near-global (if not global) solution, for anything but the simplest problems, it is always possible that there is a significantly better solution 'out there' that is worth searching for.

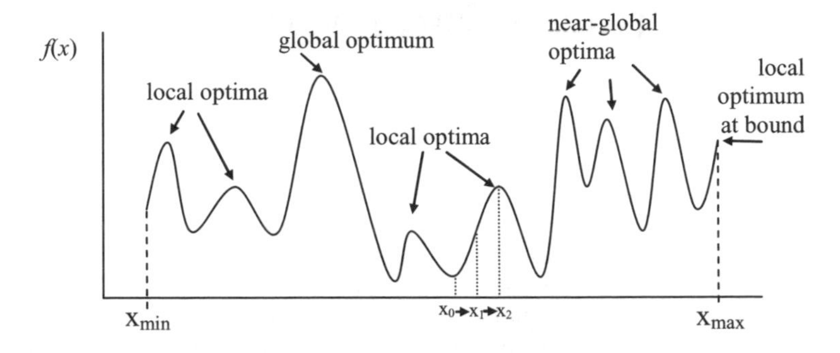


**Figure 42.1** Schematic of local, near-global and global optima of a function.

The main advantage of gradient-based local optimisers is that they are quick. Their main limitations are (i) their tendency toward local optima; (ii) the requirement of function gradients; and (iii) step-size and numerical issues similar to those of open root-finding methods.

## 42.2 THE SIMPLEX METHOD

An alternative local optimiser that does not use gradient information is the simplex method devised by Nelder and Mead (1965). This algorithm is deterministic so that the same starting point will always reach the same end-point. It is presented here as a maximiser for a function of two variables $f(x,y)$, although it is readily extended to higher dimensions or used as a minimiser. For two variables a simplex is a triangle and the function is evaluated at the three vertices of this triangle. The vertices are ordered according to their function value. Following Mathews and Fink (2004) the ranked vertices are labelled $B = (x_1,y_1)$, $G = (x_2,y_2)$ and $W = (x_3,y_3)$, corresponding to the best solution, a good solution, and the worst solution respectively. The function value at the mid-point between the best two vertices is then evaluated: $M = 0.5\,(B+G)$. A new triangle is formed based on the configuration of these points according to a variety of scenarios.

*Reflection*

The function value will increase moving along the sides of the triangle from $W$ to $B$ or from $W$ to $G$. Therefore, it is possible that the function takes on larger values for points that lie away from W but on the other side of the line between $B$ and $G$.

Consider the point $R$, shown in Figure 42.2, which is obtained as a reflection of the triangle through the mid-point: $R = M + (M-W)$ at a distance $d$.

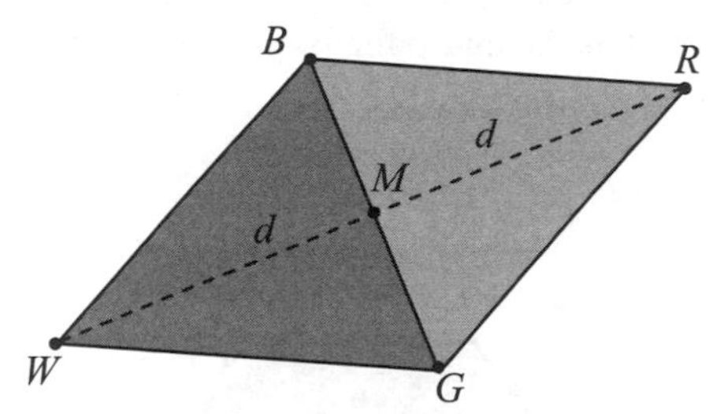


**Figure 42.2** Reflection of simplex through mid-point.

*Expansion*
If the function value at $R$ is larger than the function value at $W$, the simplex has moved in the correct direction. The line segment is then extended to the point $E$ under the possibility that the minimum lies in this direction. This forms an expanded triangle $BGE$ as shown in Figure 42.3. The point $E$ is found by moving an additional distance $d$ along the line joining $M$ and $R$ so that $E = R + (R-M)$.

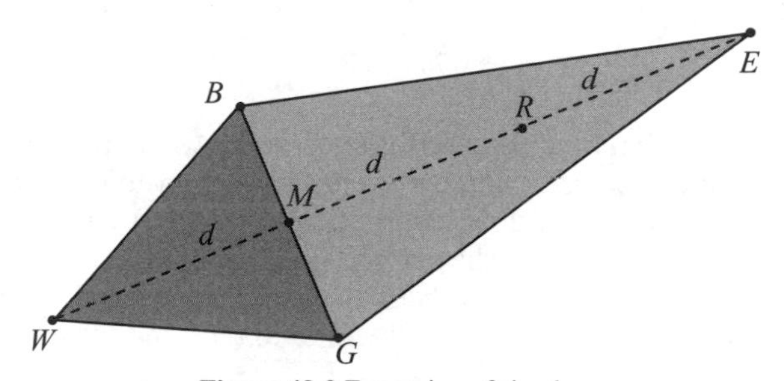


**Figure 42.3** Expansion of simplex.

*Contraction*
If the function values at $R$ and $W$ are similarly poor, another point must be tested. It is possible that the function is larger at $M$, but $M$ cannot replace $W$ since the coordinates must form a triangle. Consider the two mid-points $C_1$ and $C_2$ as shown in Figure 42.4. Both points are tested, but only the point with the larger function value is used so that the new triangle becomes $BGC$.

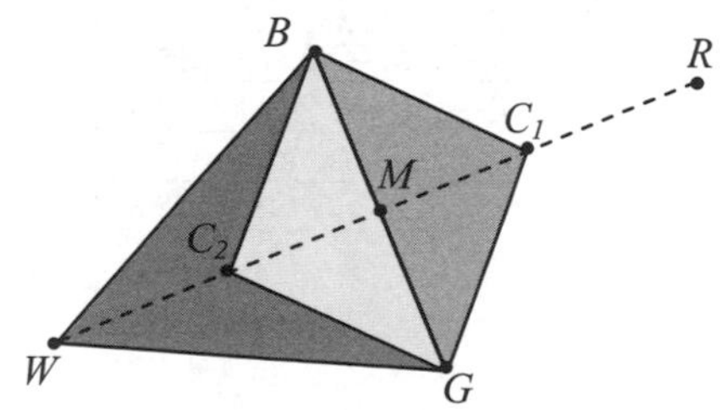


**Figure 42.4** Contraction of simplex.

*Multiple Contraction*

If the function value at $C$ is no better than the value at $W$, the points $G$ and $W$ must both be contracted toward $B$ (see Figure 42.5). The point $G$ is replaced with $M$, and $W$ is replaced with $S$, which is the mid-point of the line segment joining $B$ with $W$.

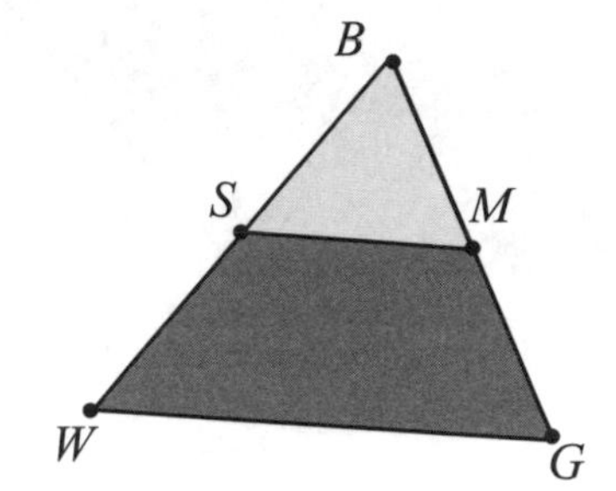


**Figure 42.5** Multiple contraction of simplex.

Depending on the shape of the function one of these scenarios will apply so that a new triangle can be generated and the algorithm can proceed. A Matlab implementation of the 2D simplex method is given in Figure 42.6 and it is applied to the Himmelblau function:

$$f(x,y) = (x^2 + y - 11)^2 + (x + y^2 - 7)^2 \qquad (42.3)$$

which has a global minimum of zero at (3,2) and is elsewhere positive. By negating the function value it can be used with the maximising algorithm in the Matlab code. Starting the optimiser from the three coordinates (–1, –4.5), (1,–4.5) and (0, –50) gives the sequence of moves as shown in Table 42.1. Figure 42.7 shows these moves graphically and demonstrates that some of them are reflections and others contractions. This figure also shows that the function has four near-global optimum values.

**Table 42.1** Moves made by simplex method over successive iterations

| Iter. | Best point | Good point | Worst point |
|---|---|---|---|
| 1 | f(–1.00, –4.50) = –360.3 | f(1.00, –4.50) = –413.3 | f(0.00, –5.00) = –580.0 |
| 2 | f(0.00, –3.50) = –237.8 | f(–1.00, –4.50) = –360.3 | f(1.00, –4.50) = –413.3 |
| 3 | f(–3.50, –3.00) = –5.3 | f(0.00, –3.50) = –237.8 | f(–1.00, –4.50) = –360.3 |
| 4 | f(–3.50, –3.00) = –5.3 | f(–2.50, –2.00) = –75.8 | f(0.00, –3.50) = –237.8 |
| 5 | f(–3.50, –3.00) = –5.3 | f(–2.50, –2.00) = –75.8 | f(–1.50, –3.00) = –138.3 |
| 6 | f(–3.50, –3.00) = –5.3 | f(–3.75, –2.25) = –33.0 | f(–2.50, –2.00) = –75.8 |
| 7 | f(–3.50, –3.00) = –5.3 | f(–4.19, –2.94) = –19.4 | f(–3.75, –2.25) = –33.0 |
| 8 | f(–3.50, –3.00) = –5.3 | f(–3.94, –3.69) = –7.7 | f(–4.19, –2.94) = –19.5 |
| 9 | f(–3.95, –3.14) = –3.4 | f(–3.50, –3.00) = –5.3 | f(–3.94 –3.69) = –7.7 |

```
function opt = simplex(B,G,W)  % Nelder-Mead simplex method
  % maximises 2D function f, % B,G,W are vectors with 2 elements
  % Code stored in file simplex.m

  % Rank B,G,W: User probably put them in a random order
  fb = f(B); fg = f(G); fw = f(W);
  if(fb<fg) % swap if needed
    temp = G; G = B; B = temp; temp = fg; fg = fb; fb = temp;
  end
  if(fg<fw)  % swap if needed
    temp = W; W = G; G = temp; temp = fw; fw = fg; fg = temp;
  end
  if(fb<fg)  % swap if needed
    temp = G; G = B; B = temp; temp = fg; fg = fb; fb = temp;
  end
  % update simplex coordinates
  for i = 1:10000 % max. iterations 10,000
    M = 0.5*(B+G); fm = f(M); % midpoint
    R = 2*M-W;     fr = f(R); % reflected value
    d = sqrt(sum((M-W).^2));  % distance
    if(fr>fg) % reflect or extend
      if(fb>fr)                              % new simplex B,R,G
        W = G; fw = fg; G = R; fg = fr;
      else
        E = R+R-M; fe = f(E); % extension
        if(fe>fb)                            % new simplex E,B,G
         W = G; fw = fg; G = B; fg = fb; B = E; fb = fe;
        else                                 % new simplex R,B,G
         W = G; fw = fg; G = B; fg = fb; B = R; fb = fr;
        end
      end
    else   % contract or shrink
      if(fw>fr); R = W; fr = fw; end % determine best side, C1 or C2
      C = 0.5*(M+R); fc = f(C); % contraction
      if(fc>fr)
         if(fc>fb)                            % new simplex C,B,G
           W = G; fw = fg; G = B; fg = fb; B = C; fb = fc;
         elseif(fc>fg)                        % new simplex B,C,G
           W = G; fw = fg;G = C; fg = fc;
         else                                 % new simplex B,G,C
           W = C; fw = fc;
         end
      else
        S = 0.5*(B+W); fs = f(S); % shrink
        if(fm>fs)                            % new simplex B,M,S
          G = M; fg = fm; W = S; fw = fs;
        else                                 % new simplex B,S,M
          G = S; fg = fs; W = M; fw = fm;
        end
      end
    end
    if(d<0.0001); break; end % exit criteria, simplex is small
  end % loop
  opt = B; % return the best point
```

```
function ans = f(x); % Himmelblau function stored in file f.m
% x is a vector having 2 elements
ans = -((x(1)^2+x(2)-11)^2+(x(1)+x(2)^2-7)^2);
```

**Figure 42.6** Matlab implementation of simplex method.

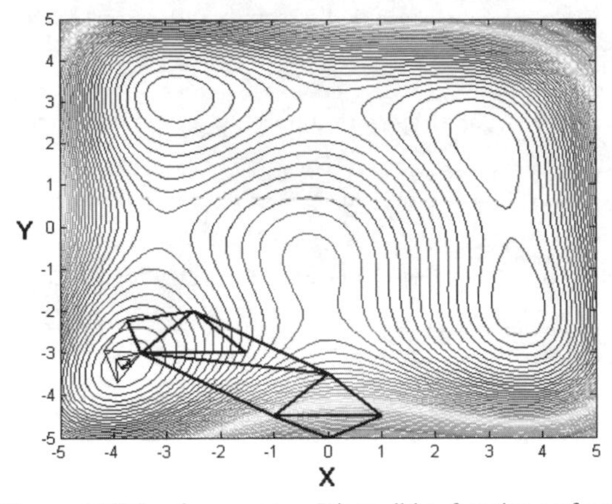


**Figure 42.7** Simplex moves on Himmelblau function surface.

## PROBLEMS

**42.1** If a local optimiser obtains a solution that is at the boundary of the search region what does this suggest?

**42.2** What are some limitations encountered using deterministic optimisation algorithms that are based on gradients of some function.

**42.3** Apply the simplex method to the Himmelblau function for different starting positions. Is the same optimal solution always found?

**42.4** Optimise the Rosenbrock function $f(x,y) = (1-x)^2+100(y-x^2)^2$.

**42.5** Plot a contour map of the Rosenbrock function on the interval. Change the code to output the simplex coordinates at each iteration. Print the contour plot and draw the simplexes onto the plot.

CHAPTER FORTY THREE

# Optimisation (Global Optimisation)

## 43.1 INTRODUCTION

Global optimisers are an alternative to local optimisers that do not share their disadvantages, but come at the cost of high computational demand. These optimisers are stochastic to encourage random exploration of the function, but also use a feedback mechanism so that promising areas are searched more thoroughly.

Many global optimisers have been inspired by real-world processes and the behaviour of these algorithms can therefore be explained by analogy. Examples include: (i) genetic algorithms that have analogy to evolutionary processes, (ii) ant colony algorithms that mimic the behaviour of ants foraging for food, and (iii) simulated annealing that has an analogy to the heating and controlled cooling of metals to increase the size of crystals. Genetic algorithms and ant colony optimisation are suited to combinatorial problems and are discussed here as a point of interest. The main focus is to present an implementation of the simulated annealing algorithm which is suitable for continuous search spaces.

## 43.2 COMBINATORIAL OPTIMISATION

Consider the problem of designing an urban water supply by choosing *n* pipes where there are *m* different options of pipe to choose from. These options may be related to the type of material used (PVC or concrete-lined iron) or may represent different diameters, since larger pipes allow for better pressure and more flow. A schematic example is given in Figure 43.1.

A solution then is a collection of pipes that is able to satisfy the demand from all customers. Smaller pipes or cheaper materials may lead to an overall lower cost but may not satisfy customer demand. Larger pipes and expensive materials may satisfy the demands easily but will be unnecessarily expensive. The optimum solution therefore is to find the cheapest solution that is also able to satisfy demand. However, this objective is not straightforward since there are $m^n$ combinations of pipe options that constitute a supply network. Designing a small network of 10 pipes with 8 pipe options therefore has $8^{10}$ possible arrangements. If a computer model took 1 second to evaluate each option this would require about 30 years of computing time to evaluate all options and find the global optimum!

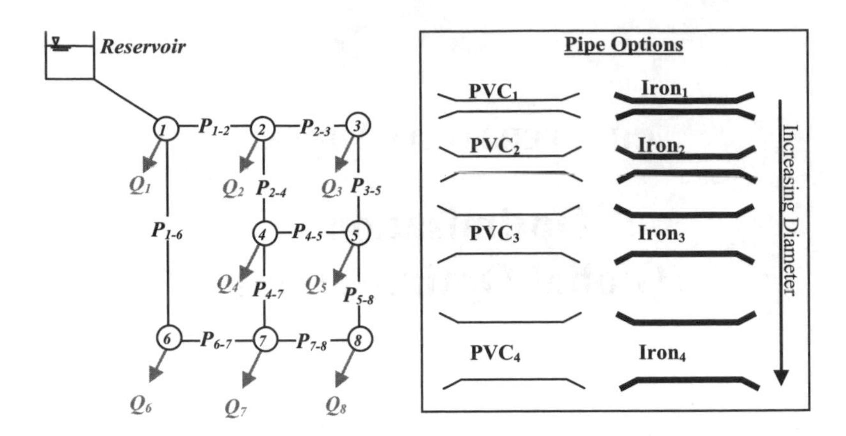


**Figure 43.1** Schematic 10-pipe network. $Q_i$ is customer demand at node *i*. $P_{i\text{-}j}$ is the pipe between nodes *i* and *j*, which can be chosen from one of 8 options that vary in material and diameter.

Genetic algorithms have been applied to the problem of optimising water distribution systems. The main concept in genetic algorithms is "survival of the fittest", where the solutions are made to compete by selecting the fittest to become parents of new solutions so that the quality of solutions evolves over time. A binary genetic algorithm requires each option to be converted into a binary number, thus the options 1 to 8 would be represented by four-digit numbers 000, 001, 010, 011, 100, 101, 110, 111. The coded options are strung together to make what is referred to as a chromosome. Thus 110100010100111110100101001000 would refer to the option set $x$={7,5,3,5,8,7,5,6,2,1} for the ten respective pipes.

The procedure starts with a population of chromosomes. The chromosomes are decoded to determine which options they represent. The "fitness" of each option is determined by the function $f(x)$ and the chromosomes are forced to compete so that only the best solutions are retained. Pairs of these solutions are used as parents to create offspring that will have similar but not identical properties. The offspring are created by a process called cross-over whereby a random point along the chromosome is chosen and the two parents swap information after this point to create two new chromosomes (Figure 43.2).

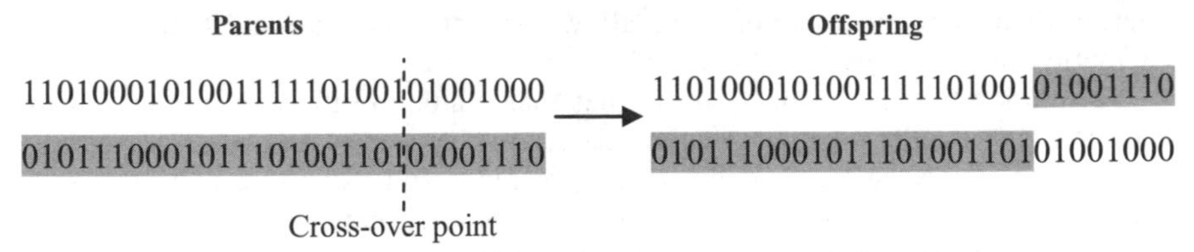


**Figure 43.2** Cross-over of two parent chromosomes to produce offspring.

Another process is mutation where, with very low probability, a digit is selected and changed from a 0 to a 1 or vice versa. Because the children are different from their parents the search will cover different regions of the function. Because the children are in some ways similar to their parents and because only the fittest are allowed to produce offspring, the outcome after numerous generations of offspring is to evolve the solutions to, hopefully, a near-global quality.

As another example consider the classic travelling salesperson problem, whereby a salesperson must complete a circuit that visits a schedule of cities and then return home. Each city is visited once, but the order in which cities are visited is not important. The problem is to find the trip that takes the shortest overall length. If there are $n$ cities, then assuming that it is possible to travel to any city from any other city, there are $(n-1)!$ possible routes that can be travelled. This becomes a large number very quickly, such that it is infeasible to consider all possible routes. One algorithm might be to travel to the next-closest city on each part of the journey, and it is aptly named the greedy algorithm. The greedy algorithm is literally near-sighted and a short-term gain does not always imply long-term good. Consider Figure 43.3, where the circles are cities to be visited. At point A the greedy algorithm chooses the closest city at C and keeps moving until point D when there is only 1 city remaining at point B. To make the journey up to B and then back to the start makes the overall journey longer than if the journey went from point A to point C via point B.

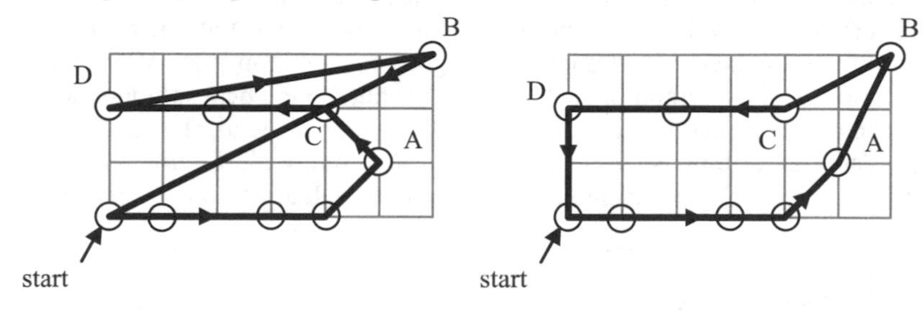


**Figure 43.3** Illustration of greedy algorithm causing an overall longer journey.

Ant colony algorithms were designed with the travelling salesperson problem in mind. Ant colonies are well known to be able to find a source of food despite the fact that the ants are blind and that they do not directly communicate with each other. Ants deposit a chemical substance, pheromone, on the paths where they walk and their antennae are able to detect the relative strengths of pheromone. If a good food source is found then a reliable pheromone trail can be laid to indicate that source to others. Consider the diagram in Figure 43.4, where there are two different routes from an ant nest to a food source, and where the problem is to find the shortest route to the food and back.

At first there may be no preference and the ants choose either path randomly. If all ants lay down a constant amount of pheromone as they move along, then when the ants at the third time step make their return journey there will be more pheromone on the shorter path and they will be more likely to take this route. As this feedback process continues, more and more ants take the shorter path and the longer path is forgotten.

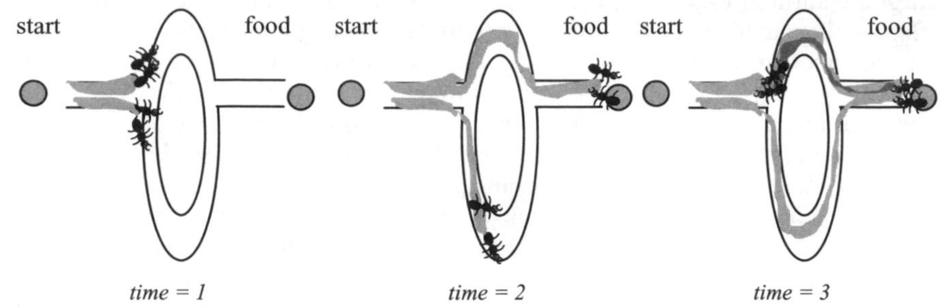


**Figure 43.4** Ants laying pheromone in their search for food.

The travelling salesperson problem can be represented similarly to the nest-food diagram depicted in Figure 43.5. In this figure there are multiple decision points along a journey and once a city is selected it is removed from the decision graph. Each ant chooses paths randomly, but in proportion to the amount of relative pheromone between the paths. Visiting all cities generates a set of options. The fitness $f(x)$ is then evaluated (shorter paths are better) and the ant puts down an amount of pheromone proportional to how good the solution was. After many ants pass over the decision graph the paths that contribute to the best solutions will have more pheromone, enabling near-global optima to be found (hopefully). While global search algorithms such as genetic and ant colony algorithms require many repetitions, this number is considerably lower than evaluating all possible solutions.

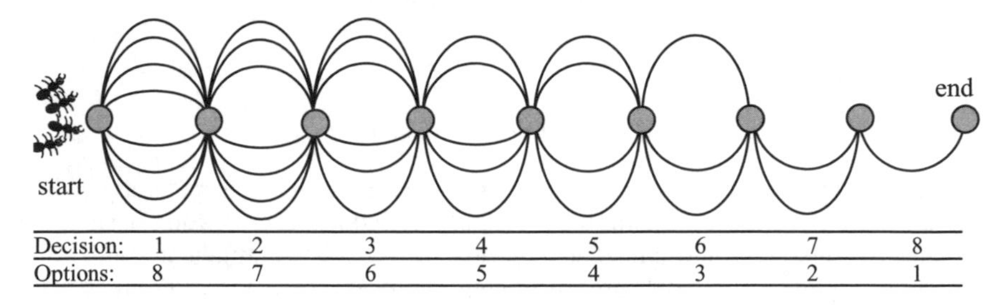


**Figure 43.5** Decision graph. Nodes are decisions, lines are options. Ants traverse the paths to decide which cities to travel to. Options reduce after a city is selected.

## 43.3 SIMULATED ANNEALING

Annealing involves the heat treatment of metals so that the structure of the metal lattice is altered. This causes changes in the metal properties such as ductility, strength and hardness. Annealing requires the metal to be heated and then gradually cooled. Heating causes atoms to vibrate, get displaced and move randomly through states of higher energy. Slow cooling allows the atoms to find stable configurations with lower internal energy than the initial state. For this reason, simulated annealing is a minimiser.

Simulated annealing, by analogy with metal atoms, causes a solution to be randomly displaced. The function value of the solution is referred to as the energy such that a decrease in energy (i.e. an improved solution) is automatically selected. If a solution has a higher energy (i.e. it is worse) it is accepted with a probability, $\alpha$, that depends on the difference in energy and on the global temperature, $T$:

$$\alpha = \exp\left(\frac{f(x_{proposed}) - f(x_{current})}{T}\right) \tag{43.1}$$

The initial temperature is set to be high so that at the start there is little discrimination if a solution is poor. The reasoning for this is that it is important initially to encourage a broad search of the area and that temporarily accepting a worse solution is like crossing a valley to get over to another hill. Gradually the temperature is reduced so that there are fewer worse solutions that are accepted. This implies that the number of successful jumps decreases and the searching gets more concentrated in a given area. Figure 43.6 shows a schematic for a two-dimension search region using a bivariate normal jump distribution centred on the current solution. At the start, the jumps are erratic and are like a random search but as the temperature decreases the jump distribution is less likely to move.

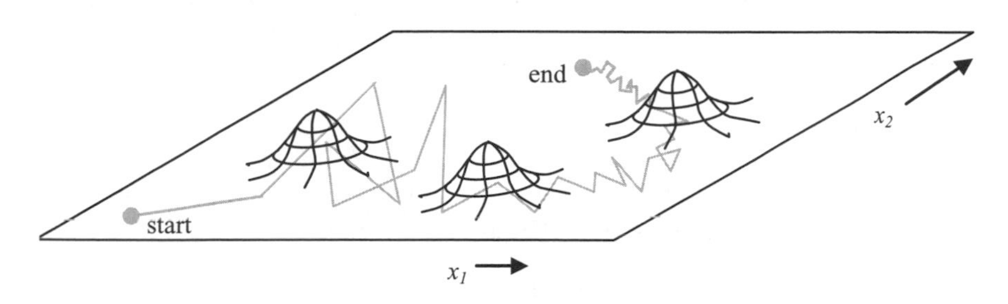


**Figure 43.6** Schematic of simulated annealing in two dimensions.

There are various cooling schedules that can be used to control the temperature. Two of the most common are shown in Figure 43.7, which shows schedules for a linear decrease in the temperature and an exponential decrease in the temperature.

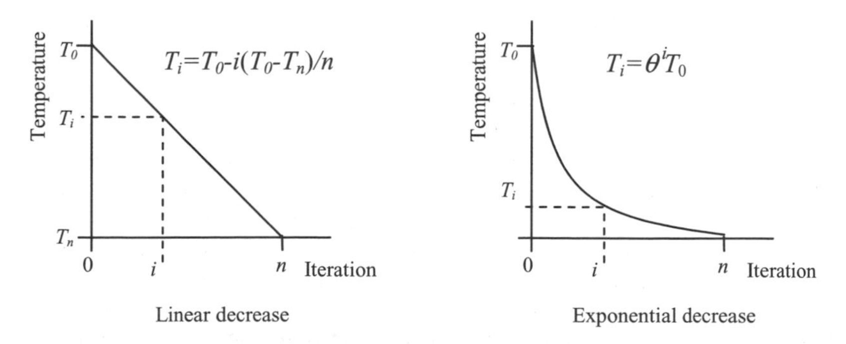


**Figure 43.7** Cooling schedules for simulated annealing.

Simulated annealing is an adaptation of the Metropolis algorithm (Chapter 36) where (i) a proposal is made, (ii) it is automatically accepted if it is better and (iii) if it is worse then it is accepted only with a certain probability. The accepted values form a Markov chain, but whereas the Metropolis algorithm is concerned with the probability distribution of this chain, simulated annealing is concerned only with the maximum value obtained along this chain.

To implement this algorithm it is necessary to specify the number of dimensions, the boundaries of each dimension, the form of the energy function, the size of the jump distribution, the number of iterations and the parameters that control the behaviour of the simulated annealing algorithm. As simulated annealing is a minimiser it may be necessary to multiply the energy function by a negative number or take a reciprocal to ensure that better solutions correspond to lower energies if a maximising algorithm is used. The parameters of the annealing algorithm are the initial temperature and the final temperature. A common method for setting the initial temperature is to make the first jump have an acceptance of, say, 0.9. If a linear cooling schedule is used, then the final temperature can be made zero. If an exponential schedule is used, then a typical setting might be $\theta = 1-5/n$, where $n$ is the number of iterations. The number of iterations to use will depend on the size of the problem and the speed at which the energy function can be evaluated. If an optimisation run does not take a long time, then it is always possible to increase the number of iterations by a factor of 10 and see if a significantly better solution is obtained.

A simulated annealing implementation is provided in Figure 43.8, where a multivariate normal jump distribution is used. Because the jump distribution is centred on the current state it is most likely to jump to a solution that is near this location. The standard deviation of the jump distribution has been set to be different in each dimension depending on the scale of that dimension. For this implementation the standard deviation has been fixed as one 100th of the distance (*upper boundary – lower boundary*). There is an issue of what to do when a jump goes out of bounds, which has been addressed here by wrapping the lower boundary with the upper boundary. Thus if a jump exceeds the upper boundary by 0.03, it is set to the lower boundary plus 0.03.

As an example implementing this code, consider the generalised Rosenbrock function:

$$f(x) = \sum_{i=1}^{m-1} \left[ (1-x_i)^2 + 100(x_{i+1} - x_i^2)^2 \right] \tag{43.2}$$

where $m$ is the number of dimensions of the problem. A dimension of 10 is used and the upper and lower bounds for all dimensions are set to ±10. Figure 43.9(a) is a histogram of 100 'optimal' solutions generated by repeating the optimisation from 100 different starting points. Figure 43.9(b) is a histogram generated using a purely random search (all solutions accepted). Comparing the two shows that 10,000 iterations of simulated annealing are capable of finding significantly better solutions than 10,000 random guesses. However, there is a high proportion of times when simulated annealing performs worse. This phenomenon is termed "sub-optimal convergence" and occurs here because the temperature cools when the algorithm is in a region of poor solutions and it becomes difficult to jump out of this region. To remedy this, the standard deviation of the jump distribution could be increased.

```
function[y]=RBrock(x)
    % stored in file RBrock.m
    % the generalised Rosenbrock function
    m = length(x);
    y = sum((1-x(1:m-1)).^2) + ...
            100*sum((x(2:m)-x(1:m-1).^2).^2);
```

```
function[opt] = SA(lb,ub,n)
  % stored in file SA.m
  % simulated annealing minimiser
  % lb is an mx1 array of lower bounds
  % ub is an mx1 array of upper bounds
  % n is the number of iterations
  theta = 1 - 5/n;   % cooling schedule
  w   = (ub-lb);     % width of search interval
  m   = length(w);   % number of decisions
  x0=w.*rand(m,1)+lb;% start at a random point
  xi  = x0;         % the current solution
  ei  = RBrock(x0); % energy of the current sol.
  opt = x0;         % the best solution
  eopt=RBrock(opt); % the best energy
  sd  = (ub-lb)/100;% jump size
  for i = 1:n     % search for n iterations
     xp = xi + randn(m,1).*sd; % a random jump
     % check out of bounds, wrap if needed
     for j = 1:m
       if (xp(j)>ub(j));xp(j)=xp(j)-w(j);end
       if (xp(j)<lb(j));xp(j)=xp(j)+w(j);end
     end
     ep = RBrock(xp); % energy at proposal
     % set initial Temp on 1st iter
     if(i==1); T0 = (ep-eopt)/log(0.9); end
     T = theta^i*T0; % exponential cooling
     % update if new optimum found
     if(ep<eopt)opt = xi; eopt = ep; end
     % determine if new location is accepted
     alpha = exp(-(ep-ei)/T); % accept ratio
     if(rand()<alpha); xi = xp; ei = ep; end
  end
```

```
% commands typed at prompt or stored in script.m
m=10;                      % set the dimension
lb = ones(m,1)*(-10);      % create lower bounds
ub = ones(m,1)*(10);       % create upper bounds
ans = SA(lb,ub,100000)     % call optimiser
RBrock(ans)                % func. value at optimum
```

**Figure 43.8** Simulated annealing code stored in three separate Matlab files.

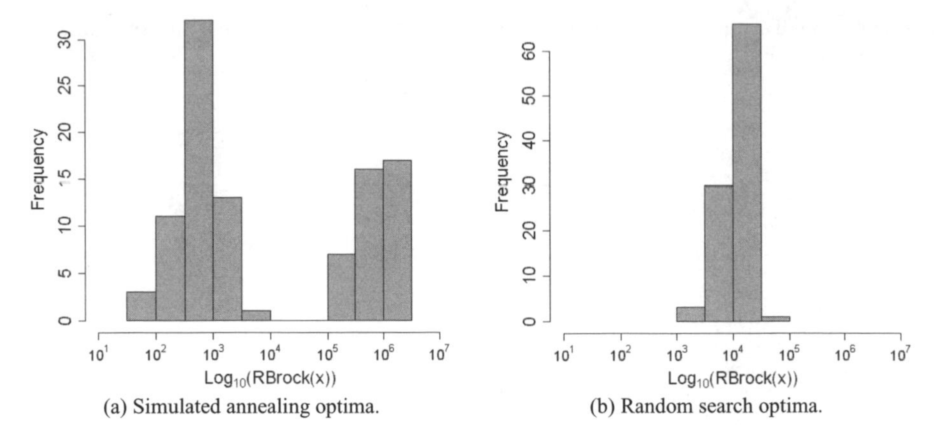


(a) Simulated annealing optima.

(b) Random search optima.

**Figure 43.9** Histograms of 100 optima for two different algorithms. The $x$ axis shows the function value at the optimum using a log scale.

## PROBLEMS

**43.1** What are the respective mechanisms in genetic algorithms and ant colony algorithms that encourage exploration? What are the feedback mechanisms that evolve improved solutions?

**43.2** Apply the simulated annealing algorithm to the generalised Rastrigin function between –10 and +10 for $m$ = 10 dimensions:

$$f(x) = 10m + \sum_{i=1}^{m} \left[x_i^2 - 10\cos(2\pi x_i)\right]$$

**43.3** Implement a linear cooling regime for the simulated annealing algorithm. Does it give better solutions? Try using a various parameter values.

**43.4** Repeat the simulated annealing algorithm for different numbers of iterations: 100; 1000; 10,000; 100,000; 1,000,000. If possible repeat this multiple times to see which number of iterations performs best.

**43.5** Change the size of the jump distribution to 1:50.

## CHAPTER FORTY FOUR

# Linear Systems and Resonance

## 44.1 INTRODUCTION

In 1831 60 soldiers were marching across a bridge in Broughton, England, when it collapsed. It was believed that the fact that the men were marching in step led to the problem and, according to Zivanovic et al. (2005), since that time notices have been placed on numerous bridges warning troops to break step when crossing. They cite an example from a railway suspension bride at Niagara Falls, USA:

*A fine of $50 to $100 will be imposed for marching over this bridge in rank and file or to music, or by keeping regular steps. Bodies of men or troops must be kept out of step when passing over this bridge. No musical band will be allowed to play while crossing except when seated in wagons or carriages.*

Although there is now considerable scepticism about the possibility of a bridge collapsing if people march in step, unacceptable levels of vibration certainly do occur. The Millennium Bridge (Figure 44.1) in London had to be closed within days of opening in 2000 due to excessive vibrations and swaying caused by the crowds eager to inspect the latest Thames crossing. There will be more on this particular example later in the chapter.

**Figure 44.1** The London Millennium Bridge. Photo by Robert Walker.

These examples highlight the importance of loads that are dynamic in nature and the need for an understanding of resonance and its possible effect on bridges, chimney stacks, buildings or even harbours. Resonance is important in civil engineering and the two key features of resonant systems are:

1. the system is free to vibrate, and has a natural frequency; and
2. the system is forced by an oscillating external force at exactly, or close to, that natural frequency.

Resonance, therefore, refers to excessive vibrations brought on by seemingly innocuous driving forces that are at a critical frequency. In the design of large chimney stacks, for example, there are two important forces that come from the wind. The first is a static force due to the drag caused by flow around the structure. This will be related to wind speed, with higher speeds leading to higher forces. A general expression for the force can be written:

$$F_D = C_D A \rho \frac{U^2}{2} \tag{44.1}$$

where $C_D$ is the drag coefficient, $A$ the area of the body, $\rho$ the density of the fluid and $U$ the fluid velocity. The second relates to the shedding of vortices from the chimney, a process that can occur at any wind speed and is dependent on the wind speed and the diameter of the chimney. The frequency of shedding, $f$, can be written:

$$f = \frac{StU}{D} \tag{44.2}$$

where $U$ is the flow velocity, $D$ the diameter of the body and $St$ is the Strouhal number. Studies have shown that over a wide range of flow conditions the Strouhal number is approximately 0.18 (Sarioglu and Yavuz, 2000). Such a process is illustrated in Figure 44.2.

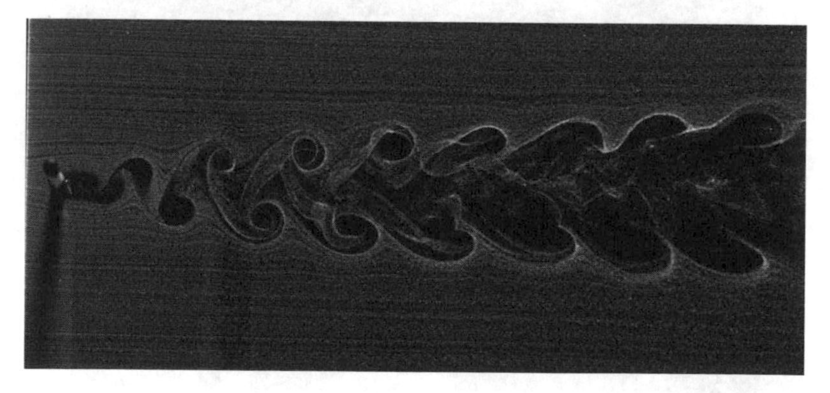

**Figure 44.2** Vortex shedding visualisation. Flow is from left to right. Photo courtesy of the School of Mechanical Engineering, the University of Adelaide.

It is evident from the visualisation that not only are eddies being generated but the direction in which they move alternates between left and right (or up and down depending on the situation). This then leads to a dynamic force with an alternating direction. To overcome the potential for the vortexes to be shed at the frequency at which the chimney would vibrate naturally, many such structures have what is referred to as helical straking. This is shown in Figure 44.3 as a line of metal that has been fixed around the outside of the pipe in a helical pattern. The purpose of this is to prevent the organised shedding of vortexes which could lead to the oscillatory force on the pipe.

**Figure 44.3** Helical straking on chimneys in Melbourne's industrial area.

Resonance is important in the design of buildings in earthquake-prone areas, and in the design of marinas, chimneys and many other structures subject to wind and water loading. And it is not just engineering structures where resonance is important. Human hearing is also based on resonance. According to Popper and Eccles (1977):

*There is a highly specialized transduction mechanism in the cochlea, where by a beautifully designed resonance mechanism, there is a frequency analysis of the complex patterns of sound waves and conversion into the discharges of neurones that project into the brain.*

As a matter of interest, the collapse of the Tacoma Rapids bridge is often cited as an example of the effects of resonance leading to a significant disaster. This is, however, wrong. The cause of the disaster was in fact dynamic instability of the bridge brought on by the wind and the bridge's shape, in much the same way that a flat ribbon suspended between two points and blown will flap wildly. The vertical motion of a suspension bridge is not well modelled as a linear system because the cables only provide a restoring force when they are under tension. However, a linear model for the transverse oscillation of the tower and cables relative to the roadway is more realistic. Analysis after the collapse was able to demonstrate that the natural frequency of the bridge was well away from the forcing frequency caused by the wind (Billah and Scanlan, 1991).

## 44.2 MATHEMATICAL DESCRIPTION OF LINEAR SYSTEMS

Many physical systems can be modelled, quite satisfactorily, as linear systems. But it should be remembered that the mathematical linear system is only an approximation of the physical system. The mathematical model is adequate if it provides an understanding of the qualitative behaviour of the physical system, and enables sufficiently accurate predictions to be made of its response to external forcing.

The defining characteristic of a linear system is that the response to a sum of forcing terms is the sum of the responses to the individual forcing terms. Consequences of this definition are that the response to a sinusoidal input is at the same frequency as the input, and that if the magnitude of the input is increased by some factor the magnitude of the response increases by the same factor.

A mass on a spring, a pendulum making small oscillations, a lightweight tubular lamp post, a cylindrical buoy, and the Salter duck wave energy device are all examples of physical systems that can be modelled as a linear system with a single natural frequency of vibration.

Consider the trolley shown in Figure 44.4. Let $y$ be the distance from the equilibrium position, where the spring is neither compressed or extended, with the positive direction being to the right. Let the mass of the trolley be $m$, and assume the spring exerts a restoring force proportional to its extension. Assume also that friction is responsible for a force, proportional to the negative velocity, which opposes the motion. Then, applying Newton's law, and writing $k$ and $c$ for the constants of proportionality for the spring and frictional force respectively gives:

$$m\ddot{y} = -c\dot{y} - ky \tag{44.3}$$

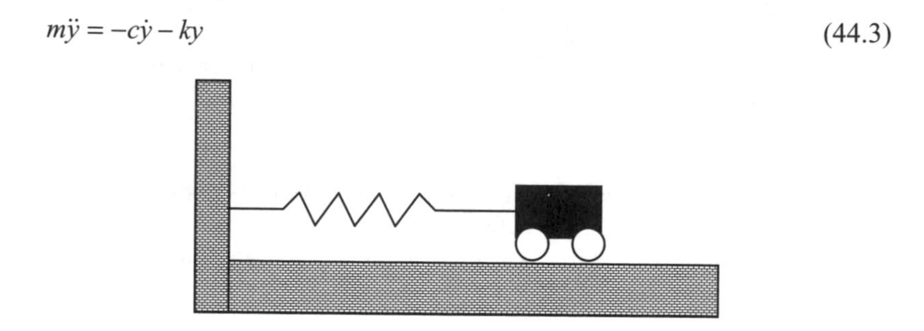

**Figure 44.4** A mass, moving on a frictionless plane under the influence of a linear spring.

For the trolley to move from its equilibrium position it needs an initial displacement, or initial velocity. It will then oscillate with decreasing amplitude until it again comes to rest at its equilibrium position. If the trolley is to keep moving it will have an external force applied to it.

The general form of the equation of motion for a linear system with a single natural frequency of vibration is a second-order linear differential equation with constant coefficients and can be written:

$$\ddot{y} + 2\varsigma\omega\dot{y} + \omega^2 y = u \tag{44.4}$$

where $y$ is a displacement, $\zeta$ is the damping coefficient, $\omega$ is the undamped natural frequency, and $u$ is a forcing term. Now, suppose there is no forcing term ($u = 0$), and assume $y$ has the form:

$$y = e^{mt} \tag{44.5}$$

Substituting Equation (44.5) into Equation (44.4) gives a quadratic equation in $m$:

$$m^2 + 2\varsigma\omega m + \omega^2 = 0 \tag{44.6}$$

The case of $\zeta = 1$ is critical damping and $\zeta$ must be less than this for an oscillatory response. If $\zeta < 1$ then the solution of Equation (44.6) is:

$$m = -\varsigma\omega \pm i\sqrt{(\omega^2(1-\varsigma^2))} \tag{44.7}$$

and the response has the form of an exponentially damped, phase-shifted sine wave (see Appendix C) with a frequency of $\omega\sqrt{(1-\varsigma^2)}$. This frequency is the natural frequency of vibration, and if the damping is light, $\zeta$ is small, and the natural frequency is only slightly less than $\omega$. The unforced response is known a transient response because it is damped and dies away. The initial amplitude of the unforced response depends on the initial displacement and initial velocity. If the initial velocity is 0 the initial amplitude is the initial displacement. A solution for particular values of $\omega$ and $\zeta$ is shown in Figure 44.5.

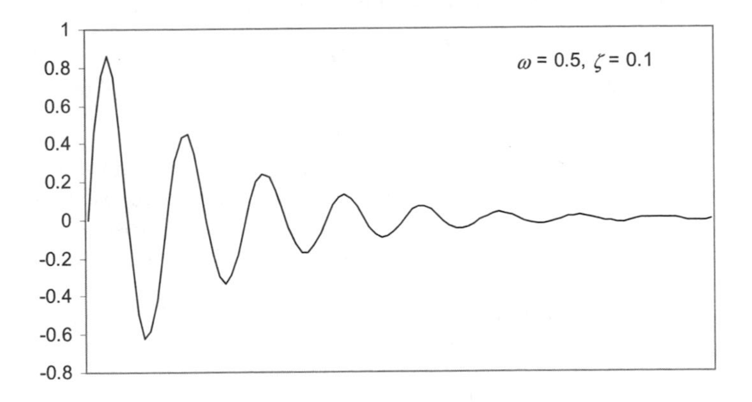


**Figure 44.5** Displacement of a mass for a particular value of damping.

Now consider a periodic forcing term:

$$u = e^{i\Omega t} \tag{44.8}$$

and assume a steady-state response at the same frequency but with a change of amplitude and phase, that is:

$$y = Ae^{i(\Omega t+\phi)} \tag{44.9}$$

The relative amplitude of the response is known as the gain, and both it and the phase shift $\phi$ depend on the forcing frequency $\Omega$. They can be calculated as:

$$A = \frac{1}{\sqrt{((\omega^2 - \Omega^2) + 4\varsigma^2\omega^2\Omega^2)}} \tag{44.10}$$

$$\tan(\phi) = \frac{2\varsigma\omega\Omega}{\omega^2 - \Omega^2} \tag{44.11}$$

If the damping is light, the amplitude will become large if the forcing frequency is close to the natural frequency, and this is phenomenon is known as resonance. In the case of zero damping (admittedly unrealistic) the equation of motion can be written:

$$\ddot{y} + \omega^2 y = a\cos\Omega t \tag{44.12}$$

where $a$ is $F_0/m$ with $F_0$ being the amplitude of the force and $m$ being the mass of the object under consideration; $\Omega$ is the frequency of the driving force. According to Acheson (1997) this equation, known as simple harmonic motion, was studied by Euler in 1739. (The analytic solution is straightforward for any driving frequency.) For the situation where the driving frequency is identical to the natural frequency of oscillation there is an exact analytical solution that can be written:

$$y = \frac{a}{2\omega} t\sin\omega t \tag{44.13}$$

This shows a signal that increases in amplitude linearly with time. In reality, the assumptions inherent in this solution (no damping, small oscillations) soon break down and the actual solution would be limited in its final oscillations.

Solution of the governing equation using a fourth-order Runge-Kutta scheme, provides a check of the Runge-Kutta scheme against the known analytic answer. The sensitivity of the system to the frequency of the driving force can be illustrated by changing the driving frequency. For example, in Figure 44.6 the driving frequency is 80% of the natural period of oscillation. Note the occasional changes in the amplitude of vibration, brought on through the interaction of the driving frequency and the natural frequency of oscillation.

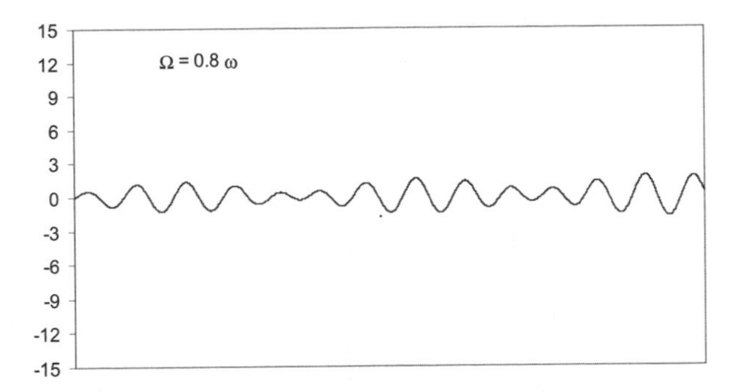


**Figure 44.6** Oscillations that result from a system being driven at 80% of its natural frequency of vibration. The vertical axis shows displacement and the horizontal axis time.

## 44.3 RESONANCE IN THE COASTAL ENVIRONMENT

The design of coastal features such as marinas requires consideration of the effects of resonance. The water in harbours and marinas can oscillate although the period of oscillation tends to be terms of tens of seconds or minutes. In any event, if the basin is subjected to an oscillating driving force that is at or extremely close to its natural frequency, large displacements are possible, and these can have a devastating effect on ships that are moored there. One aspect of marina design, therefore, is to calculate the natural period of oscillation and ensure that it is well removed from any waves of that same period in the coastal environment.

As an illustration of marina oscillation, although not resonance, results from a field study are described. In 2003 Holdfast marina in Glenelg, South Australia, was instrumented to observe sea wave penetration. In addition to the short-period sea waves some longer-period oscillations were observed. A part of a record is shown in Figure 44.7.

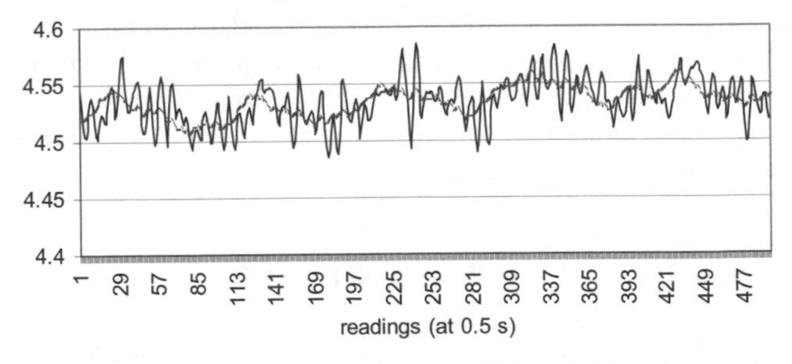


**Figure 44.7** A 250-second section of wave record from 28 June 2004. A 10-second moving average has been calculated and plotted. The period of oscillation is approximately 50 seconds.

If the water in the marina were to oscillate as a whole it would be possible to estimate the period of oscillation using simple resonance theory. In this case the period of resonance for a closed basin can be calculated as:

$$T = \frac{2L}{\sqrt{gd}} \tag{44.14}$$

where $L$ is the length of the basin, $g$ is the gravitational constant (9.8 m/s$^2$), and $d$ is the water depth. In the case given, water depth was approximately 4.5 metres and the basin length approximately 200 metres. This would predict a natural period of oscillation of 60 seconds, which is very close to the observed value shown in Figure 44.7. This period is significantly longer than any significant energy arriving from the local coastal environment so, although there is some observed oscillation, resonance is avoided at this particular site.

Tides were one feature of the natural environmental that defied proper description despite much early observation and study. Even once the basic driving forces were understood, the ability of tides to be so different at locations that were so close left many early mariners perplexed. Burling et al. (2003) illustrate one of the causes of such behaviour in their study of Shark Bay in Western Australia. The tides on the open coast at the site are a mixture of 12-hour and 24-hour periods due mainly to the effects of the sun and moon acting on a rotating earth. However, the tides in Hopeless Reach are mainly semi-diurnal (12-hour period) and those on the Freycinet Reach are mainly diurnal (24-hour period). The researchers were able to show that the length (134 km) and depth (15 m) of Hopeless Reach were such that there was a near-quarter wavelength resonance (associated with an open driving boundary) of the semi-diurnal component leading to a factor-of-two increase in the amplitude of that tidal component. Hence the quite different behaviours of two sites driven by the same tidal signal. The quarter-wave resonance condition for a basin open at one end can be written:

$$T = \frac{4L}{\sqrt{gd}} \tag{44.15}$$

where the terms are as defined previously. There are a number of other locations where resonance is important for tidal dynamics. For example, van Maren and Hoekstra (2004) give an example of a location in Vietnam where two of the once-a-day tidal components ($K_1$ with period of 23.9 hours and $O_1$ with a period of 25.8 hours) resonate and therefore dominate the tidal signal leading to diurnal tides (one high water and one low water per day).

## 44.4 RESONANCE IN STRUCTURAL ENGINEERING

The design of buildings requires that the applied loads be properly assessed. Loads are generally sub-divided into live loads and dead loads, where the live loads include all those loads that are due to externally applied forces that may vary in their application and the dead loads tend to be those associated with the weight of the structure itself. It is important to assess not only the magnitude of the live load but also its frequency characteristics. This, however, may not be as easy as it sounds, particularly in the case of human-induced loading. According to Dallard et al. (2001) once the number of pedestrians on a bridge reaches a certain level there is the potential for any small lateral movements to cause the people to start walking in step in order to deal with the effect that these movements have on their walking comfort. In this way they achieve a frequency "lock-in" and a positive feedback results. This is particularly so if the natural frequency of the motion is around 1 Hz (Wilkinson and Knapton, 2006). It is believed that this is what happened with London's Millennium Bridge at its opening. Under light pedestrian loads everything may have been fine, but the sheer number of people using the bridge led to the situation where the lateral motion became excessive. The solution was to increase the damping, at a cost of £5 million (Wilkinson and Knapton, 2006).

According to Salvadori (1980), the World Trade Center buildings, when standing, had a natural period of oscillation of 10 seconds. Salvadori suggested that "if a long series of wind gusts with a relatively slow period of ten seconds were to hit the towers, the swing of the towers would slowly increase until the structure of the building might sway so widely as to collapse."

Earthquakes can be very destructive, and their effect can be magnified due to resonance. One recent case was the collapse of a 1.2-km section of a double-decker highway (the Cypress Structure) in the San Francisco area at Loma Prieta in 1989. The quake lasted only 5 to 7 seconds but it was enough to set up violent oscillations with a period of approximately that of the bridge supports. According to Salvadori (1980), "In the case of the Cypress Structure, it turns out that the principal vertical frequency was almost identical to the frequency imposed by the earthquake at the site of the collapsed section. This coincidence led to a reinforcement, or amplification of the movements – i.e. resonance."

**Mexico Earthquake, 1985**

In September 1985, Mexico City was hit by an earthquake that caused significant damage. Jain (1990) reports that although the firm-soil earth movements, caused by the undersea quake, were relatively mild (less than 0.04g), resonance in the soft sedimentary materials that much of Mexico City sits on amplified this to between 0.05g and 0.20g. The period of the oscillation was quite steady at around 2 seconds. Much damage was caused to multi-storey buildings where the undamaged natural period of vibration ranged from 0.75 to 2 seconds. As the quake progressed, damage in the upper storeys moved the natural periods closer to the period of forcing, leading to resonance and, in many cases, collapse.

## PROBLEMS

**44.1** Determine the range of frequencies likely to be generated by wind passing a large flagpole. Assume a diameter of 150 mm and a range of wind speeds up to 30 knots (1 knot = 0.5144 m/s). If possible, make observations of flag poles being driven to vibration by strong winds and determine if there is agreement between the natural frequency of vibration and the driving frequency based on the frequency of shedding given in Equation (44.2).

**44.2** Set up an Excel spreadsheet and reproduce the results shown in Figure 44.5 by solving Equations (44.4) to (44.6). Investigate the effect of varying the damping on the behaviour.

**44.3** Verify that the length and depth of Hopeless Reach in Western Australia are likely to cause a quarter-wave resonance with the 12-hour tidal signal.

**44.4** Investigate the use of tuned dampers in structural engineering. Where are they used most commonly and what is their purpose?

**44.5** Cars with poorly balanced front wheels can develop an annoying steering wheel vibration around a speed of 80 km/h. If this is the case, estimate the natural frequency of vibration of the steering system. Make what assumptions are necessary to determine a solution.

CHAPTER FORTY FIVE

# Spectral Analysis (Introduction)

## 45.1 INTRODUCTION

If an undamped linear system is driven by an oscillation close to the system's natural frequency the response will exhibit some variation in its response. An example of this is shown in Figure 45.1, where the driving force is at 1.25 times the natural frequency of the system.

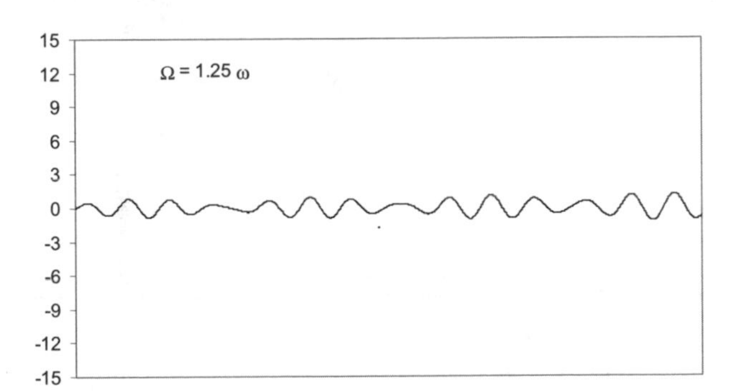


**Figure 45.1** Oscillations that result from a system being driven at 125% of its natural frequency of vibration. Displacement on vertical axis, time on horizontal.

The question is: how to determine these if only the final signal is available for analysis? This, and many other problems, fall within the topic that deals with spectral analysis: the determination of frequency information from a complicated signal.

Spectral analysis is a most important tool in engineering. It finds its way into structures, hydraulics and many other aspects of engineering practice. Spectral analysis is concerned with analysing data and determining from it not only the magnitude of disturbances, displacements or forces, but also information on the frequencies of the processes. This information is vital for design, as in many cases it is the frequency of disturbances as well as their amplitude that must be considered. This is because resonance must generally be avoided. Some examples where resonance may be important are:

- waves in harbours (Bellotti, 2007; Losada et al., 2008);
- design of tall chimneys;
- offshore structures (Dong et al., 2002);
- the behaviour of buildings during earthquakes;
- vibration in pipes (Lee et al., 2005);
- measuring and removing the hum in the earth's vibration (Coontz, 1999);
- design of bridges (Briseghella and Zordan, 2002; Dallard et al., 2001).

## 45.2 FREQUENCY AND TIME DOMAINS

Plots of waves or any time-varying signal are most commonly shown in the time or space domain. For example, a plot of:

$$f(x) = sin(x) \;\text{ or }\; f(t) = sin(t) \tag{45.1}$$

is shown in Figure 45.2(a) with time or space on the horizontal axis and amplitude on the vertical. But what of the frequency domain? Here a plot shows frequency on the $x$ axis and amplitude on the $y$ axis. A sine wave of frequency $\omega$ and unit amplitude would be plotted as a single line at $\omega$ as shown in Figure 45.2(b).

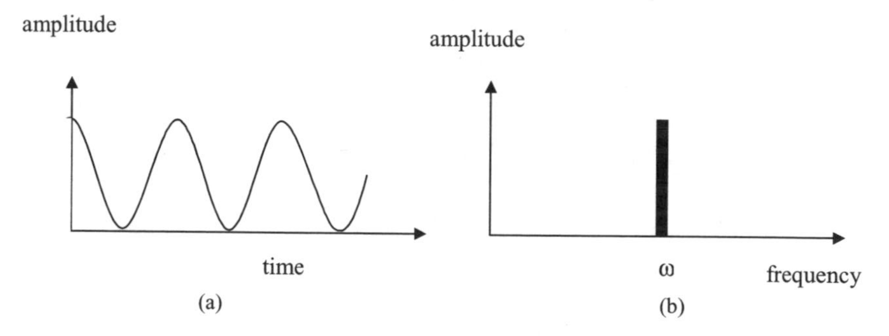


**Figure 45.2** A pure and simple sine wave plotted in (a) time and (b) frequency domains.

This gives almost the same amount of information as the time plot except that there is no information about the phase of the signal. So the representation of the two diagrams, the time and frequency plots, gives virtually the same information but from a quite different viewpoint.

For a simple sine wave the advantage of the frequency plot is perhaps not evident. But consider for a moment a signal such as that shown in Figure 45.1 that is composed of multiple frequencies (and therefore potentially represented by multiple sine waves). In this case the plot in the time domain gives little real information other than the fact that more than one frequency is involved. However, if spectral analysis is carried out the frequency plot is as shown in Figure 45.3. Now the two frequencies in the system are clearly evident. These are: the natural frequency of the system (3.162 rads/s) and the other driving frequency (3.953 rads/s). The frequency representation is clearly superior in terms of extracting information about the signals in this case.

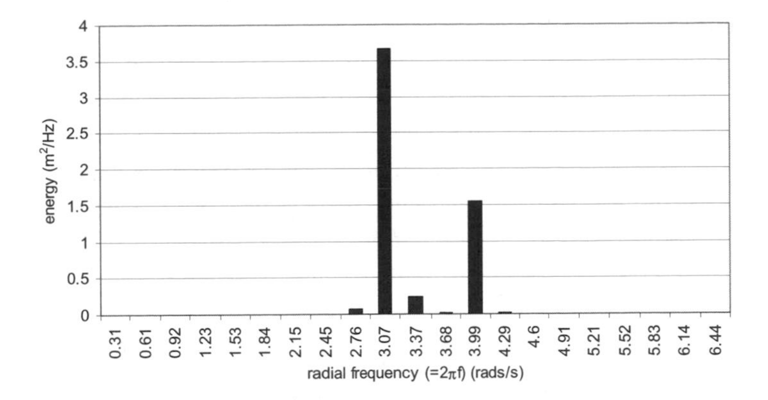


**Figure 45.3** Spectrum of data shown in Figure 45.1. For this case the driving frequencies of the system were 3.162 rads/s and 3.953 rads/s. Both are evident in the plot.

## 45.3 HISTORY – FOURIER

The Fourier transform is named after a Frenchman Jean Baptiste Fourier (1768–1830), a physicist and mathematician. Regarding another topic, Fourier was the first to suggest the greenhouse analogy when discussing how the earth manages to be kept warm by its atmosphere (Weart, 2003). This came about following calculations that he carried out on the heat balance of the earth. He found that without some mechanism to retain heat the planet would be much colder than observed. He also spent some time studying heat conduction in metal conductors and devised the Fourier series to help develop solutions to the problem. Although not strictly relevant to the topic, the general heat equation can be written:

$$\frac{\partial u}{\partial t} = \alpha^2 \frac{\partial^2 u}{\partial x^2} \tag{45.2}$$

where the general solution, obtained using separation of variables, is:

$$u(x,t) = F(x).G(t) \tag{45.3}$$

where

$$F_n(x) = \sin\frac{n\pi x}{L} \qquad \text{and} \qquad G_n(t) = B_n e^{-\lambda^2 t} \tag{45.4}$$

To find specific solutions Fourier realised it was necessary for any general initial temperature distribution to be able to be represented by a sum of sinusoidal terms. He found that this was the case where the function was “reasonably friendly” (Farlow, 1993). Before looking at some of the details of Fourier analysis it is worth taking note of the fact that Kreyszig (1983) states that the theory is complicated, but the application simple.

## 45.4 THE FOURIER TRANSFORM

Based on the work of Fourier it is possible to represent any continuous periodic function as an infinite series of sine and cosine terms (with a particular set of amplitudes, frequencies and phases). Such a function is shown in Figure 45.4.

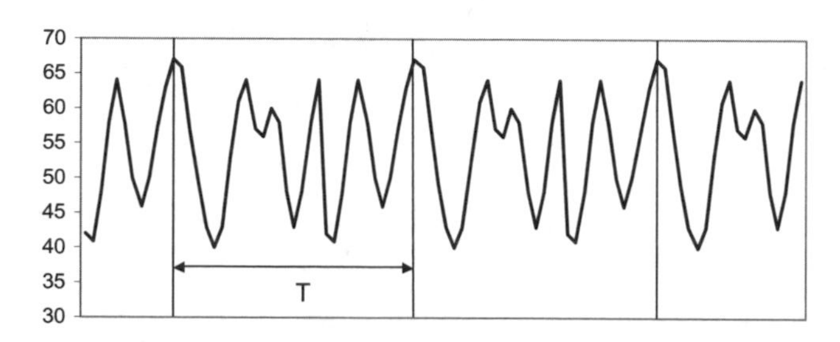


**Figure 45.4** Example of a continuous periodic function. Fourier showed that any arbitrary continuous function can be represented by an infinite series of sinusoids.

Formally, any function $f(t)$ can be written as:

$$f(t) = a_0 + a_1 \cos(\omega_0 t) + b_1 \sin(\omega_0 t) + a_2 \cos(2\omega_0 t) + b_2 \sin(2\omega_0 t) + ... \quad (45.5)$$

or more concisely:

$$f(t) = \frac{a_0}{2} + \sum_{k=1}^{\infty} [a_k \cos(kw_0 t) + b_k \sin(kw_0 t)] \quad (45.6)$$

where $\omega_0$ is the fundamental frequency and the other frequencies that are whole multiples of it are called harmonics. The determination of the constants can be expressed:

$$a_k = \frac{2}{T} \int_0^T f(t) \cos(k\omega_0 t) dt \quad \text{for k=0,1,2,3, ...} \quad (45.7)$$

$$b_k = \frac{2}{T} \int_0^T f(t) \sin(k\omega_0 t) dt \quad \text{for k=1,2,3, ...} \quad (45.8)$$

As a demonstration of this, consider the description of a square wave using the Fourier method. Following the mathematics (not shown here) it is possible to describe the function as an infinite series of cosine waves. This can be written:

$$f(t) = \frac{4}{\pi}\cos(\omega_0 t) - \frac{4}{3\pi}\cos(3\omega_0 t) + \frac{4}{5\pi}\cos(5\omega_0 t) - + ... \qquad (45.9)$$

If *f(t)* is plotted (Figure 45.5) it can be seen that by increasing the number of terms a better approximation to the square wave can be achieved. There will seldom be time to plot an infinite number of terms but after 27 the results are quite good.

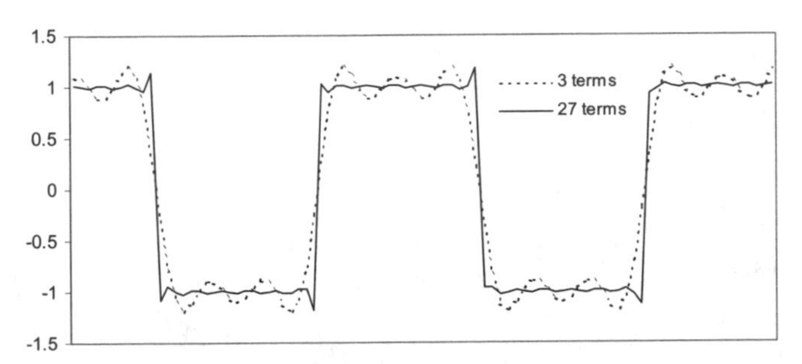


**Figure 45.5** Effect of increasing the number of components in Fourier representation of a square wave.

Therefore, a square wave can be thought of as a series of cosine waves of gradually increasing frequency. This is rather artificial because there is a perfectly good definition of a square wave, but if one can fit cosine waves to a square wave it demonstrates the potential for almost any periodic function to be represented as a sum of sine or cosine functions.

Incidentally, the spectrum for the signal shown in Figure 45.5 can be derived straight from the frequencies and amplitudes shown in Equation (45.9). It is plotted in Figure 45.6, where the frequencies are shown on the horizontal axis and the vertical axis shows the absolute magnitude of the terms.

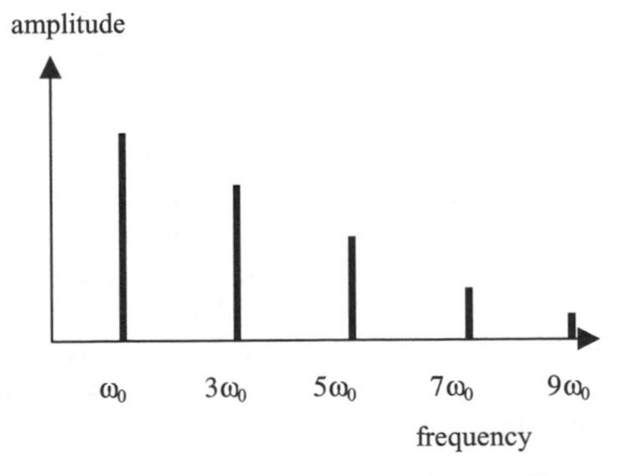


**Figure 45.6** Spectrum of square wave, shown to five terms.

The absolute value is shown since there is no difference between $x(t) = a \sin(\omega t)$ and $x(t) = -a \sin(\omega t)$ other than a phase difference. In the diagram, each spectral line represents one of the components of the square wave. It is important to note that all components are multiples of the basic frequency, $\omega_0$.

## 45.5 CASE STUDY 1: EARTHQUAKE ANALYSIS

In regions of high earthquake activity data on the accelerations, velocities and motions that are experienced are now routinely collected. In Pakistan, for example, a single recorder installed near Islamabad in 1969 became part of a network of 13 recorders in 1972 with the installation of a further 12 analogue accelerograms (Rizwan et al., 2008). The output from the recorders is generally a trace of acceleration: a typical one is shown in Figure 45.7.

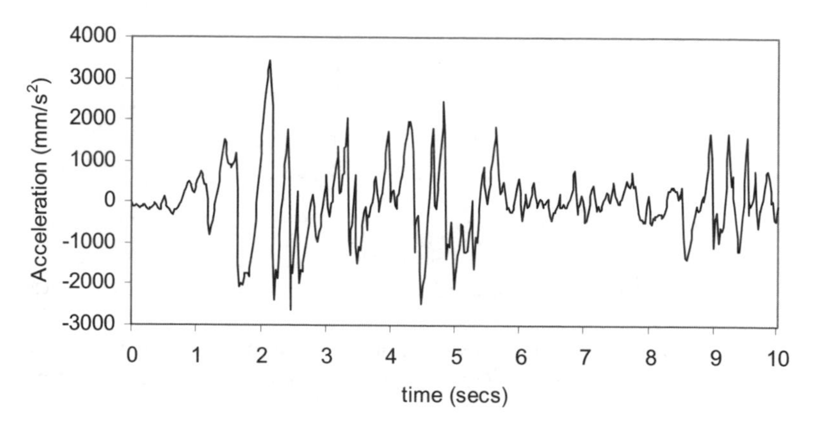


**Figure 45.7** Acceleration measured at El Centro, Imperial Valley Irrigation District, 18 May 1940.

Although a number of key features, such as peak acceleration, can be read directly from the record, the frequency mix in the record is a key feature of the record and this can be recovered most conveniently using spectral analysis. Exactly how this is done will be set out in the following chapters.

## 45.6 CASE STUDY 2: PINK NOISE

Gribben (2004), in a discussion of earthquakes, frozen potatoes being smashed, and the size of cities demonstrates how many quite separate phenomena can be explained in terms of a power law where, generally, the size of an event and its likelihood are related in some way that can be explained. That is, the size of the event is inversely proportional to the likelihood raised to some power. A convenient way to show this is with the spectrum which, when plotted on log paper, should show a characteristic straight line that reduces with increasing frequency. An example is shown in Figure 45.8.

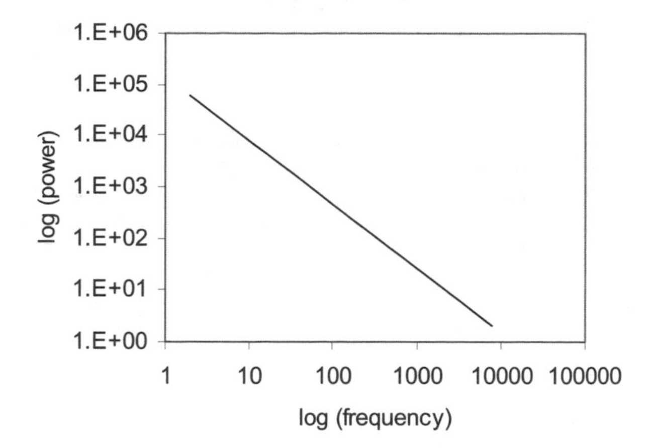


**Figure 45.8** Pink noise spectrum showing classic 1/f power law.

This particular shape of spectrum is referred to as pink noise, in contrast to white noise, which has a flat spectrum, and red noise that has more power at the lower frequencies.

## 45.7 CASE STUDY 3: RED NOISE

Torrence and Compo (1998) describe how a Markov process of the form:

$$x_n = \alpha x_{n-1} + z_n \tag{45.10}$$

where $x_n$ are the data, $\alpha$ is a constant and $z_n$ is a Gaussian noise term, should give a red spectrum. Generating a set of data with $\alpha = 0.8$ and $x_0 = 0.0$ gave a dataset which, when analysed using Excel's Fourier transform (under Tools – Data Analysis – Fourier Transform) gave the data and spectrum shown in Figure 45.9.

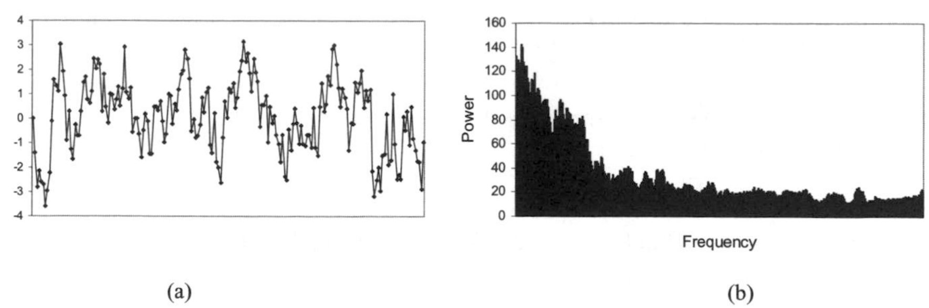


(a) (b)

**Figure 45.9** (a) The raw data and (b) the red noise spectrum based on a Markov process. The spectrum is considered red because of the relatively high energies at the low frequency (red) end of the spectrum.

**Data Encryption**

A scientist, Lou Pecora, who was working on encryption using artificially generated chaos believed he had come up with a foolproof system. He showed it off to another expert:

*Pecora, in his naïveté, bet one of them a beer that he could hide a sine wave in the chaos, and challenged the visitor to extract it. The visitor ran the circuits for a minute, measured the voltage waveforms, then did a computation called a fast Fourier transform to measure the strengths of all the component frequencies being transmitted. The sine wave stood out nakedly as a spike in the spectrum. Pecora realized then that he had a lot to learn about encryption.* Quoted in Strogatz (2003).

## 45.8 WORKED EXAMPLE

This problem is listed to show the mathematical approach to the derivation of the Fourier transform. It is not the approach that engineers would take when analysing data: this will be covered in the next chapter. It is, however, valuable for showing an analytical derivation of the spectrum.

*Problem*

Derive the Fourier transform for the continuous function of a square wave where the definition is given as:

$$f(t) = \begin{cases} -1 & -T/2 < t < -T/4 \\ 1 & -T/4 < t < +T/4 \\ -1 & +T/4 < t < +T/2 \end{cases} \tag{45.11}$$

*Solution*

To evaluate the $a$ and $b$ terms Equations (45.8) to (45.10) are applied. It is convenient to use the period of interest from $-T/2$ to $T/2$ rather than 0 to $T$ so:

$$a_k = \frac{2}{T} \int_{-T/2}^{T/2} f(t) \cos(k\omega_0 t) \tag{45.12}$$

If the integration is split into the three different sections based on its definition, it is possible to write:

$$a_k = \frac{2}{T}\left( - \int_{-T/2}^{-T/4} \cos(k\omega_0 t)\, dt + \int_{-T/4}^{T/4} \cos(k\omega_0 t)\, dt - \int_{T/4}^{T/2} \cos(k\omega_0 t)\, dt \right) \tag{45.13}$$

where $k$ can take on a range of values. By considering the properties of the sine and cosine functions it is possible to show that it is not necessary to evaluate all the terms individually. The first term in the series, where $k = 1$, can be written:

$$a_1 = \frac{2}{T}\left( -\int_{-T/2}^{-T/4} \cos(\omega_0 t)dt + \int_{-T/4}^{T/4} \cos(\omega_0 t)dt - \int_{T/4}^{T/2} \cos(\omega_0 t)dt \right) \qquad (45.14)$$

By using the fact that $sin(x) = -sin(-x)$, and $\omega_0 = 2\pi/T$ it is possible to show that $a_1 = 4/\pi$. Subsequent terms follow in much the same way. The final pattern of the solutions is:

$$a_k = \frac{4}{k\pi} \text{ for } k\text{=1,5,9,...,}$$

$$a_k = -\frac{4}{k\pi} \text{ for } k\text{=3,7,1,...,}$$

$$a_k = 0 \text{ for all } k = \text{even number.}$$

It can also be shown that all the $b$ terms are zero. The square wave is an even function since $f(t) = f(-t)$. In this case it can be shown that the $b$ terms will always be zero. For odd functions where $f(t) = -f(-t)$ all the $a$ terms will be zero. It is also possible to show that $a_0 = 0$.

## PROBLEMS

**45.1** Assume the temperature at some place on earth is made up of an annual cycle (based on the seasons) and a daily cycle (based on the rising and setting of the sun). Assume further that both forcing mechanisms can be described by simple sine waves. Without detailed calculations, sketch the signal that might be expected over a year and also the spectrum for this time series.

**45.2** A sine wave with a period of 8 seconds, $f(t) = \sin(2\pi t/8)$, will show a spectrum with a single spike of energy at a frequency of 0.125 Hz. What would the spectrum look like for the function $f(t) = \sin^2(2\pi t/8)$? Attempt an answer without any formal calculations. A sketch of the function may assist.

**45.3** Sketch the spectrum that might arise from a random number generator. Assume a uniform distribution of numbers.

**45.4** Figure 45.10 shows data recorded on a patient at a local hospital. Like much data collected where the subject is some sort of complex system, it shows a mix of amplitudes and periods (or frequencies). In many situations, this mix can point to potential problems and behavioural characteristics that may be useful as a diagnostic tool. As a prelude to more detailed analysis, and using a minimum of calculations, estimate the key frequencies that are in the signal and their relative magnitudes.

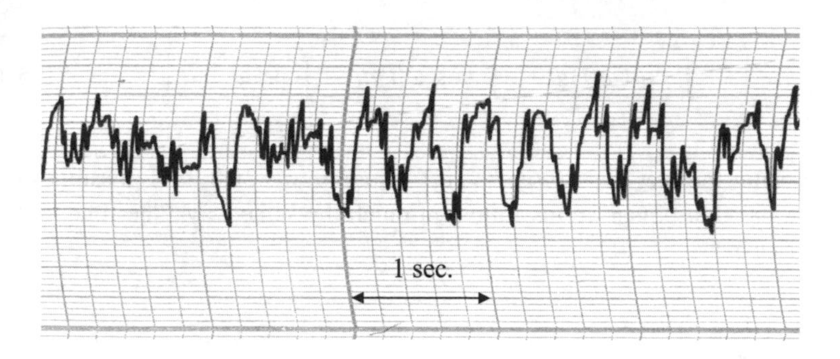


**Figure 45.10** Trace of a patient's vital signs.

**45.5** Derive the Fourier transform of the sawtooth wave:

$f(t) = t$ for $-T/2 < t < T/2$

**45.6** Confirm the calculations from the previous question by plotting in Excel the representation of the sawtooth waveform based on an increasing number of terms.

**45.7** Assume a continuous function has been obtained that related to slow variations in the frequency and amplitude of the 240V alternating current supplied to a home. If the Fourier transform was calculated:

(a) what would it look like?
(b) how would a change in frequency be identified?
(c) how would a change in voltage be identified?

**45.8** Consider tide data collected each hour for a number of months. Tides are known to be made up of a number of components, the major ones being the semi-diurnal (12-hour period) and diurnal (24-hour period). In Adelaide the two components are approximately equal. Draw the spectrum that you might expect from the tide data.

# CHAPTER FORTY SIX

# Spectral Analysis (Discrete Fourier Transform)

## 46.1 INTRODUCTION

The original definition of the Fourier transform was based on continuous functions, possibly well-defined ones such as a polynomial, or perhaps a general function such as that shown in Figure 46.1.

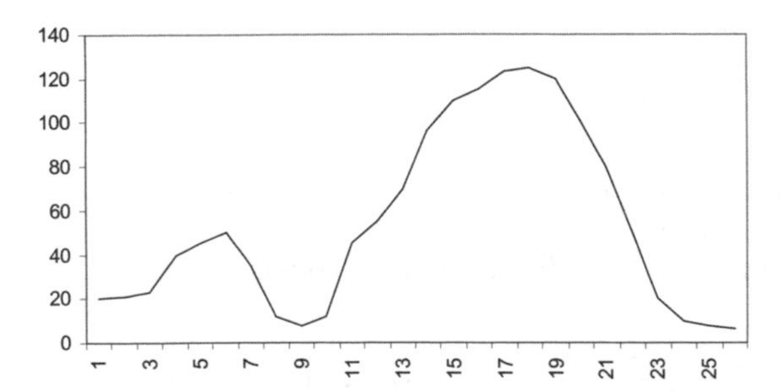


**Figure 46.1** Example of a continuous function.

In engineering it is more likely that a series of data points will have been collected rather than a continuous representation. The data are also often taken at a constant rate and is therefore evenly spaced in time. Such a representation is shown in Figure 46.2.

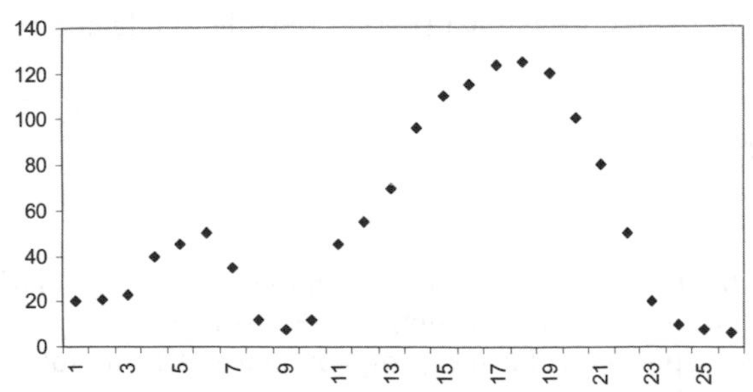


**Figure 46.2** Example of discrete data. The data are taken at a regular time interval, normally called $\delta t$. The sampling frequency is $1/\delta t$ and measured in cycles per second or Hertz (Hz.).

## 46.2 THE FOURIER TRANSFORM OF DISCRETE DATA

It has previously been shown that any well-behaved periodic function $f(t)$ can be written:

$$f(t)=\frac{a_0}{2}+\sum_{k=1}^{\infty}\left[a_k \cos(k\omega_0 t)+b_k \sin(k\omega_0 t)\right] \tag{46.1}$$

The coefficients for a discrete set of data points can be determined by replacing the integral in Equations (45.7) to (45.8) with a summation:

$$a_k = \frac{1}{N}\sum_{m=0}^{N-1} f(m)\cos(m\omega_0 k) \quad \text{for k=1,2,3, ...} \tag{46.2}$$

$$b_k = \frac{1}{N}\sum_{m=0}^{N-1} f(m)\sin(m\omega_0 k) \quad \text{for k=1,2,3, ...} \tag{46.3}$$

Now using the Euler identity:

$$e^{ia} = cos(a) + i\, sin(a) \tag{46.4}$$

it is possible to write the two formulae more compactly as:

$$F_k = \frac{1}{N}\sum_{m=0}^{N-1} f_m e^{-im\omega_0 nk} \quad \text{for k = 0 to N–1} \tag{46.5}$$

where $N$ is the number of data in the set, $\omega_0 = 2\pi/N$, $F_k$ is the $k^{th}$ component of the spectrum, and $f_m$ is the $m^{th}$ data point from the set being analysed where the data are numbered $f_0$ to $f_{N-1}$.

This is in fact a key result as it shows how time series data ($f_m$) can be transformed into its frequency data ($F_k$). In this, the fundamental frequency can be calculated as:

$$\omega_0 = \frac{2\pi}{N} \tag{46.6}$$

It is possible to write a simple algorithm to calculate the Fourier transform of a data set, $f_n$, and this is shown in Figure 46.3. Notice that the real and imaginary parts have been split to make the programming easier to follow and that those arrays must be defined from the $0^{th}$ element. The input data are the array $f()$, the number of data, $N$. Note that $f_i$ is the $i^{th}$ data value in the data series, and $w_0=2\pi/N$. The output are the arrays containing the real and imaginary parts of the transform.

```
do k = 0 to N-1
  do m = 0 to N-1
    angle = k*w0*m
    r(k)  = r(k)  + f(m)*cos(angle)/N
    im(k) = im(k) - f(m)*sin(angle)/N
  end do
end do
```

**Figure 46.3** Fortran code segment to calculate the discrete Fourier transform.

Since there are two nested loops each going from 0 to $N$–1 it takes $N^2$ operations to compute the discrete Fourier transform (DFT) of a data set. The frequencies are calculated from the formula:

$$freq_k = \frac{k}{T} \qquad (46.7)$$

where $T$ (=$N\delta t$) is the total length of the records. The total energy at any particular frequency can be calculated by summing the real and imaginary components of the transform since they relate to the same frequency, but differ in that one is a sine wave and the other a cosine wave.

$$e_k = \text{real}_k^2 + \text{imaginary}_k^2 \qquad (46.8)$$

The results obtained from the DFT contains information on the zero frequency or mean value of the series (Index 0), the positive frequencies (Index 1 .. $N/2$) and the negative frequencies (Index $N/2+1$ .. $N-1$). The different indices are associated with different frequencies from 0 up to $1/2\Delta t$, which is called the Nyquist frequency.

$$f_{NQ} = \frac{1}{2\Delta t} \qquad (46.9)$$

The result of this is that it is not possible to determine any information about frequencies higher than those that are represented by at least two points per cycle in the raw data. This is explained further in the section of the sampling theorem.

## 46.3 WORKED EXAMPLE

As an example, consider a set of 32 data points that have been generated with a time-step set ($\delta t$) of 0.01 seconds from a unit amplitude cosine function with a frequency of 12.5 Hz., that is: $f(t) = \cos(2\pi(12.5)t)$. If this is put through a DFT, with appropriate post-processing to determine the frequencies and energies of the various components, the results are as listed in Table 46.1. It is important to note that the output includes a set of real and imaginary components that are associated with a set of negative frequencies. This is in fact a mirror image of the first half of the output, which is associated with the positive frequencies. It is usual to ignore the second half, but since half the energy is effectively also ignored the energies for the positive frequencies must be doubled, that is:

$$e_k = 2(\text{real}_k^2 + \text{imaginary}_k^2) \quad \text{if the positive frequencies are used.} \qquad (46.10)$$

The energy at $k = 4$, therefore is $2(0.5^2 + 0^2) = 0.50$.

**Table 46.1** Results of DFT on a simple cosine wave with frequency of 12.5 Hertz. The values in light font are associated with the negative frequencies and are essentially ignored.

| k | $f_i$ | real | imaginary | freq (Hz.) | energy ($m^2$) |
|---|---|---|---|---|---|
| 0 | 1.000 | 0 | 0 | 0.00 | 0 |
| 1 | 0.707 | 0 | 0 | 3.13 | 0 |
| 2 | 0.000 | 0 | 0 | 6.25 | 0 |
| 3 | –0.707 | 0 | 0 | 9.38 | 0 |
| 4 | –1.000 | 0.50 | 0 | 12.50 | 0.50 |
| 5 | –0.707 | 0 | 0 | 15.63 | 0 |
| 6 | 0.000 | 0 | 0 | 18.75 | 0 |
| 7 | 0.707 | 0 | 0 | 21.88 | 0 |
| 8 | 1.000 | 0 | 0 | 25.00 | 0 |
| 9 | 0.707 | 0 | 0 | 28.13 | 0 |
| 10 | 0.000 | 0 | 0 | 31.25 | 0 |
| 11 | –0.707 | 0 | 0 | 34.38 | 0 |
| 12 | –1.000 | 0 | 0 | 37.50 | 0 |
| 13 | –0.707 | 0 | 0 | 40.63 | 0 |
| 14 | 0.000 | 0 | 0 | 43.75 | 0 |
| 15 | 0.707 | 0 | 0 | 46.88 | 0 |
| 16 | 1.000 | 0 | 0 | 50.00 | 0 |
| 17 | 0.707 | 0 | 0 | | |
| 18 | 0.000 | 0 | 0 | | |
| 19 | –0.707 | 0 | 0 | | |
| 20 | –1.000 | 0 | 0 | | |
| 21 | –0.707 | 0 | 0 | | |
| 22 | 0.000 | 0 | 0 | | |
| 23 | 0.707 | 0 | 0 | | |
| 24 | 1.000 | 0 | 0 | | |
| 25 | 0.707 | 0 | 0 | | |
| 26 | 0.000 | 0 | 0 | | |
| 27 | –0.707 | 0 | 0 | | |
| 28 | –1.000 | 0.50 | 0 | | |
| 29 | –0.707 | 0 | 0 | | |
| 30 | 0.000 | 0 | 0 | | |
| 31 | 0.707 | 0 | 0 | | |

Referring to Table 46.1 it is evident that most of the components of the real and imaginary arrays have zero values. The exceptions are for $k = 4$ and $k = 28$ where the real array has a value of 0.5 in each case. First, ignore the $k = 28$ value because it is associated with the negative frequencies and is in fact a duplicate of the first half of the output. In all Fourier transforms both positive and negative frequencies are calculated and the spectrum has a mirror point about $k = N/2$ ($k = 16$ in this case). Therefore the only component with any energy is at $k = 4$ which has a frequency of 12.5 Hz. as expected. Therefore the DFT has taken a raw data set and, by performing a Fourier transform, been able to identify the frequency component in the raw data. The results are plotted in Figure 46.4 where both the input and output are shown.

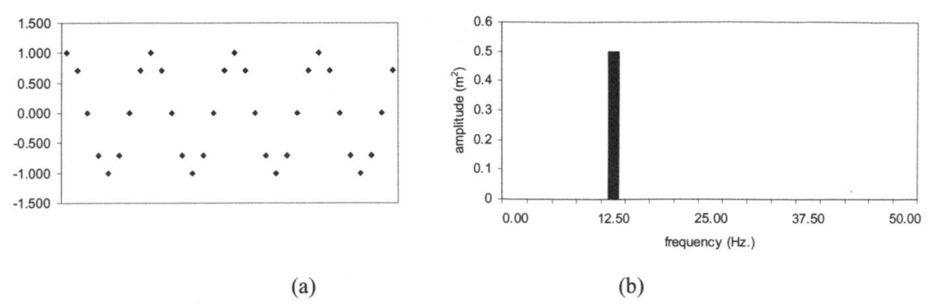


(a) (b)

**Figure 46.4** (a) Raw data and (b) Fourier transform plotted as spectral energy against frequency.

## 46.4 THE FAST FOURIER TRANSFORM IN MATLAB

Matlab includes a fast Fourier transform command as part of its standard repertoire of analysis capabilities. To reproduce the results in the previous section the following code would be required. Note that the FFT does not carry out the divide by the number of points in the data set, hence the need to explicitly carry out that aspect of the calculations. The Matlab file is shown in Figure 46.5 and the output is shown in Figure 46.6.

```
N=32;                    % number of points
dt=0.01;                 % time between data
T=N*dt;                  % total record duration
freq=12.5;               % 12.5 Hz frequency
t=(dt:dt:N*dt);          % generate time points
f=sin(2*pi*freq*t);      % generate signal
F=fft(f)/N;              % fft and scale
w=(0:1/T:1/(2*dt));      % frequency bands
subplot(1,2,1);          % plot input signal
plot(t,f);
axis([0 0.32 -1.5 1.5]);
title('Input signal');
xlabel('time (secs)');
ylabel('amplitude');
subplot(1,2,2);          % plot output transform
bar(w,abs(F(1:17)));     % only the first N/2 + 1 are useful
axis([0 50 0 1]);
title('Fourier transform');
xlabel('frequency (Hz.)');
ylabel('amplitude');
%end
```

**Figure 46.5** Matlab example of a Fourier transform of a simple sine wave.

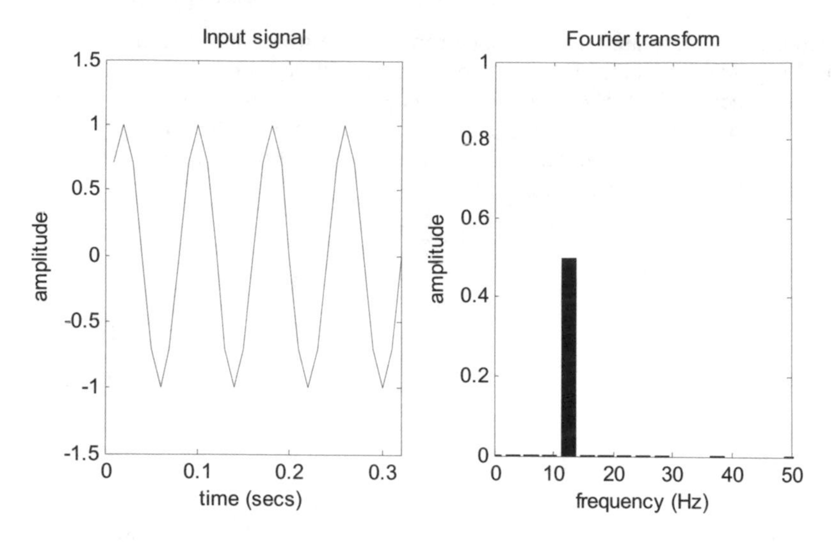


**Figure 46.6** Matlab output from a Fourier transform of a simple sine wave.

## 46.5 THE FAST FOURIER TRANSFORM AND EXCEL

The algorithm for the discrete Fourier transform has been shown to involve of the order of $N^2$ computations for its implementation. For small data sets this does not cause a problem but once $N$ gets large the time taken can quickly become an issue. In the mid 1960s work by Cooley, Tukey and Sande led to the development of an algorithm, the fast Fourier transform (FFT), that allowed the calculation of the Fourier transform in the order of $N\log_2 N$ operations: a much faster result, particularly for large $N$. No attempt will be made to describe the operation of the FFT here; suffice it to say the algorithm works by dividing the data set into two halves, then each of those halves are also halved and so on. For this reason the data set must have a power-of-2 number of points (2, 4, 8, 16, 32, 64, ..., 1024, 2048, etc.). Most spectral analysis packages nowadays would use a version of the FFT.

In many ways the fast Fourier transform that is provided in Excel (Tools – Data Analysis – Fourier Analysis) is an example of how specialist tools should be provided: it works but has some features that are a little different. The end result is that one has to know what is going on to make correct use of it.

The main issue is that when calculating the real and imaginary components the algorithm does not include the divide by $N$ in the calculations. Therefore, transforming a 16-point pure sine wave results in an imaginary component of 8 at the appropriate frequency, a 32-point transform gives a value of 16 and so on. To reduce the real and imaginary components to what is expected from the transform requires the user to divide estimates by $N$.

The Excel algorithm makes use of complex number notation so the transform is a single complex (real and imaginary components) number. To extract the real part, use the function `IMREAL`. To extract the imaginary component use the function `IMAGINARY`. To combine the two to determine the magnitude (in most cases where a signal is made up of both sine and cosine components) use the function `IMABS`. The Excel algorithm also forces the user to determine the appropriate frequency values for the N transform elements. No assistance is given with this task and it is necessary to know that the first element is the zero frequency component (the mean value of the series) and that the next $N/2$ elements have frequencies that can be calculated from Equation (46.7).

## 46.6 DISCRETE FOURIER TRANSFORM SUMMARY

The result of a Fourier transform on discrete data can be summarised quite succinctly. Firstly, assume $N$ data have been collected at a time spacing of $\Delta t$ seconds and a discrete Fourier transform has been calculated.

- The total length of the data set $T$ can be calculated as: $T = N\Delta t$.
- With $N$ data there will be $1 + N/2$ useful spectral components. The first component is associated with the mean of the data series. The remaining components represent the negative frequencies and are, in fact, a mirror image of the positive frequency components.
- The Nyquist frequency for the spectrum (the highest frequency in the spectrum) can be calculated as: $f_{NQ} = 1/2\Delta t$.
- The frequencies of the $1 + N/2$ spectral components will range from 0 Hz to $f_{NQ}$ Hz where $f_{NQ}$ is the Nyquist frequency.
- The width of each spectral component (or the spacing between them, to put it another way) will be $1/T$ Hz (assuming $\Delta t$ is measured in seconds; if this is not the case the frequency will be in terms of cycles per minute or cycles per hour as appropriate to the $\Delta t$ value).

## 46.7 THE SAMPLING THEOREM

Suppose that a set of data have been collected from a signal with the aim of determining the characteristics of that signal. To begin with a data point was sampled every 5 seconds over a period of 20 seconds. The data are plotted in Figure 46.7. The key question is: what can be deduced about the underling process that produced those data? It would appear that the process is oscillating with a gradually increasing amplitude.

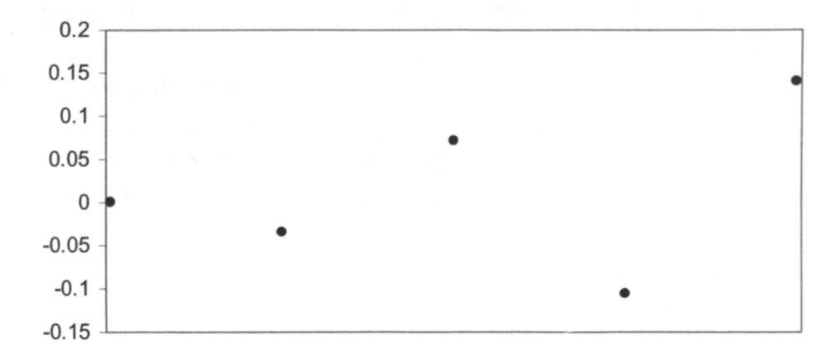


**Figure 46.7** Data sampled at 5-second intervals for a total of 20 seconds.

Now suppose that the data collection interval is reduced to 3 seconds and the process is still sampled for 20 seconds. The result is shown in Figure 46.8. The data seem to confirm the previous conclusion.

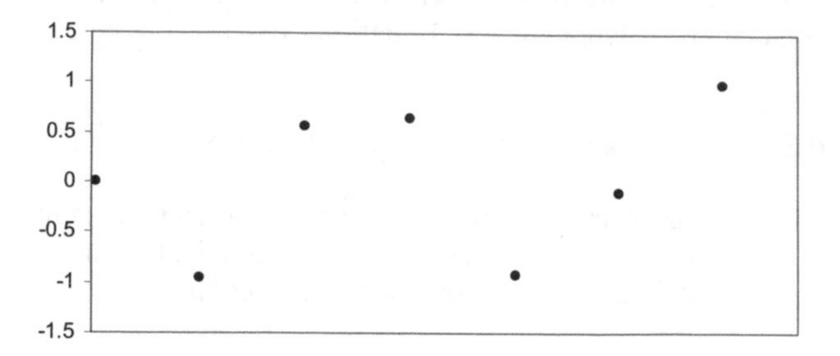


**Figure 46.8** Data sampled at 3-second intervals for a total of 20 seconds.

Once again, the sampling interval is reduced – this time to 1.5 seconds. The result is shown in Figure 46.9. The record now contains many more points and the trend of increasing amplitude as time goes by appears to have disappeared. It may just be some sort of oscillatory behaviour although the period seems to be varying through the record.

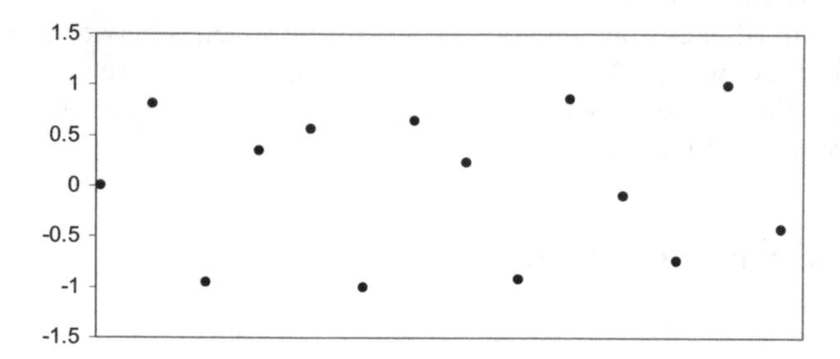


**Figure 46.9** Data sampled at 1.5-second intervals for a total of 20 seconds.

Finally, the data are sampled each 0.25 seconds and then plotted. This is shown in Figure 46.10. Finally the true pattern of the signal is revealed. It is a pure sine wave. The period is not a multiple of 0.25 seconds and the sampled points are out of synchronisation with the underlying sine wave.

The point to be made here is that if one does not sample a signal quickly enough, then not only will the underlying properties of the signal be missed completely, but a completely false set of properties will be suggested. Look back again to the case where the signal was sampled at 5-second intervals. The truth of the underlying signal has been missed but a perfectly reasonable alternative has been presented. Unless the signal is sampled quickly enough the truth will be hidden and replaced by a plausible alternative. This is the basis of the sampling theorem.

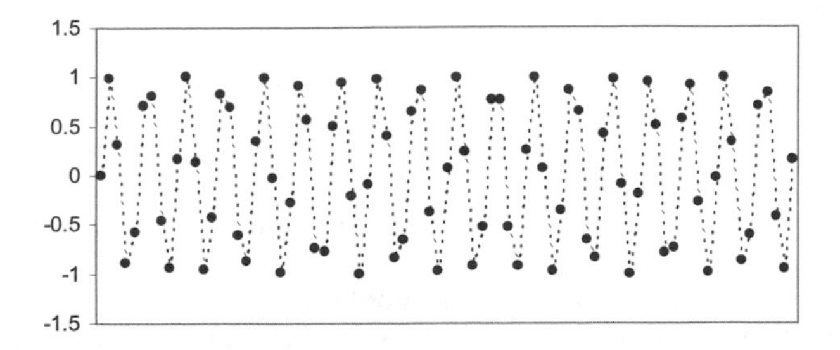


**Figure 46.10** Data sampled at 0.25-second intervals for a total of 20 seconds.

In summary: the collection of data puts restrictions on the analysis that can be carried out subsequently. It is summarised in the sampling theorem, which states that it is not possible to determine any information about frequencies higher than the Nyquist frequency in any data set. Of more consequence is the fact that the presence of such frequencies will contaminate information in the lower frequencies as well. This is because high frequencies are misinterpreted as low frequencies, and the contamination is referred to as aliasing. As a consequence the sampling frequency must be set such that it is at least twice the frequency of the highest frequency that is present in the phenomenon being sampled. This is generally achieved either by sampling at a sufficient rate or by filtering out higher frequencies in the original continuous signal using analogue electronics. It is not possible to filter out higher frequencies once the data have been sampled.

## 46.8 EXAMPLES FROM DATA COLLECTION STUDIES

In a study of sediment flux in Swedish and Swiss lakes Weyhenmeyer and Bloesch (2001) installed sediment traps and sampled them every two weeks. They were trying to investigate the seasonal variation in sediment production but found that by sampling too infrequently they were missing the large variation that might be expected to occur over a day, and the results of their analysis were therefore suspect. They concluded that their main conclusions were "partly caused by sampling frequency".

A similar problem was encountered by researchers investigating currents off the California coast (Di Lorenzo, 2003). Although a data set going back 60 years was available its spatial resolution (80 km) and temporal resolution (readings every 3 months) left the author with "an incomplete understanding of the rich mesoscale oceanic structure". So much data, so little use!

Horn (2002) reviewed two decades of research into the interaction between coastal tides and waves and groundwater. She suggested, based on a knowledge of the frequency of the likely changes, that if groundwater levels were being measured some distance inland from the coast "watertable elevations only need to [be] obtained every 15–20 minutes".

## PROBLEMS

**46.1** Write a program to carry out the Fourier transform and reproduce the results shown in the example problem in Section 46.3. Use the program to investigate the behaviour of the spectrum in a variety of situations. For example:

(a) What happens to the real and imaginary components if the amplitude of the signal doubles?
(b) What happens to the energy if the amplitude of the signal doubles?
(c) It has been stated that the Fourier transform of a pure sinc wave should be a single spike of energy at the appropriate energy, but investigate what happens if the frequency of the input signal changes slightly from the 12.5 Hz to (say) 13 Hz Explain the result.
(d) As the amplitude and frequency of the signal changes, this brings associated changes to the spectrum. However, the frequency bands (0.0, 3.13, 6.25 Hz) in the spectrum remain constant in value. Why is this?
(e) Determine the spectrum for a pure sine wave with a frequency this is mid-way between two of the frequencies. Try, for example, a sine wave with a frequency of 26.56 Hz Explain the resulting spectrum.

**46.2** Modify the program written in Problem 46.1 to have an input time series that is the sum of two sine waves of different frequencies and amplitudes. Determine the output spectrum and verify that the two waves have been identified.

**46.3** The trace in Figure 46.11 shows data recorded on a patient at a local hospital. If it is required to digitise the data and carry out spectral analysis, what is the key factor in deciding the digitisation frequency (sampling rate)? Select and justify a digitisation frequency.

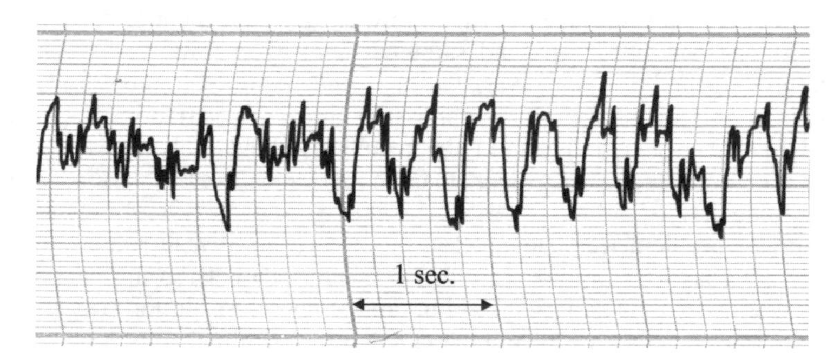


**Figure 46.11** Trace of a patient's vital signs.

**46.4** In Section 46.8 it was stated that in a study of sediment dynamics in a set of lakes, the sampling frequency of a reading every two weeks was insufficient. Suggest and justify a better value for the inter-record duration.

# CHAPTER FORTY SEVEN

# Spectral Analysis (Practical Aspects I)

## 47.1 INTRODUCTION

There are a number of practical issues with the Fourier transform that must be understood if it is to be used correctly. In some instances these may seem minor, but they can be crucial to the proper operation of any spectral analysis programs. These will be outlined in the following two chapters.

## 47.2 DATA PRE-PROCESSING

In the outline of the theory of the Fourier transform it was mentioned that the equations assumed that the function being transformed was periodic. Not much was made of this at the time, but it turns out that this is crucial to a proper analysis of the results. If a data set is being analysed the assumption of periodicity means that, in effect, what is actually being analysed is a repeating pattern made up of the finite length data set (see Figure 47.1).

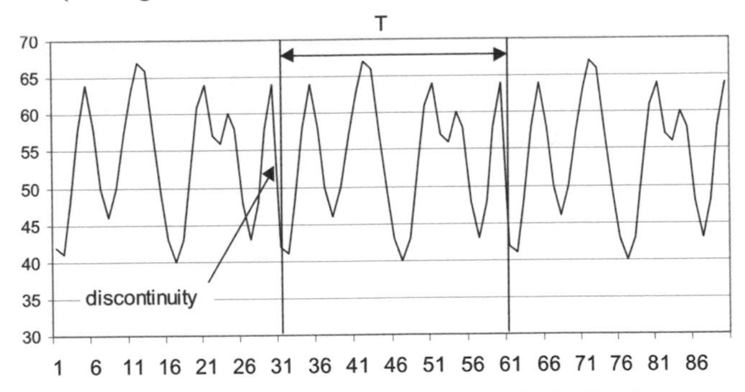


**Figure 47.1** A data set as it would be represented for Fourier analysis where it is repeated to infinity.

This may seem to cause few problems, but consider for a moment what happens at the point where the data set finishes and is about to repeat itself: there is likely to be a discontinuity in the signal from the last point of the data set to the first point. The situation would be worse for a data set where there was a trend in the data as this would lead to a more significant discontinuity in the data representation. To overcome these issues a number of pre-processing tasks are normally required to obtain an accurate Fourier transform.

*1. Remove any mean from the data.*
This task is accomplished by determining the mean and subtracting it from the data.

*2. Remove any trend in the data or any wavelengths longer than the data set.*
A Fortran subroutine to remove a linear trend is given in Figure 47.2.

```
subroutine detrend(y)
  ! takes out any mean trend in data
  implicit none
  real :: y(:) ! data
  integer :: i,n  ! data index, size
  real    :: alpha, beta ! parameters
  real    :: num,den ! temporary
  real    :: my,mx    ! mean values
  n = size(y)
  mx = real(n)/2.0
  my = sum(y)/real(n)
  do i = 0,n-1
    num = num+(i-mx)*(y(i+1)-my)
    den = den+(i-mx)**2.0
  end do
  beta  = num/den
  alpha = my - beta*mx
  do i = 0,n-1
    y(i+1)  = y(i+1)-alpha-beta*i
  end do
end subroutine
```

**Figure 47.2** A Fortran subroutine to remove a linear trend from a data set.

*3. Smooth the ends of the data with a window that removes discontinuities.*
The discontinuity at the start and end of the data series can lead to false energy components and one method of reducing this effect is to gradually ramp up the data by multiplying the raw data by a suitable function such as the a sine or cosine wave (see Figure 47.3).

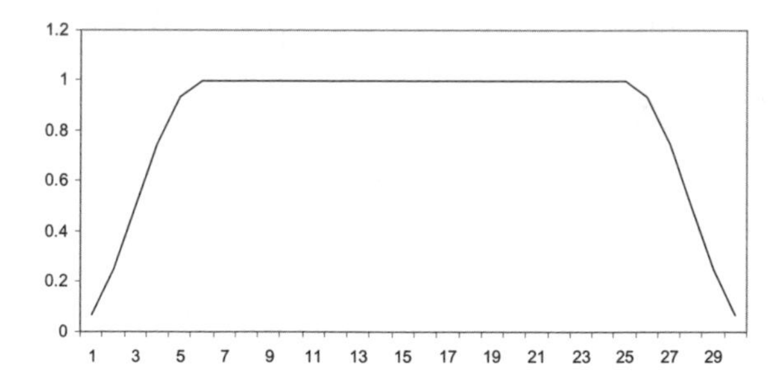


**Figure 47.3** Cosine bell window for data transformation.

A routine (listed in Figure 47.4) to multiply the raw data with a cosine bell for the first and last 10% is given below. As this reduces the strength of the signal for part of it, a scaling factor is applied to the rest of the data or to the final spectrum. The effect on the raw data of the window is shown in Figure 47.5.

```
subroutine modulate(x)
 ! modulates data with half cosine
 ! bell, 10% at start and finish and increases raw
 ! data values to compensate for lost power
  implicit none
  real :: x(:) ! data
  real, parameter :: pi=3.1415926
  integer :: i,j,k,n
  n = size(x)
  k=n/10 ! integer truncation
  do i=1,k
   x(i)=0.5*(1.0-cos(pi*real(i-1)/real(k)))*x(i)
   j = n-k+i
   x(j)=0.5*(1+cos(pi*real(j-n+k)/k))*x(j)
  end do
  x(:)=x(:)*1.078 !correct energy level
end subroutine
```

**Figure 47.4** A Fortran subroutine to apply a cosine bell to a data set.

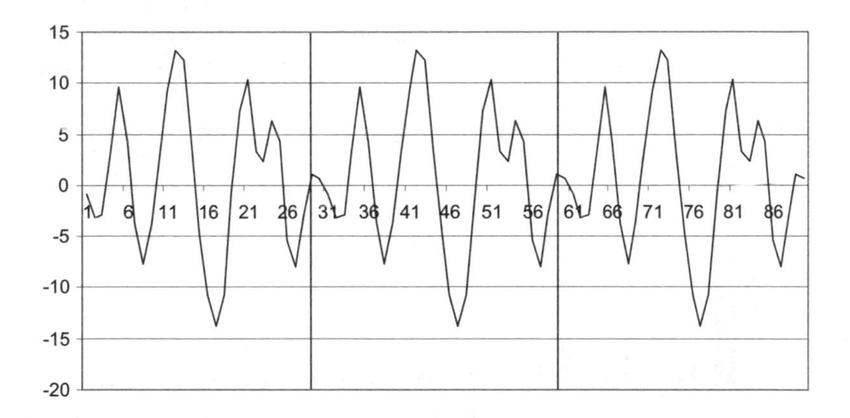


**Figure 47.5** The transformed data that has had the mean removed and had a cosine bell transformation applied to it. Note that the data now join up at the start and end of each set.

Other window shapes and methods of treating the data are available in standard texts. Rodriguez et al. (1999) carried out some comparisons with a range of these and found that there was little variation when the total energy under the spectrum was concerned, but the peak of the spectrum did show some sensitivity to the window chosen. A description of these other windows is beyond the scope of this text.

## 47.3 WORKED EXAMPLE: WATER TEMPERATURE SERIES

Temperature data collected at a station on the Torrens Lake were subjected to spectral analysis. The data were taken from records collected during the last week in March 2002 and the raw data are shown in Figure 47.6.

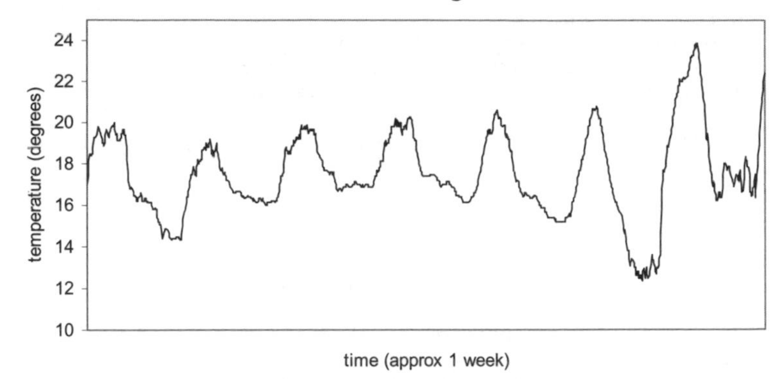


**Figure 47.6** Raw temperature data from Torrens Lake, March 2002. Data were collected at 10-minute intervals, with 1024 points plotted.

Excel was used to determine the spectrum based on the raw data (Menu – Tools – Analysis – Fourier Transform). The resulting spectrum is shown in Figure 47.7. There is certainly a major spike at a period of approximately 24 hours, but notice also that there are spikes of low frequency that have similar or larger energy.

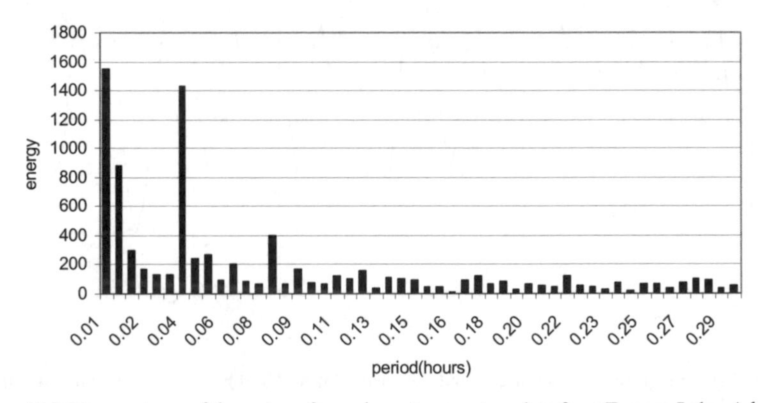


**Figure 47.7** The spectrum of the untransformed raw temperature data from Torrens Lake. A large spike at 0 frequency has been removed to allow the other detail to be seen.

The data had their mean removed and were then transformed using a cosine bell. The resulting spectrum is shown in Figure 47.8. The effect of removing the mean and applying the cosine bell to the data is quite clearly evident. The large spike at 0 frequency has gone, as have the spurious spikes of 'energy' at the low frequencies leaving the 24-hour period spike. While it is true that there are still some low-frequency energy spikes, reference to the data indicates that with the range of the daily temperatures increasing through the record there should be some longer period variation to be accounted for. The higher-frequency oscillations in the raw data are also seen in the spectrum in Figure 47.8.

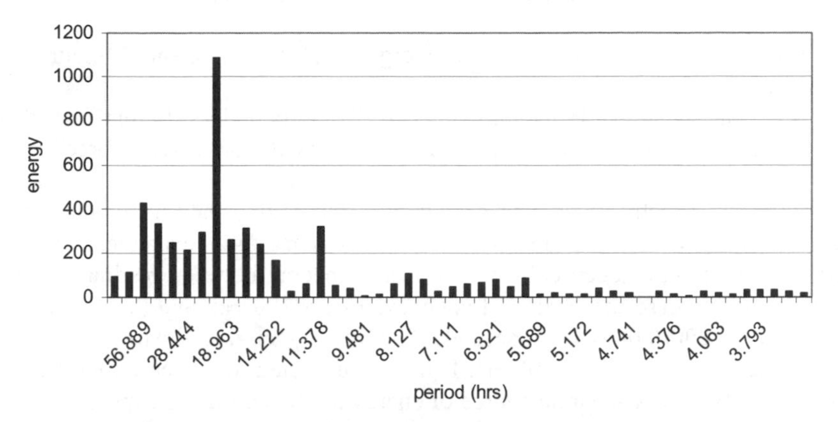


**Figure 47.8** The spectrum of the transformed temperature data from Torrens Lake. The energy at 24 hours is clearly evident.

## 47.4 SPECTRAL ENERGY

When a spectrum is plotted it is usual to plot the spectral energy density against the frequency. In this case, the energy density can be determined from the raw Fourier transform components by:

$$E(f) = T|F(f)|^2 \tag{47.1}$$

where $E(f)$ is the energy spectral density, $T$ is the total length of the record being transformed and $F(f)$ is the absolute value of the Fourier components. There is occasionally a factor of 2 in the equation to ensure that the sum of the energy under the spectrum is equal to the variance of the record (Young, 1999), but this may be due to failing to ignore the negative energies in the spectrum as noted in Chapter 46. The variance of the record can be determined as:

$$\sigma^2 = \sum_{i=1}^{N/2} E(f)\Delta f \tag{47.2}$$

where $\sigma^2$ is the variance of the data set, $N$ is the number of data (meaning there are $N/2$ valid spectral components), $E(f)$ is the energy density and $\Delta f$ is the width of the spectral bands ($\Delta f = 1/T$), $T$ is the total record length ($T = N\Delta t$) and $\Delta t$ is the time between data points. For example, where the interval between subsequent raw data readings has been measured in seconds and the data are in metres, the frequency will be in Hertz (cycles per second) and the energy in $m^2$/Hz or $m^2$s.

## 47.5 FREQUENCY RESOLUTION

As a first step, consider the spectral plot based on temperature data collected in the Torrens Lake (Figure 47.6). There are a few points to ponder:

1. How sure can one be about the energy level shown in the 24-hour period energy band?
2. Is the energy shown in the period of 11.378 hours really different from that in the adjacent band or are they part of the same oscillatory driving force?

With $N/2$ spectral estimates spread over the frequencies up to the Nyquist frequency the raw estimates are spaced at $1/T$ Hz apart. As an example, assume that 200 data points have been collected at a time spacing between readings of 0.005 seconds. In this case the Nyquist frequency, $F_{NQ}$ = 100 Hz, and the length of the record, $T$ = 200(0.005) = 1 second. The raw spectral estimates would have frequencies of 0, 1, 2, 3, ... 100 Hz. If it was suspected in this case that there was energy at 5 Hz and a separate source of energy at 10 Hz, then the spectrum should be able to show that there are 4 components (6 Hz to 9 Hz) spaced between the two frequencies of interest. If, however, it was suspected that there was energy at 5 Hz and a separate source at 5.5 Hz the raw estimates would be spaced too close together to be able to separate this. In this case it would be desirable to have no greater than a 0.5 Hz spacing, meaning that 200 spectral components and, therefore, 400 data points were required. A larger number of data collected at the same sampling frequency would further enhance the resolution capability of the analysis.

## 47.6 SPECTRAL CONFIDENCE

The output from an FFT is based on the data that have been chosen, and the energies that are calculated are in fact just estimates of the actual energy at each frequency. It has been shown (e.g. Newland, 1984; Young, 1999) that the distribution of spectral estimates follows a chi-squared distribution where the degrees of freedom parameter, $d$, can be approximated as $d = 2\,p$ where $p$ is the number of raw spectral bands that were averaged together to form the spectrum. The chi-squared distribution with $d$ degrees of freedom is the distribution of the sum of $d$ independent squared standard normal variables. The raw spectral estimates (of which there are $N/2$, where $N$ is the number of data points analysed) have no averaging and therefore have 2 degrees of freedom. Table 47.1 sets out the values that can be used to determine the confidence limits that apply to the spectral estimates.

Given a spectral estimate of $m$, the best estimate for the actual value, $e$, is in a range of values for a particular confidence interval $i$ with $d$ degrees of freedom that can be calculated from:

$$\frac{m}{\chi^2(X_{\frac{1-i}{2}}, d)} > e > \frac{m}{\chi^2(X_{\frac{i+1}{2}}, d)} \tag{47.3}$$

For example, the 80% confidence limits for 10 degrees of freedom are:

$$\frac{m}{\chi^2(X_{0.1}, 10)} > e > \frac{m}{\chi^2(X_{0.9}, 10)} = \frac{m}{0.487} > e > \frac{m}{1.599} = 2.05m > e > 0.625m$$

**Table 47.1** $\chi^2$ values for a range of confidence limits. Generated in Excel using CHIINV().

| DoF | $X_{0.01}$ | $X_{0.02}$ | $X_{0.05}$ | $X_{0.10}$ | $X_{0.90}$ | $X_{0.95}$ | $X_{0.98}$ | $X_{0.99}$ |
|---|---|---|---|---|---|---|---|---|
| 2 | 0.010 | 0.020 | 0.051 | 0.105 | 2.303 | 2.996 | 3.912 | 4.605 |
| 4 | 0.074 | 0.107 | 0.178 | 0.266 | 1.945 | 2.372 | 2.917 | 3.319 |
| 6 | 0.145 | 0.189 | 0.273 | 0.367 | 1.774 | 2.099 | 2.506 | 2.802 |
| 8 | 0.206 | 0.254 | 0.342 | 0.436 | 1.670 | 1.938 | 2.271 | 2.511 |
| 10 | 0.256 | 0.306 | 0.394 | 0.487 | 1.599 | 1.831 | 2.116 | 2.321 |
| 12 | 0.298 | 0.348 | 0.436 | 0.525 | 1.546 | 1.752 | 2.004 | 2.185 |
| 14 | 0.333 | 0.383 | 0.469 | 0.556 | 1.505 | 1.692 | 1.919 | 2.082 |
| 16 | 0.363 | 0.413 | 0.498 | 0.582 | 1.471 | 1.644 | 1.852 | 2.000 |
| 18 | 0.390 | 0.439 | 0.522 | 0.604 | 1.444 | 1.604 | 1.797 | 1.934 |
| 20 | 0.413 | 0.462 | 0.543 | 0.622 | 1.421 | 1.571 | 1.751 | 1.878 |
| 22 | 0.434 | 0.482 | 0.561 | 0.638 | 1.401 | 1.542 | 1.712 | 1.831 |
| 24 | 0.452 | 0.500 | 0.577 | 0.652 | 1.383 | 1.517 | 1.678 | 1.791 |
| 26 | 0.469 | 0.516 | 0.592 | 0.665 | 1.368 | 1.496 | 1.648 | 1.755 |
| 28 | 0.484 | 0.530 | 0.605 | 0.676 | 1.354 | 1.476 | 1.622 | 1.724 |
| 30 | 0.498 | 0.544 | 0.616 | 0.687 | 1.342 | 1.459 | 1.599 | 1.696 |
| 32 | 0.511 | 0.556 | 0.627 | 0.696 | 1.331 | 1.444 | 1.578 | 1.671 |
| 34 | 0.523 | 0.567 | 0.637 | 0.704 | 1.321 | 1.429 | 1.559 | 1.649 |
| 36 | 0.534 | 0.577 | 0.646 | 0.712 | 1.311 | 1.417 | 1.541 | 1.628 |
| 38 | 0.545 | 0.587 | 0.655 | 0.720 | 1.303 | 1.405 | 1.525 | 1.610 |
| 40 | 0.554 | 0.596 | 0.663 | 0.726 | 1.295 | 1.394 | 1.511 | 1.592 |
| 50 | 0.594 | 0.633 | 0.695 | 0.754 | 1.263 | 1.350 | 1.452 | 1.523 |
| 60 | 0.625 | 0.662 | 0.720 | 0.774 | 1.240 | 1.318 | 1.410 | 1.473 |
| 70 | 0.649 | 0.684 | 0.739 | 0.790 | 1.222 | 1.293 | 1.377 | 1.435 |
| 80 | 0.669 | 0.703 | 0.755 | 0.803 | 1.207 | 1.273 | 1.351 | 1.404 |
| 90 | 0.686 | 0.718 | 0.768 | 0.814 | 1.195 | 1.257 | 1.329 | 1.379 |
| 100 | 0.701 | 0.731 | 0.779 | 0.824 | 1.185 | 1.243 | 1.311 | 1.358 |

*Example*
Determine the 90% confidence limits for the raw spectral estimate which has a value of 10 $m^2$/Hz.

*Solution*
Degrees of freedom = 2 × 1 = 2, and for 90% confidence limits require area under curve between 5% and 90%. Therefore $X_{95}$ = 2.996, $X_{05}$ = 0.051. The 90% confidence limits are therefore:

$$\frac{10}{2.996} < x < \frac{10}{0.051}$$

so, while the estimate is 10, the range of values to be 90% confident is from 3.34 to 196.1 $m^2$/Hz. This is an excellent illustration of why the raw spectral estimates are rarely quoted. One can have very poor confidence in each estimate. For 100 degrees of freedom, achieved by averaging over 50 raw spectral estimates, the range of values for 90% confidence are: 8.05 $m^2/Hz < x < 12.84$ $m^2$/Hz.

## PROBLEMS

**47.1** When testing Fourier transform routines with pure sine waves, why would it not be necessary to use a cosine bell, or similar, to transform the input data for the situation $N = 50$, $\Delta t = 2.0$ seconds and $f = 0.1$ Hz but it would be necessary if the frequency of the input signal was $f = 0.15$ Hz?

**47.2** Explain the problem that occurs with spectral analysis when attempting to analyse a data set where there is a linear upward trend in the record.

**47.3** The variance of a unit amplitude sine wave can be shown to be 0.50. Based on the Fourier transform of a simple sine wave, show that the energy under the spectrum will be equal to the variance of the signal.

**47.4** A spectral analysis has been carried out on a record of 1024 data values. How many raw spectral estimates would this generate? If averaging was carried out over 15 adjacent bands, what are the 90% confidence limits for any particular spectral estimate? If the original data had been collected with $\Delta t = 0.1$ seconds, what would be the width of the raw spectral estimates and the width of the spectral bands in the final spectrum?

**47.5** It has been suggested that there are advantages in padding a data set with 0s to make the number up to a power of 2 rather than discarding points. For example, with 8000 data points it might be a good idea to add 96 zeros to the start and end to make the total up to 8192, which is $2^{13}$. Comment on that idea, its advantages and disadvantages.

## CHAPTER FORTY EIGHT

# Spectral Analysis (Practical Aspects II)

### 48.1 INTRODUCTION

To overcome the problems associated with poor confidence levels in the raw spectral estimates, the most common approach is to average over adjacent energy bands. It will be seen that while this improves the confidence it also reduces the frequency resolution, so that once the size of the data set has been determined there is a trade-off between resolution and confidence. The effect of different effective bandwidths is illustrated for an ocean wave data set containing 2048 values. Figure 48.1 shows the full spectrum without any smoothing. It is evident from the plot that while the peak energy is associated with a frequency of approximately 0.12 Hz, there are also a number of apparent spikes of energy in nearby frequency bands. They key question is: how confident could one be in each of the individual energy spikes? The quick answer is: not very. In fact, the 90% confidence interval for the raw spectral estimates indicate that the height of any particular energy spike could be between 0.33 and 19 times the height shown (see Chapter 47).

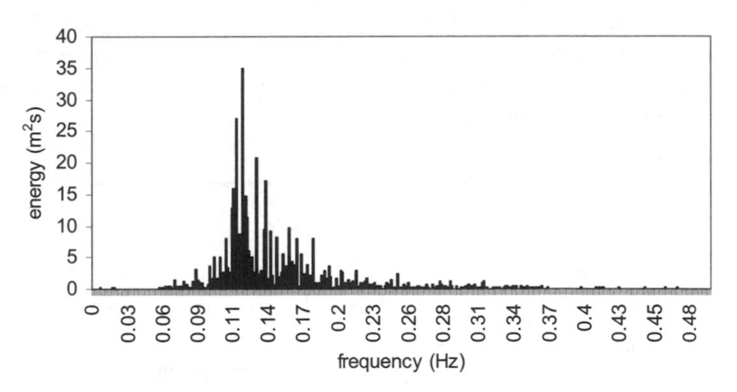


**Figure 48.1** A wave energy spectrum where 2048 data led to 1024 spectral components.

In the next plot, Figure 48.2, averaging over 16 adjacent frequency bands has been carried out to reduce the total number of spectral components to 64. This results in 32 degrees of freedom in the $\chi^2$ distribution and means that the 90%

confidence interval for any particular energy spike is between (approximately) 70% and 160% of the value shown. While much of the resolution has been lost, the lack of confidence in the raw spectrum means that few reliable conclusions could be drawn from it anyway.

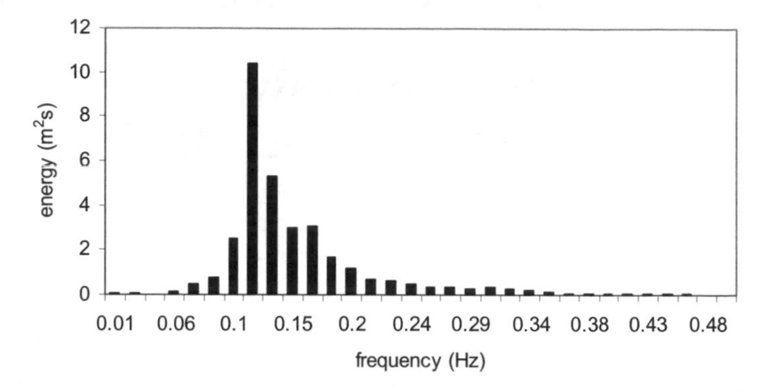


**Figure 48.2** A wave energy spectrum where 2048 data led to 64 spectral components.

## 48.2 EXCEL EXAMPLE

To illustrate the procedure of determining a spectrum a time series was generated in Excel as the sum of two sin waves and a random component. The equation for the data can be written:

$$f_t = A_1 \sin(2\pi t / T_1 + \theta_1) + A_2 \sin(2\pi t / T_2 + \theta_2) + R_t \tag{48.1}$$

where, for this example, $A_1 = 1.0$, $T_1 = 4.0$, $\theta_1 = 0.0$, $A_2 = 0.5$, $T_2 = 7$, $\theta_2 = \pi/6$ and $R$ is a random number of uniform distribution. In total, 32 data were generated with a time spacing ($\Delta t$) of 0.5 seconds, giving a total record length ($T$) of 16 seconds. The data are plotted in Figure 48.3.

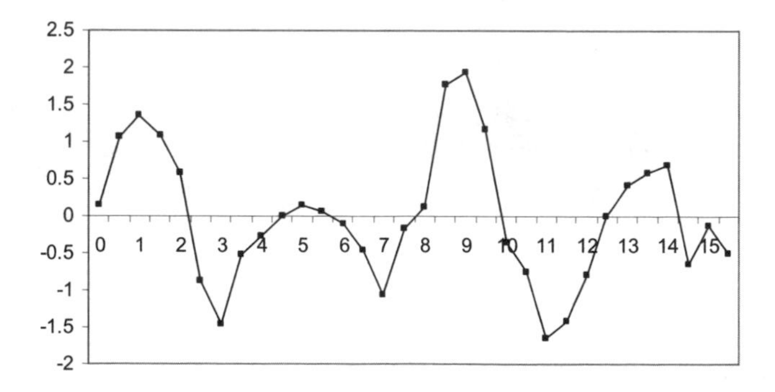


**Figure 48.3** A raw data set comprising two sine waves and a random component.

Selecting the Fourier Analysis option from Tools – Data Analysis and using the 32 data points described earlier gave an output of 32 complex numbers. Recalling that the second half of the output is a mirror image of the first and therefore ignoring it left 17 (1 + *N*/2) values. These complex numbers were separated into real and imaginary components and divided by N to give the correct values. The energy was determined ($e_k = r_k^2 + i_k^2$) and the energy per spectral band width was calculated by dividing by 1/*T* (1/16 in this case). The output is shown in Table 48.1. The total energy was 0.784 $m^2$, which was very close, as expected, to the variance of the raw data (0.808 $m^2$). The spectrum is plotted in Figure 48.4.

**Table 48.1** Output from Fourier analysis.

| k | real | imaginary | real/N | imaginary/N | energy | Freq | energy/Hz |
|---|---|---|---|---|---|---|---|
| 0 | 0.000 | 0.000 | 0.000 | 0.000 | 0.000 | 0.000 | 0.000 |
| 1 | 0.879 | –0.307 | 0.027 | –0.010 | 0.001 | 0.063 | 0.014 |
| 2 | 9.211 | –1.827 | 0.288 | –0.057 | 0.086 | 0.125 | 1.378 |
| 3 | –3.219 | 2.787 | –0.101 | 0.087 | 0.018 | 0.188 | 0.283 |
| 4 | –1.554 | –15.950 | –0.049 | –0.498 | 0.251 | 0.250 | 4.013 |
| 5 | –0.398 | 0.110 | –0.012 | 0.003 | 0.000 | 0.313 | 0.003 |
| 6 | –2.364 | –2.002 | –0.074 | –0.063 | 0.009 | 0.375 | 0.150 |
| 7 | 0.108 | 2.264 | 0.003 | 0.071 | 0.005 | 0.438 | 0.080 |
| 8 | 0.482 | 0.201 | 0.015 | 0.006 | 0.000 | 0.500 | 0.004 |
| 9 | 1.304 | –0.303 | 0.041 | –0.009 | 0.002 | 0.563 | 0.028 |
| 10 | –1.528 | –0.099 | –0.048 | –0.003 | 0.002 | 0.625 | 0.037 |
| 11 | –1.374 | –1.764 | –0.043 | –0.055 | 0.005 | 0.688 | 0.078 |
| 12 | –1.648 | 0.359 | –0.051 | 0.011 | 0.003 | 0.750 | 0.044 |
| 13 | 0.191 | –0.812 | 0.006 | –0.025 | 0.001 | 0.813 | 0.011 |
| 14 | 0.029 | –1.357 | 0.001 | –0.042 | 0.002 | 0.875 | 0.029 |
| 15 | 2.590 | –0.367 | 0.081 | –0.011 | 0.007 | 0.938 | 0.107 |
| 16 | –0.815 | 0.000 | –0.025 | 0.000 | 0.001 | 1.000 | 0.010 |

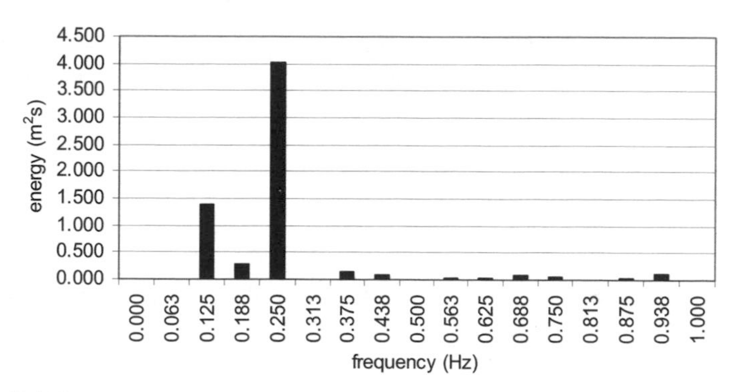


**Figure 48.4** The spectrum of the data set comprising two sine waves and a random component.

Inspection of Figure 48.4 shows a number of interesting features:

1. The analysis has identified quite clearly that there are two dominant period waves in the data and that the frequencies are closest to 0.125 Hz and 0.250 Hz.
2. In addition to the waves there is some noise in the signal contributing small energies to the higher frequencies.
3. The 4-second period wave has been identified correctly with the peak at 0.250 Hz.
4. The 7-second period wave, which has a frequency of 0.143 Hz, has caused energy to be allocated to the two nearest frequency bands (0.125 Hz and 0.188 Hz). This is a reflection of the relatively poor resolution of the analysis brought on by the wide spectral bands, which are in turn due to the very short time series that was analysed. If a longer data set was generated and analysed a more accurate result could be expected. For example, generating 1024 data points gives the spectrum shown in Figure 48.5, where the 7-second wave has been identified in the frequency band of 0.143 Hz Note that, although the energy density is much higher, the total energy remains the same since it is the product of the density and the individual spectral bandwidth, which is much smaller in this case (being equal to $1/T$). The variances of the raw data and the total energy are now virtually identical at 0.720 $m^2$.

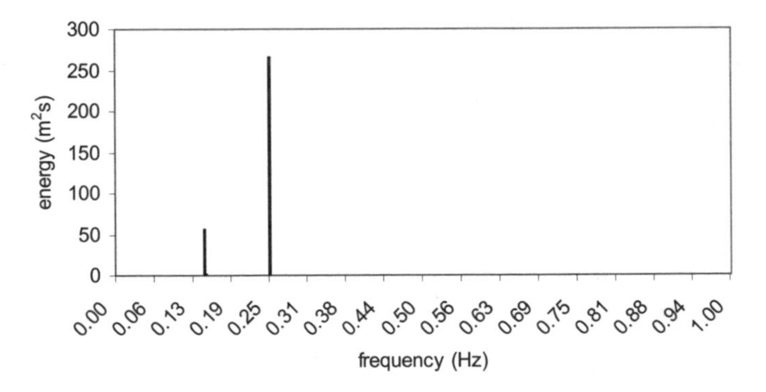


**Figure 48.5** The spectrum of the 1024-point data set comprising two sine waves and a random component.

## 48.3 PRACTICAL ASPECTS OF SPECTRAL ANALYSIS

From the brief look so far a number of features are clear:

1. Sampling must be at a frequency at least double the highest frequency present in the signal being investigated.

2. The maximum frequency that can be determined (the Nyquist frequency) is calculated from the sampling frequency; $f_{NQ} = 1/(2\Delta t)$ where $\Delta t$ is the time interval between readings.
3. Since there are $N/2$ spectral estimates produced by a DFT (or FFT) the frequency resolution (bandwidth) is $f_{NQ}/(N/2) = 1/T$ where $T$ is the total length of the data set ($T = N\Delta t$). Therefore the greater $N$ the better the frequency resolution possible.
4. Although there are $N/2$ spectral estimates it is advisable to average over several adjacent ones to reduce the number of estimates and increase the bandwidth. The advantage of this is that it increases the confidence in the estimates.
5. Averaging increases the effective bandwidth, which improves the confidence in the estimates that result. The drawback is that it reduces the resolution – that is, the ability to differentiate between energy at frequencies that are close to one another.

It is important, therefore, that if useful analysis is to be undertaken, the experiment should be correctly set up.

## 48.4 DATA COLLECTION FOR SPECTRAL ANALYSIS

In setting up any data collection where spectral analysis is to be performed, there are a number of decisions that must be made. These decisions determine the performance of the spectral analysis. The decisions are:

- sampling frequency (and therefore $\Delta t$);
- length of the data set ($N$);
- required frequency resolution;
- required level of confidence in each estimate.

*Problem*
Design a data collection strategy for monitoring vibration in a building. Assume that it is known that the highest frequency present in the building is of the order of 50 Hz. A frequency resolution of 1 Hz is required. Assume 50 degrees of freedom are required to give reasonable confidence in the final estimates.

How long should the sample be? At what frequency should the data be collected? What averaging must be employed in the spectral estimates to achieve the desired confidence?

*Solution*
According to the sampling theorem the sampling frequency must be at least double the highest frequency present. Therefore, select a sampling frequency of 100 Hz giving $\Delta t = 0.01$ s. This gives the Nyquist frequency $= 1/2\Delta t = 50$ Hz. Therefore the spectrum will be able to show the full range of frequencies that are believed to

be able to be present in any data set. Any lower frequency sampling would lead to a reduced Nyquist frequency and the potential for aliasing in the spectrum to spoil the results.

If 50 degrees of freedom are required it will be necessary to average over 25 adjacent bands (since d.o.f. = $2p$). If it is going to be possible to resolve energy that is 1 Hz apart, it will be necessary for the adjacent frequency bands to be approximately half this value apart. Therefore the final effective bandwidth of the estimates should be 0.5 Hz. This 0.5-Hz width will be made up by averaging over 25 raw spectral bands so the raw bands must be 0.5/25 Hz wide or 0.02 Hz. With a Nyquist frequency of 50 Hz, there will need to be at least 50/0.02 = 2,500 raw spectral bands in the spectrum. Since there are $N/2$ raw spectral bands the number of data will have to be at least 5,000. Given this number of data it would be wise to use an FFT, so a power of 2 is required. The next power of 2 greater than 5,000 is $2^{13} = 8192$. Sampling at 100 Hz means that this will involve a data set of duration no shorter than 81.92 seconds. It would be prudent to collect a little longer so that if there are issues at the start or end of the data set some values can be discarded. For this reason select 90 seconds of data.

Therefore, to achieve the objectives, sample for 90 seconds at 100 Hz and, following the Fourier transform, average the raw spectrum over 25 adjacent frequency bands.

## PROBLEMS

**48.1** Design a data collection program to measure the frequency spectrum of the motion of a footbridge under loading conditions due to normal foot traffic. Assume that the highest frequency of movement is likely to be 2 Hz. The data should be suitable for analysis using an FFT and the frequency resolution should be such that there is a spacing of at least 0.1 Hz between adjacent spectral estimates. How long, and at what rate, should one collect data if you require 50 degrees of freedom for the estimates.

**48.2** The South Australian Department of Marine and Harbours operated a wave-recording device at various sites around the South Australian coastline. The device consisted of a float on a vertical pole which was free to move and follow the variations in water surface. The mass of the float and the friction in the vertical pole meant that waves of period less than 1 second (frequency greater than 1 Hz) were effectively filtered out of the recorded motion. Data were collected from the device with records lasting 35 minutes with data being recorded every 0.5 seconds.

(a) If the data are to be analysed using an FFT routine, what length of data set would be chosen ? Why ?
(b) Following some preliminary analysis it is found that the energy is concentrated between frequencies of 0.1 Hz and 0.4 Hz. If it is required to have approximately ±30% confidence in the spectral estimates, what

frequency resolution could be expected? How many spectral bands would have to be averaged over to achieve this resolution ?

(c) To get more detailed information in the frequencies of interest it is suggested that either the sampling interval be halved to 0.25 seconds (for the 35-minute duration) or the recording period be increased to 70 minutes (while retaining the 0.5-second data interval). Comment on these two options. (Note: they are quite different and one of the two is effectively useless.)

## APPENDIX A

# Taylor Series

### A.1 INTRODUCTION

An understanding of the behaviour of functions in the neighbourhood of a point leads to a result that is central to many mathematical developments, including the numerical evaluation of the basic trigonometric and logarithmic functions. What is now known as a Taylor expansion was first published in 1715 by Brook Taylor, although it was known and used much earlier by, among others, Newton. The Taylor expansion for a function *f*() about a point *a* requires that the function be continuous and differentiable at *a* and can be written:

$$f(a+h)=f(a)+h.f'(a)+\frac{h^2}{2!}f''(a)+\frac{h^3}{3!}f'''(a)+\ldots \qquad (A.1)$$

where *f(a)* is the function evaluated at the point *a*, $f'(a)$ is the first derivative evaluated at the point *a*, $f''(a)$ is the second derivative evaluated at the point *a*, and so on, and *h* is a small increment. The expansion is justified by assuming a finite polynomial approximation

$$f(a+h)=c_0+c_1h+c_2h^2+\cdots+c_nh^n \qquad (A.2)$$

and then requiring that the function and *n* derivatives on both sides of the equation be equal when $h = 0$. Thus $c_0 = f(a)$. Differentiating both sides gives:

$$f'(a+h)=c_1+2c_2h+\cdots+nc_nh^{n-1} \qquad (A.3)$$

Substituting h=0 gives $c_1 = f'(a)$. Continuing, the *n*th derivative:

$$f^{(n)}(a+h)=n!c_n \qquad (A.4)$$

and $c_n = f^{(n)}(a)/n!$ where $f^{(n)}$ represents the *n*th derivative.

If the point a is taken as 0 the Taylor series is often referred to as a Maclaurin series. The Maclaurin series is equivalent to, rather than just a special case of, a Taylor series because it is possible to define $g(h)=f(a+h)$.

The preceding argument ignores the crucial question of whether or not the right-hand side converges to the function as the number of terms $n$ increases. In general, the Taylor series will converge provided $|h|$ is less than some number, known as the radius of convergence, which depends on both the function and the point $a$. However, the Taylor series for exponential, sine, and cosine converge for all values of $h$. Kreysig (1983) provides more detail.

*Example*
Calculate the value of $e^{1.1}$ given that $e^{1} = 2.718281828$.

*Solution*
If each term of the expansion is computed individually and then a sum generated ,the individual values and the progressing sum are as shown in Table A.1. It is evident that for this situation, where $h = 0.1$, the required number of terms to get quite an accurate answer (better than 99%) is only 2.

**Table A.1** Calculation of $e^{1.1}$ based on expansion of terms at $e^{1.0}$.

| i | term | value | sum | % error |
|---|---|---|---|---|
| 1 | $f(x)$ | 2.718282 | 2.71828 | 9.516258 |
| 2 | $h\,f'(x)$ | 0.271828 | 2.99011 | 0.467884 |
| 3 | $h^2/2!\,f''(x)$ | 0.013591 | 3.0037 | 0.015465 |
| 4 | ... | 0.000453 | 3.00415 | 0.000385 |
| 5 | ... | 1.13E–05 | 3.00416 | 7.67E–06 |
| exact | | | 3.004166024 | |

If instead a value of $h = 0.2$ is used to calculate $e^{1.1}$ from a known value of $e^{0.9}$, then the summation proceeds as shown in Table A.2. Note that the errors are higher and that more terms are required to reach the same accuracy, but that the solution does converge to the correct value, again in a relatively small number of steps.

**Table A.2** Calculation of $e^{1.1}$ based on expansion of terms at $e^{0.9}$.

| term | value | sum | % error |
|---|---|---|---|
| 1 | 2.459603 | 2.459603 | 18.12692 |
| 2 | 0.491921 | 2.951524 | 1.75231 |
| 3 | 0.049192 | 3.000716 | 0.114848 |
| 4 | 0.003279 | 3.003995 | 0.005684 |
| 5 | 0.000164 | 3.004159 | 0.000226 |
| 6 | 6.56E–06 | 3.004166 | 7.49E–06 |
| 7 | 2.19E–07 | 3.004166 | 2.13E–07 |
| exact | | 3.004166024 | |

## A.2 FUNDAMENTAL TAYLOR (MACLAURIN) SERIES

Taylor series are used for calculating the numerical values of common functions. If $a = 0$, Equation (A.4) is referred to as a Maclaurin series and it is usual to write $x$ rather than $h$ for the increment from 0. An essential example is $f(x) = e^x$. In this case all the derivatives (first, second, etc.) are the same as the function itself and Equation (A.4) can be written:

$$e^x = 1 + x + \frac{x^2}{2!} + \frac{x^3}{3!} + \ldots \tag{A.5}$$

For example, $e^1 = 1 + 1 + 1/2! + 1/3! + \ldots = 2.5$ after 3 terms, 2.716666667 after 5 terms, and 2.718253968 after 7 terms (a 0.001% error). Two other fundamental Maclaurin series are:

$$\sin x = x - \frac{x^3}{3!} + \frac{x^5}{5!} - \frac{x^7}{7!} + - \ldots \tag{A.6}$$

$$\cos x = 1 - \frac{x^2}{2!} + \frac{x^4}{4!} - \frac{x^6}{6!} + - \ldots \tag{A.7}$$

The Maclaurin series for exp($x$), sin($x$), cos($x$) are all convergent for all values of $x$. The following fundamental Maclaurin series are only convergent within the given domain of $x$:

$$(1+x)^r = 1 + rx + \frac{r(r-1)}{2!}x^2 + \frac{r(r-1)(r-2)}{3!}x^3 + \cdots \quad \text{for } -1 < x < 1 \tag{A.8}$$

$$\ln(1+x) = x - \frac{x^2}{2} + \frac{x^3}{3} - \cdots \quad \text{for } -1 < x \leq 1 \tag{A.9}$$

Equation (A.5) is known as the generalised binomial expansion. It is possible to calculate logarithms of any positive number using Equation (A.6) by noting that any positive number can be written as $(1+x)/(1-x)$ for some $x$ and then expanding $\ln((1+x)/(1-x))$.

The calculation of the *sin(x)* function using the Maclaurin series representation of it and the first 9 terms is shown in Figure A.1. Again, it is evident that for small $x$ the approximation is very good and it is only as $x$ increases that more terms are required to give a sufficiently accurate answer.

## A.3 GEOMETRIC SERIES AND CONVERGENCE

The Maclaurin expansion for arctan($x$) can be evaluated at $x$=1 to give an infinite series representation for $\pi$. However convergence is very slow, more than 300 terms are need to get 2-decimal-place accuracy, and it is not a very practical way of approximating $\pi$.

$$\frac{\pi}{4}=1-\frac{1}{3}+\frac{1}{5}-\frac{1}{7}+-... \tag{A.10}$$

Furthermore, computing finite sums can give a misleading impression that a series converges when it does not. A striking example is the harmonic series:

$$1+\frac{1}{2}+\frac{1}{3}+\frac{1}{4}+\frac{1}{5}+\frac{1}{6}+\frac{1}{7}+\frac{1}{8}+\frac{1}{9}+\cdots\geq 1+\frac{1}{2}+\frac{1}{2}+\frac{1}{2}+\cdots$$

which diverges, but very slowly.

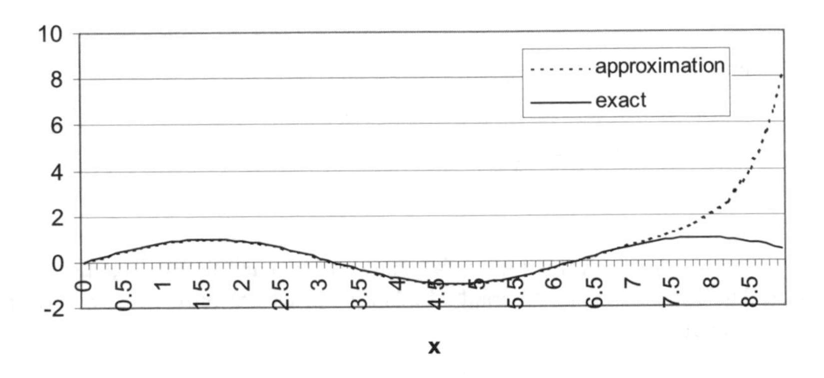


**Figure A.1** Generation of the sine function using 9 terms in the Maclaurin series. The accuracy falls away quickly around an x value of 7.

The convergence of infinite series is often proved by demonstrating that the terms are less in absolute magnitude than some convergent geometric series. The sum to infinity of a geometric series is an important result that we have used at several places in the book. Write $S_n$ for the sum of the first $n$ terms in a geometric progression with a first term $a$ and a common ratio $r$.

$$\begin{aligned} S_n &= a+ar+ar^2+ar^3+ar^4+\cdots+ar^{n-1} \\ rS_n &= ar+ar^2+ar^3+ar^4+ar^5+\cdots+ar^n \\ S_n-rS_n &= a-ar^n \\ S_n &= \frac{a(1-r^n)}{1-r} \end{aligned} \tag{A.11}$$

Now provided $|r|<1$ the sum of the infinite series is:

$$S_\infty=\frac{a}{1-r} \tag{A.12}$$

## A.4 CODING IN FORTRAN

As an illustration of computing techniques, the Taylor series expansion for the exponential function given in Equation (A.5) is coded in Fortran as a function. For the coding note that, independent of the *x* value, the series starts at value 1 and that each term in the series is of the form $x^i/i!$. Summing of the series continues until the individual terms get so small that their value adds little to the total. Note that this last point is possibly an important one and ultimately limits the accuracy of the calculation. The function is shown in Figure A.2. This code requires a function, in fact, to compute the factorial. Be wary of factorials since it is very easy to generate a numerical overflow condition when integers are being used.

```
real function my_exp(x)
implicit none
real:: x,sum,tol,term
integer:: i
sum = 1.0              ! progressive sum
tol = 0.001            ! accuracy parameter
i = 0                  ! term number
term = 10              ! dummy value to start loop
do while(term > tol) ! keep going while terms > tol
 i = i+1               ! next term
 term=x**i/fact(i)
 sum = sum+term        ! increment total sum
end do
my_exp = sum           ! return value in function name
end
```

**Figure A.2** Fortran function to determine exp(x).

## A.5 CODING IN EXCEL

Following the method in the previous section, the Taylor series can also be calculated in an Excel VBA function. It is listed in Figure A.3. The similarity with the Fortran function is striking.

```
Function my_exp(x)
' start value, tolerance
my_exp = 1
tol = 0.0001
' start with dummy value to enter loop
term = 1
' counter for term number
i = 1
' loop and sum until term < tolerance required
Do While (term > tol)
 term = x ^ i / Application.Fact(i)
 my_exp = my_exp + term
 i = i + 1
Loop
End Function
```

**Figure A.3** Excel Visual Basic function to determine exp(x).

## A.6 FINITE DIFFERENCE APPROXIMATION

An important reason for being familiar with Taylor series as an engineer is in the derivation of finite difference approximations to partial differential equations. These are covered in a sequence of earlier chapters. There, it was seen that the standard Taylor expansion, about a point $x$, can be re-arranged into the form:

$$f'(x) = \frac{f(x+h) - f(x)}{h} + \frac{h}{2!} f''(x) + \frac{h^2}{3!} f'''(x) + \ldots \qquad \text{(A.13)}$$

which allows an approximation to be determined for the first derivative of a function in terms of neighbouring points. The expansion (A.11), which converges to the exact derivative, is often truncated to:

$$f'(x) = \frac{f(x+h) - f(x)}{h} + O(h) \qquad \text{(A.14)}$$

where the error is based on the power to which $h$ was raised in the largest term that has been neglected, in this case 1. This gives what is referred to as a first-order error. With some manipulation it is possible to derive a more accurate approximation to the first derivative:

$$f'(x) = \frac{f(x+h) - f(x-h)}{2h} + O(h^2) \qquad \text{(A.15)}$$

which is now second-order accurate. Although it is generally unwise to rely on intuition in matters mathematical, it is not surprising that an approximation to the derivative at a point that uses information from both sides of the point rather than from one side only is more accurate. The Taylor expansion allows this notion to be quantified.

Given that much of engineering data are collected at a series of discrete points in time or space, and that most of the modelling also deals with representing phenomena at a number of discrete points in time and/or space, the importance of the expansion cannot be over-estimated.

## A.7 MATRIX EXPONENTIAL

If A is a square matrix, the matrix exponential of A is defined as:

$$\exp(At) = I + At + A^2t^2/2! + A^3t^3/3! + \cdots \qquad \text{(A.16)}$$

It follows from this definition that:

$$\frac{d}{dt}\exp(At) = A\exp(At) \qquad \text{(A.17)}$$

In applications A is often diagonisable – that is:

$$A = M\Lambda M^{-1} \tag{A.18}$$

where $\Lambda$ is a diagonal matrix of eigenvalues, and M contains the corresponding eigenvectors. Then:

$$\exp(At) = M \exp(\Lambda t) M^{-1} \tag{A.19}$$

Note that unless a matrix is diagonal, the matrix exponential is not obtained by taking exponential of the elements of the matrix. The importance of the matrix exponential is that it is the solution for a linear system expressed in state space form.

APPENDIX B

# Error Function and Gamma Function

## B.1 ERROR FUNCTION

The error function, which comes up frequently in engineering modelling and analysis, is defined as:

$$erf(x) = \frac{2}{\sqrt{\pi}} \int_0^x e^{-w^2} dw \tag{B.1}$$

It is the CDF of a folded normal distribution, and arises in many physical applications. Examples include the effect of the infiltration trenches in raising the water table (Walton, 1970), pollutant dispersion (Gunnerson and French, 1996), the distribution of velocities inside the viscous sublayer (Levi, 1995), and nearshore coastal depth profiles (Kobayashi, 1987).

The integral cannot be expressed in terms of the elementary mathematical functions using the usual elementary tricks of calculus. It can, however, be evaluated using a finite approximation to the infinite sum:

$$erf(x) = \frac{2}{\sqrt{\pi}} \left( x - \frac{x^3}{1!(3)} + \frac{x^5}{2!(5)} - \frac{x^7}{3!(7)} + - ... \right) \tag{B.2}$$

and is shown plotted in Figure B.1 around the area of $x$=0 where its behaviour is most varied. Carslaw and Jaeger (1959) give more efficient formulae for the complementary error function, *erfc(x)*, and the value of *erf(x)* follows from the relation: $erf(x) = 1 - erfc(x)$. Their formulae are:

$$erfc(x) = 1 - \frac{2}{\sqrt{\pi}} \sum_{n=0}^{10} \frac{(-1)^n x^{2n+1}}{(2n+1)n!} \qquad \text{for } x \leq 1.65 \tag{B.3}$$

else

$$erfc(x) = \pi^{-1/2} \exp(-x^2) \left( \frac{1}{x} - \frac{1}{2x^3} + \frac{1x3}{2^2 x^5} - \frac{1x3x5}{2^3 x^7} \right) \tag{B.4}$$

The error function, and complementary error function, are defined in Excel for positive values of $x$. Negative values can be obtained from the relation:

$$erf(-x) = -erf(x) \tag{B.5}$$

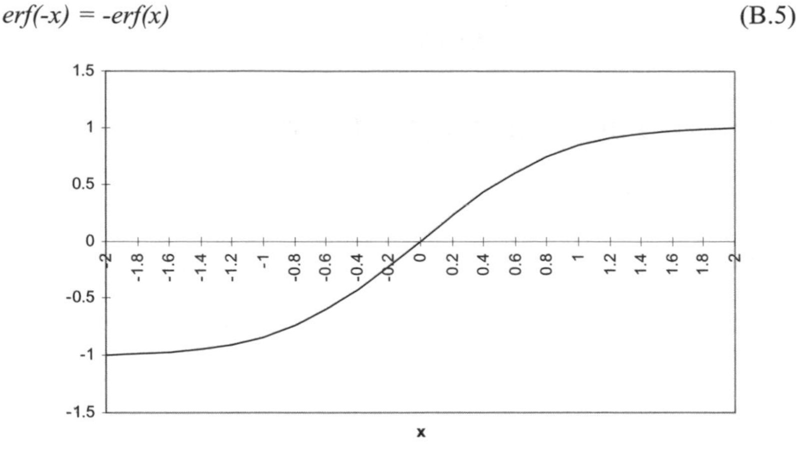


**Figure B.1** The Error function near the origin, $x = 0$.

## B.2 GAMMA FUNCTION

The gamma function is a generalisation of the factorial function. It is defined as:

$$\Gamma(\alpha) = \int_0^\infty t^{\alpha-1} e^{-t} dt \qquad 0 < \alpha \tag{B.6}$$

The function is defined for non-integer negative α, but these values are less often needed. Straightforward integration by parts leads to the result:

$$\Gamma(\alpha) = (\alpha - 1)! \qquad \Gamma(1) = 1 \tag{B.7}$$

This explains why 0! = 1. Kreyszig (1983) provides a table of $\Gamma(\alpha)$ for $1 \leq \alpha \leq 2$. In Excel, the function `GAMMALN()` returns the natural logarithm of the gamma function. In Matlab the gamma function is evaluated by gamma().

The incomplete gamma function is a function of the upper limit of the integral when α is fixed:

$$F(x) = \int_0^x t^{\alpha-1} e^{-t} dt \qquad 0 \leq x \tag{B.8}$$

It is the CDF of a gamma distribution, except for the normalising constant, which is $\Gamma(\alpha)$.

# APPENDIX C

# Complex Sinusoid

## C.1 COMPLEX EXPONENTIAL

The complex exponential $e^{i\omega t}$, also written $\exp(i\omega t)$, facilitates algebraic operations concerned with system inputs and outputs. It is a generalisation of sinusoidal function. The essential result is:

$$\exp(i\omega t) = \cos(\omega t) + i\sin(\omega t) \tag{C.1}$$

and it can be proved in several ways. It follows directly from the Taylor series expansions if it is assumed they remain valid for complex arguments. Thus:

$$\begin{aligned} e^{i\omega t} &= 1 + i\omega t + \frac{(i\omega t)^2}{2!} + \frac{(i\omega t)^3}{3!} + \frac{(iwt)^4}{4!} + \frac{(i\omega t)^5}{5!} + \cdots \\ &= 1 - \frac{(\omega t)^2}{2!} + \frac{(\omega t)^4}{4!} - \cdots + i\left(\omega t - \frac{(\omega t)^3}{3!} + \frac{(\omega t)^5}{5!} - \cdots\right) \\ &= \cos(\omega t) + i\sin(\omega t) \end{aligned} \tag{C.2}$$

A general point, $z$, on the unit circle in an Argand diagram with centre at 0, is defined by $z = e^{i\omega t}$, where $\omega t$ is the angle in radians, measured anti-clockwise from the positive real axis, $\omega$ is the angular velocity in radians per unit of time and $t$ is time. This is shown in Figure C.1. Substitution of $i\pi$ for $\omega t$ gives the celebrated formula named after Leonhard Euler:

$$e^{i\pi} = -1 \tag{C.3}$$

As a point of interest, this formula is celebrated because it summarises four fundamental developments in the history of mathematics: the exponential number, imaginary numbers, $\pi$ and negative numbers.

## C.2 CODING IN EXCEL AND MATLAB

A complex number such as 6+8i can be held in an Excel cell as `COMPLEX(6,8)`. Arithmetic is performed with the functions `IMSUM(,)`, `IMPRODUCT(,)` and others with the IM prefix. In the Fourier transform tool in Excel, for example, the output is as a complex number and the overall magnitude can be calculated using IMABS(), the real part can be isolated using `IMREAL()` and the imaginary part isolated by `IMAGINARY()`.

In Matlab a variable can be assigned a complex number, such as 6+8i, by `z=6+8i`. Arithmetic operations in Matlab work with complex arguments without the need for any special treatment.

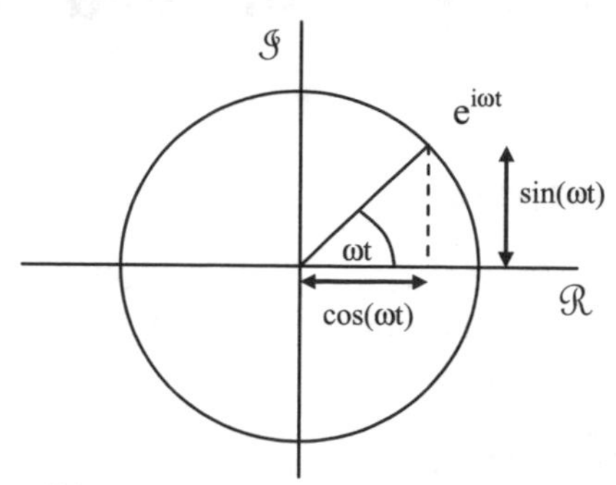


**Figure C.1** Argand diagram showing the general position of *z*.

## C.3 SINGLE MODE OF VIBRATION

As an example of the power of a complex representation in modelling, consider the differential equation representing a linear system with a single mode of vibration, such as a mass on a spring.

$$\ddot{y} + 2\zeta\Omega\dot{y} + \Omega^2 y = x \tag{C.3}$$

in which $\zeta$ is the damping coefficient, which is less than 1 for an oscillatory response, and $\Omega$ is the un-damped natural frequency. Now let the input $x$ be $x = e^{i\omega t}$ and the response be $y = Ax = Ae^{i\omega t}$, where $A$ is a complex number. Substitution into the equation and simple rearrangement gives:

$$A = \frac{1}{-\omega^2 + 2\zeta\Omega\omega i + \Omega^2} \tag{C.4}$$

The gain of the system is:

$$|A| = \sqrt{\frac{1}{(2\zeta\Omega\omega)^2 + (\Omega^2 - \omega^2)^2}} \tag{C.5}$$

The phase lag $\phi$ satisfies the equation:

$$A = |A| e^{-i\phi} \tag{C.6}$$

# APPENDIX D

# Open-Source Software

## D.1 INTRODUCTION

The examples provided in this book have used commercial packages such as Matlab and Microsoft Excel. This has been done under the assumption that most universities and engineering workplaces will have licenses to these packages. This, however, is not universally true and some places prefer to use open-source software packages.

The open source initiative refers to software that has been developed to be freely distributed. The GNU foundation states that the "free" is as in "free speech" rather than "free beer":

*Free software is a matter of the users' freedom to run, copy, distribute, study, change and improve the software.* (www.gnu.org)

In addition to providing binary executable files, open-source packages provide access to the source code so that nothing is hidden from a user and they may compile the program themselves. Open-source programs are typically developed from a community of programmers.

There are several open-source alternatives for the programming tools discussed in this book. These programs are of a high standard and are recommended by the authors. Owing to the model of community development, the expectation from open-source software should not be as high. Thus, there may be less technical support or fewer features in the program since large sums have not been charged to provide it. This, however, is not always the case, since an active project may have good support and as there are no license restrictions it may be possible to obtain access to licensed features that are prohibitively expensive in commercial programs.

## D.2 OPEN-SOURCE NUMERICAL SOFTWARE

Some examples of useful open source software are listed below along with a brief discussion:

- **Gnumeric**, **Calc** and **KSpread** are spreadsheet programs that are similar in style and functionality to Excel. These programs are able to import Excel spreadsheets, but beyond numbers and formulas it is possible that not all items are able to be imported. The basic statistical manipulations and graphics they provide are easy to use. For more advanced users who use Visual Basic scripts, there are scripting options for each of these platforms, but they are not identical to Visual Basic.

- **Octave** is an open-source numerical program that is comparable to Matlab. Octave has high-level functionality for manipulating matrices, running scripts and generating plots. Octave has many in-built functions, but the functionality is not as extensive as Matlab when the Matlab toolboxes are considered.

- **R** is a programming language that mimics S+. S+ was a statistical programming language conceived at the Bell laboratories, and is therefore a distant cousin to C. R uses a command prompt environment that is similar to Matlab and is capable of basic data manipulation and detailed graphics. The high-level functionality in R is mostly focused on statistical applications and there is an active community that regularly contribute packages of extended functionality.

## D.3 FREELY AVAILABLE COMPILERS

There are numerous compilers that are freely available for a variety of languages. Only compilers for the C and Fortran languages are listed here, since these languages are most commonly used in engineering. There are many other languages, but these often produce programs that run considerably slower. It is important to note that compilers often do not adhere strictly to the standards of the programming language. Be aware that code that can be built using one compiler may use some non-standard features that are not compatible with another compiler.

- **g++** and **g95** are two open-source GNU compilers for the Fortran 95 and C++ programming languages respectively. They can be used under a Microsoft Windows platform if the MinGW environment is installed. The compilers are equally as capable as their commercial equivalents. To invoke the compiler it is necessary to use a command prompt, which may be different from the experience of those who are familiar with integrated development environments (IDEs). The main limitation, therefore, is that convenient text formatting and debugging features are not provided by these programs.

- **MinGW Developer Studio** is an open-source development environment for C++. At time of writing there were no free (easy-to-use) environments for Fortran 95.

- **Silverfrost FTN95** is a Fortran 95 development environment that is freely available *for personal use*. That is, the compiler is not open-source, but the distributors allow a single copy to be used at home. It is listed here, since it this may be all that is required for students of this book wishing to implement Fortran code.

# References


Acheson, D. (1997) *From Calculus to Chaos.* Oxford University Press, 269pp.

Adamson, P. T. (1994) *Flood Risk Analysis in Extremely Large Catchments: Studies in the La Plata Basin.* Burderop Park, Swindon, Sir William Halcrow and Partners Ltd.

Aitkins P., (2005) Fattening Children or Fattening Farmers? School Milk in Britain, 1921–1941. *Economic History Review*, 58(1), 57–78.

Andrews, D. F. and Herzberg, A. M. (1985) *Data: A Collection of Problems from Many Fields for the Student and Research Worker.* Springer-Verlag, New York.

Ang, A. H. S. and Tang, W. H. (1975) *Probability Concepts in Engineering Planning and Design, Volume 1 – Basic Principles.* John Wiley and Sons.

Arega, F. and Sanders, B. F. (2004) Dispersion Model for Tidal Wetlands. *Journal of Hydraulic Engineering*, ASCE, 130(8), 739–745.

Arnold, B. C. (1983) *Pareto Distributions.* International Co-operative Publishing House, 326pp.

Australian Bureau of Meteorology (2008) Southern Oscillation Index. Viewed 18 March 2008, at http://www.bom.gov.au/climate/current/soi2.shtml.

Bagnold, R. A. (1979) Acceptance (of Sorby Medal) by Brigadier Ralph A. Bagnold. *Sedimentology*, 26, 159–160.

Balas, C. E. and Balas, L. (2002) Risk Assessment of Some Revetments in Southwest Wales, United Kingdom. *Journal of Waterway, Port, Coastal and Ocean Engineering*, ASCE, 128(5), 216–223.

Barrow, J. D. (1999) It's All Platonic Pi in the Sky. In: *Between Inner Space and Outer Space*. Oxford University Press, 99–101.

Beasley, J. D. and Springer, S. G. (1977). Algorithm AS 111: The Percentage Points of the Normal Distribution. *Applied Statistics*, 26, 118–121.

Bellman, R. and Cooke, K. L. (1995) *Modern Elementary Differential Equations.* Dover Publications, 228pp.

Bellotti, G. (2007) Transient Response of Harbours to Long Waves under Resonance Conditions. *Coastal Engineering*, 54, 680–693.

Bendat, J. S. and Piersol, A. G. (1980) *Engineering Applications of Correlation and Spectral Analysis*. Wiley–Interscience, 302pp.

Bhallamudi, S. M. and Chaudhry, M. H. (1991) Numerical Modeling of Aggradation and Degradation in Alluvial Channels. *Journal of Hydraulic Engineering*, ASCE, 117(9), 1145–1164.

Billah, K. Y. and Scanlan, R. H. (1991) Resonance, Tacoma Narrows Bridge Failure, and Undergraduate Physics Textbooks. *American Journal of Physics*, 59(2), 118–124.

Birkhoff, G. (1990) Fluid Dynamics, Reactor Computations, and Surface Representation. In: Nash, S. G. (ed.) *A History of Scientific Computing*. ACM Press, 358pp.

Bowles, J. E. (1988) *Foundation Analysis and Design*, 4th ed., McGraw-Hill, Singapore.

Briseghella, B. and Zordan, T. (2002) Design and Analysis of a Variable Stiffness Movable Footbridge. *IABSE Symposium*, Melbourne, CD-ROM Proceedings, 10pp.

Brooks, E. B. (2001) *Statistics, The Poisson Distribution*. Viewed 22 January 2008 at http://www.umass.edu/wsp/statistics/lessons/poisson/index.html.

Burling, M. C., Pattiaratchi, C. B. and Ivey, G. N. (2003) The Tidal Regime of Shark Bay, Western Australia. *Estuarine, Coastal and Shelf Sciences*, 57, 725–735.

Camilloni I. and Barros V. (1997). On the Urban Heat Island Effect Dependence on Temperature Trends. *Climatic Change,* 37, 665–681.

Carslaw, H. S. and Jaeger, J. C. (1959) *Conduction of Heat in Solids*. Oxford University Press, New York.

Chambers, J. M., Cleveland, W. S., Kleiner, B. and Tukey, P. A. (1983) *Graphical Methods for Data Analysis*. Wadsworth, Belmont, CA.

Chapra, S. C. and Canale, R. P. (1989) *Numerical Methods for Engineers*, 2nd ed., McGraw-Hill.

Chatfield, C. (1984) *The Analysis of Time Series, An Introduction*. 3rd ed. Chapman and Hall, 286pp.

Colucci, J. M. and Begeman, C. R. (1971) Carcinogenic Air Pollutants in Relation to Automobile Traffic in New York City, *Environmental Science and Technology*, 5(2), 145–150.

Coontz, R. (1999) The Planet that Hums. *New Scientist*, 163(2203), 30–33.

Corotis, R. B., Sigl, A. B. and Klein, J. (1978) Probability Models of Wind Velocity Magnitude and Persistence. *Solar Energy*, 20, 483–493.

Dallard, P., Fitzpatrick, T., Flint, A., Low, A., Ridsdill Smith, R., Willford, M. and Roche, M. (2001) London Millennium Bridge: Pedestrian-Induced Lateral Vibration. *Journal of Bridge Engineering*, ASCE, 6(6), 412–417.

Dandy, G., Walker, D., Daniell, T. and Warner, R. (2008) *Planning and Design of Engineering Systems.* Taylor and Francis, 403 pp.

Day, K. W. (1999) *Concrete Mix Design, Quality Control and Specification*, Taylor and Francis.

Di Lorenzo, E. (2003) Seasonal Dynamics of the Surface Circulation in the Southern California Current System. *Deep–Sea Research II*, 50, 2371–2388.

Dong, S., Li, H., Lu, M. and Takayama, T. (2002) Experimental Study of the Effectiveness of TLDs under Wave Loading. *Journal of Ocean University of Qingdao* (English ed.), 1(1), 80–86.

Duggan, B. (1999) Y2K: What You Don't Know Can Hurt You. *Journal of Emergency Nursing*, 25, 41–42.

Dunn P. (2008), *Datasets for Statistical Analysis.* Viewed April 2008 at http://jamsb.austms.org.au/staff/dunn/Datasets.

Eilon, S. (1982) Comment on "In Praise of Unicorns" by Patrick Rivett. *The Journal of the Operational Research Society*, 33(5), 485–486.

Engineers Australia (2004) *Coastal Engineering Guidelines for Working with the Australian Coast in an Ecologically Sustainable Way*. The National Committee on Coastal and Ocean Engineering, 128pp.

Ezekiel, M. and Fox, K. A. (1959) *Methods of Correlation and Regression Analysis*. John Wiley and sons, New York, 256pp.

Farlow, S. J. (1993) *Partial Differential Equations for Scientists and Engineers*. Dover Publications, 414pp.

Fox, J. (1997) *Applied Regression, Linear Models, and Related Methods*. Sage.

Gear, C. W. and Skeel, R. D. (1990) The Development of ODE Methods: A Symbiosis between Hardware and Numerical Analysis. In: *A History of Scientific Computing*, Ed. S. G. Nash, ACM Press, 88-105.

Gell-Mann, M. (1994) *The Quark and the Jaguar*. Abacus, 392pp.

Gigerenzer, G. and Goldstein, D. G. (1996) Reasoning the Fast and Frugal Way: Models of Bounded Rationality. *Psychological Review*, 103(4), 650–669.

Gleick, J. (1987) *Chaos*. Cardinal, 352pp.

Goldstine, H. H. (1990) Remembrance of Things Past. In: Nash, S. G. (ed.) *A History of Scientific Computing*. ACM Press, 358pp.

Gómez–Pujol, L., Orfila, A., Cañellas, B., Alvarez-Ellacuria, A., Méndez, F. J., Medina, R. and Tintoré, J. (2007) Morphodynamic Classification of Sandy Beaches in Low Energetic Marine Environment. *Marine Geology*, 242, 235–246.

Good, I. J. (1986) Some Statistical Applications of Poisson's Work. *Statistical Science*, 1(2), 157–170.

Gould, S. J. (1991) *Bully for Brontosaurus*. Penguin.

Gribben, J. (2004) *Deep Simplicity*. Penguin, 251pp.

Gunnerson, C. G. and French, J. A. (1996) *Wastewater Management for Coastal Cities: The Ocean Disposal Option*. 2nd ed. Springer.

Gy, P. (2004) Sampling of Discrete Materials, III: Quantitative Approach – Sampling of One Dimensional Objects. *Chemometrics and Intelligent Laboratory Systems*, 74, 39–47.

Hardy, T. A., McConochie, J. D. and Mason, L. B. (2003) Modeling Tropical Cyclone Wave Population of the Great Barrier Reef. *Journal of Waterway, Port, Coastal and Ocean Engineering*, ASCE, 129(3), 104–113.

Håkanson, L. (2000) The Role of Characteristic Coefficients of Variation in Uncertainty and Sensitivity Analyses, with Examples Related to the Structuring of Lake Eutrophication Models. *Ecological Modelling*, 131, 1–20.

Hall, J. W., Meadowcroft, I. C., Lee, E. M. and van Gelder, P. H. A. J. M. (2002) Stochastic Simulation of Episodic Soft Coastal Cliff Erosion. *Coastal Engineering*, 46, 159–174.

Hargrove, W. W., Gardner, R. H., Turner, M. G., Romme, W. H. and Despain, D. G. (2000) Simulating Fire Patterns in Heterogeneous Landscapes. *Ecological Modelling*, 135, 243–263.

He, J. (1994) Analogue Simulation of Commonly Encountered Nonlinearity for Dynamic Structures. *Computers & Structures*, 51(6), 661–669.

Heaney, J. P. and Wright, L. T. (1997) On Integrating Continuous Simulation and Statistical Methods for Evaluating Urban Stormwater Systems. In: James, W. (ed.) Modelling the Management of Stormwater Impacts, *Computational Hydraulics International*, 5, 45–76.

Hesslein, R. H. (1980) In Situ Measurements of Pore Water Diffusion Coefficients Using Tritiated Water. *Canadian Journal of Fisheries and Aquatic Sciences*, 37, 545–551.

Hinwood, J. B., McLean, E. J. and Lewis, H. B. (1998) Time and Frequency Domain Analysis of Long Period Waves in Port Phillip Bay. *Proceedings 13th Australasian Fluid Mechanics Conference*, Monash University, Melbourne.

Hipel, K. W. and McLeod, A. I. (1994) *Time Series Modelling of Water Resources and Environmental Systems*. Elsevier, Amsterdam, 1013pp.

Hoffman, P. (1999) *The Man Who Loved Only Numbers*. Fourth Estate, 302pp.

Holmgren, A. J. And Molin, S. (2006) Using Disturbance Data to Assess Vulnerability of Electric Power Delivery Systems. *Journal of Infrastructure Systems*, ASCE, 12(4), 243–251.

Horn, B. K. P. (1983) The Curve of Least Energy. *ACM Transactions on Mathematical Software*, 9(4), 441–460.

Horn, D. P. (2002) Beach Groundwater Dynamics. *Geomorphology*, 48, 121–146.

IPCC (2007) *Climate Change 2007: 4th Assessment Report*. The Intergovernmental Panel on Climate Change. Available at http://www.ipcc.ch/.

Izquierdo, J., Perez, R. and Iglesias, P. L. (2004) Mathematical Models and Methods in the Water Industry. *Mathematical and Computer Modelling*, 39, 1353–1374.

Jain, A. K. (1990) Inelastic Response of Reinforced Concrete Frames Subjected to the 1985 Mexico Earthquake. *Computers & Structures*, 34(3), 445–454.

Joyner, W. B., Boore, D. M. and Porcella, R. D. (1981). *Peak Horizontal Acceleration and Velocity from Strong-motion Records Including Records from the 1979 Imperial Valley, California Earthquake*. USGS Open File report 81–365. Menlo Park, CA.

Kobayashi, N. (1987) Analytical Solution for Dune Erosion by Storms. *Journal of Waterway, Port, Coastal, and Ocean Engineering*, ASCE, 113(4), 401–418.

Kreyszig, E. (1983) *Advanced Engineering Mathematics*. Wiley, 5th ed.

Kullenberg, G. (1971) Vertical Diffusion in Shallow Waters. *Tellus*, 23, 129–135.

Kunzig, R. (2000) *Mapping the Deep*. Sort of Books.

Kuo, J.-T., Yen, B.-C., Shsu, Y.-C. and Lin, H.-F. (2007) Risk Analysis for Dam Overtopping: Feitsui Reservoir as a Case Study. *Journal of Hydraulic Engineering*, ASCE, 133(8), 955–963.

Law, A. W. K. (2000) Taylor Dispersion of Contaminants due to Surface Waves. Journal of Hydraulic Research, Vol. 38, No. 1, 41-48

Lee, P. J., Vitkovsky, J. P., Lambert, M. F., Simpson, A. R. and Liggett, J. A. (2005) Frequency Domain Analysis for Detecting Pipeline Leaks. *Journal of Hydraulic Engineering*, ASCE, 131(7), 596–604.

Leuliette, E. W, Nerem, R. S. and Mitchum, G. T. (2004) Calibration of TOPEX/Poseidon and Jason altimeter data to construct a continuous record of mean sea level change. *Marine Geodesy*, 27(1–2), 79–94.

Levi, E. (1995) *The Science of Water: The Foundation of Modern Hydraulics*. ASCE Press.

Levy, M. and Salvadori, M. (1994) *Why Buildings Fall Down*. Norton, 346pp.

Lorenz, E. N. (1993) *The Essence of Chaos*. University of Washington Press, 227pp.

Losada, I. J., Gonzalex–Ondina, J. M., Diaz-Hernandez, G. and Gonzalez, E. M. (2008) Numerical Modeling of Nonlinear Resonance of Semi-Enclosed Water Bodies: Description and Experimental Validation. *Coastal Engineering*, 55, 21–34.

Luckenbach, M. W., Coen, L. D., Ross, P. G. Jr. and Stephen, J. A. (2005) Oyster Reef Habitat Restoration: Relationships Between Oyster Abundance and Community Development Based on Two Studies in Virginia and South Carolina. *Journal of Coastal Research*, Special Issue 40, 64–78.

McIntyre, N., Jackson, B., Wheater, H. and Chapra, S. (2004) Numerical Efficiency in Monte Carlo Simulations: Case Study of a River Thermodynamic Model. *Journal of Environmental Engineering*, ASCE, 130(4), 456–464.

McMahon, T. A. and Mein, R. G. (1986): *River and Reservoir Yield*, Water Resources Publications, Colorado, USA, 368pp.

Malamud, B. D. and Turcotte, D. L. (2006) The Applicability of Power-law Frequency Statistics to Floods. *Journal of Hydrology*, 322(1–4), 168–180.

Mallayachari, V. and Sundar, V. (1996) Wave Transformation over Submerged Obstacles in Finite Water Depths. *Journal of Coastal Research*, 12(2), 477–483.

Maloof, S. and Protopapas, A. L. (2001) New Parameters for Solution of Two-Well Dispersion Problem. *Journal of Hydrologic Engineering*, ASCE, 6(2), 167–171.

Mandel, J. (1964) *The Statistical Analysis of Experimental Data*. Dover, 410pp.

Marcy, G. W., Butler, R. P. and Williams, E. (1997) The Planet Around 51 Pegasi, *The Astrophysical Journal*, 481, 926–935.

Masselink, G. and Pattiaratchi, C. (2001) Characteristics of the Sea Breeze System in Perth, Western Australia, and its Effect on the Nearshore. *Journal of Coastal Research*, 17(1), 173–187.

Mathematica (2007) *The Bivariate Normal and Conditional Distributions*. From The Wolfram Demonstrations Project, Accessed Feb 2007 at http://demonstrations.wolfram.com/TheBivariateNormalAndConditionalDistributions/.

Mathews, J. H. and Fink, K. K. (2004) *Numerical Methods Using Matlab*, 4th ed. Prentice Hall Inc., New Jersey, USA.

Metcalfe, A. V. (1997) *Statistics in Civil Engineering*. Arnold Publishers, London.

Metropolis, N. and Ulam, S. (1949) The Monte Carlo Method. Journal of the American Statistical Association, 44(247), 335–341.

Myers, R. H. (1994) *Classical and Modern Regression with Applications*. 2nd ed. Duxbury Press, Boston, MA, USA.

Nahin, P. J. (1998) *An Imaginary Tale. The Story of* $\sqrt{-1}$. Princeton University Press, 267pp.

Nelder, J. A. and Mead, R. (1965) A Simplex Method for Function Minimization. *Computer Journal*, 7, 308–313.

Nielsen, P. (1990) Tidal Dynamics of the Water Table in Beaches. *Water Resources Research*, 26(9), 2127–2134.

Newland, D. E. (1984) *An Introduction to Random Vibrations and Spectral Analysis*, 2nd ed. Longman, 377pp.

Nordin, C. F. Jr. (1971) Statistical Properties of Dune Profiles. In: *Sediment Transport in Alluvial Channels*. United States Geological Survey Professional Paper 562-F, 41pp.

O'Hanlon, L. (2000) The Time Travelling Mountain. *New Scientist*, 167(2245), 30–33.

Pacheco, A., Vila-Concejo, A., Ferreira, O. and Dias, J. A. (2008) Assessment of Tidal Inlet Evolution and Stability using Sediment Budget Computations and Hydraulic Parameter Analysis. *Marine Geology*, 247, 104–127.

Papanicolaou, A. N., Elhakeem, M., Krallis, G., Prakash, S. and Edinger, J. (2008) Sediment Transport Modelling Review: Current and Future Developments. *Journal of Hydraulic Engineering*, ASCE, 134(1), 1–14.

Perlin, A. and Kit, E. (2002) Apparent Roughness in Wave-Current Flow: Implication for Coastal Studies. *Journal of Hydraulic Research*, 128, 729–741.

Peterson, I. (1998) *The Jungles of Randomness: Mathematics on the Edge of Certainty*. Penguin Books.

Phillimore, J. and Davison, A. (2002) A Precautionary Tale: Y2K and the Politics of Foresight. *Futures*, 34, 147–157.

Pickering, S. (1999) Y2K Leaves its Mark. *III–Vs Review*, 12(5), 39–40.

Popper, K. and Eccles, J. C. (1977) *The Self and Its Brain*. Routledge, 597pp.

Press, W. H., Teukolsky, S. A., Vetterling, W. T. and Flannery, B. P. (1992) *Numerical Recipes in Fortran: The Art of Scientific Computing*. Cambridge University Press.

Priest, S. D. (2004) Determination of Discontinuity Size Distributions from Scanline Data. *Rock Mechanics and Rock Engineering*, 37(5), 347–368.

Qu, R. and Ye, J. (2000) Approximation of Minimum Energy Curves. *Applied Mathematics and Computation*, 108, 153-166.

Raubenheimer, B., Guza, R. T. and Elgar, S. (1999) Tidal Water Table Fluctuations in a Sandy Ocean Beach. *Water Resources Research*, 35(8), 2313-2320.

Ravetz, J. (2000) How I Got it Wrong. *Futures*, 32, 937–939.

Rebeiz, K. S., Rosett, J. W., Nesbit, S. M. and Craft, A. P. (1996) Tensile Properties of Polyester Mortar Using PET and Fly Ash Wastes. *Journal of Materials Science Letters*, 15, 1273–1275.

Reif, F. (1985) *Fundamentals of Statistical and Thermal Physics*. McGraw-Hill International Editions, 651pp.

Rivett, P. (1981) In Praise of Unicorns. *The Journal of the Operational Research Society*, 32(12), 1051–1059.

Rizwan, M., Ilyas, M., Masood, A. and Towhata, I. (2008) An Approach to Digitize Analog Form of Accelerograms Recorded at Tarbela, Pakistan. *Soil Dynamics and Earthquake Engineering*, 28, 328–332.

Roache, P. J. (1982) *Computational Fluid Dynamics*. Hermosa Publishers, 446pp.

Rodriguez, G., Guedes Soares, C. and Machado, U. (1999) Uncertainty from the Sea State Parameters Resulting from the Methods of Spectral Estimation. *Ocean Engineering*, 26, 991-1002.

Rozanov, Y. A. (1969) *Probability Theory: A Concise Course*. Dover Publications, 148pp.

Salvadori, M. (1980) *Why Buildings Stand Up*. W.W. Norton and Co., 323pp.

Sarioglu, M. and Yavuz, T. (2000) Vortex Shedding from Circular and Rectangular Cylinders Placed Horizontally in a Turbulent Flow. *Turkish Journal of Engineering and Environmental Sciences*, 24, 217–228.

Scheibehenne, B., Miesler, L. and Todd, P. M. (2007) Fast and Frugal Food Choices: Uncovering Individual Decision Heuristics. *Appetite*, 49, 578–589.

Singpurwalla, N. D., Castillino, V. F. and Goldschen, D. Y. (1975). Inference From Accelerated Life Tests Using Eyring Typed Reparameterizations. *Naval Research Logistics Quarterly*, 22(2), 289–296.

Sobey, R. J. (1978) Statistical Prediction of Extreme Waves behind the Great Barrier Reef. *Proceedings of the Fourth Australian Conference on Coastal and Ocean Engineering*, Adelaide, 39–43.

Steedman, R. K. (1987) A method for estimating the joint probability of occurrence of extreme storm water levels with application to Exmouth gulf. *8th Australasian Conference on Coastal and Ocean Engineering*, 28–32.

Strogatz, S. (2003) *Sync: The Emerging Science of Spontaneous Order*. Penguin Books, 338pp.

Swade, D. (2000) *The Cogwheel Brain*. Abacus, 342pp.

Swamee, P. K., Pathak, S. K. and Sohrab, M. (2000) Empirical Relations for Longitudinal Dispersion in Streams. *Journal of Environmental Engineering*, ASCE, 126(11), 1056–1062.

Tabatabaei, J. and Zoppou, C. (2000) Remedial Works on Cotter Dam and the Risk to these from Floods. *ANCOLD Bulletin*, 114, 97–111.

Taylor, A. L. III (1984) The Wizard Inside The Machine, *Time Magazine*, 16 April 1984. Downloaded 13 March, 2008 from http://www.time.com/time/printout/0,8816,954266,00.html.

Torrence, C. and Compo, G. P. (1998) A Practical Guide to Wavelet Analysis. *Bulletin of the American Meteorological Society*, 79(1), 61–78.

Turner, I. L., Aarninkhof, S. G. J. and Holman, R. A. (2006) Coastal Imaging Applications and Research in Australia. *Journal of Coastal Research*, 22(1), 37–48.

US Navy (2007) *Joint Typhoon Warning Centre Tropical Cyclone Best Track Data Site*. Viewed 15 March 2008 at https://metocph.nmci.navy.mil/jtwc/best_tracks/.

van Maren, D. S. and Hoekstra, P. (2004) Seasonal Variation of Hydrodynamics and Sediment Dynamics in a Shallow Subtropical Estuary: The Ba Lat River, Vietnam. *Estuarine, Coastal and Shelf Science*, 60, 529–540.

Viney, N., and Bates, B. (2004). It Never Rains on Sunday: The Prevalence and Implications of Untagged Multi-Day Rainfall Accumulations in the Australian High Quality Data Set. *International Journal of Climatology*, 24, 1171–1192.

Walpole, R. E. and Myers, R. H. (1989) *Probability and Statistics for Engineers and Scientists*. 4th ed. Macmillan.

Walton W. C. (1970) *Groundwater Resource Evaluation*. McGraw-Hill.

Wang, G.-T., Chen, S., Barber, M. E. and Yonge, D. R. (2004) Modeling Flow and Pollutant Removal of Wet Detention Pond Treating Stormwater Runoff. *Journal of Environmental Engineering*, ASCE, 130(11), 1315–1321.

Warner, R. F. and Kabaila, A. P. (1968) Monte Carlo Study of Structural Safety. *Journal of the Structural Division, Proceedings of the ASCE*, 94(ST12), 2847–2859.

Weart, S. R. (2003) *The Discovery of Global Warming*. Harvard University Press, 228pp.

Webb, T. and Mirfenderesk, H. (1997) Coastal Pollution Dispersion Models for a Sloping Beach. *Combined Australasian Coastal Engineering and Ports Conference*, Christchurch, 655–660.

Weyhenmeyer, G. A. and Bloesch, J. (2001) The Pattern of Particle Flux Variability in Swedish and Swiss Lakes. *The Science of the Total Environment*, 266, 69–78.

Wichura, M. J. (1988). Algorithm AS241: The Percentage Points of the Normal Distribution. *Applied Statistics,* 37, 477–484.

Wilkinson, S. M. and Knapton, J. (2006) Analysis and Solution to Human-Induced Lateral Vibrations on a Historic Footbridge. *Journal of Bridge Engineering*, ASCE, 11(1), 4–12.

Willis, N. R. T. and Thethi, K. S. (1999) Stride DIP: Steel Risers in Deepwater Environments – Progress Summary. *1999 Offshore Technology Conference*, Houston.

Woo, S.-B. and Liu, P.L.-F. (2004) Finite-Element Model for Modified Boussinesq Equations, I: Model Development. *Journal of Waterway, Port, Coastal and Ocean Engineering*, ASCE, 130(1), 1–16.

Young, I. R. (1999) *Wind Generated Ocean Waves*. Elsevier Ocean Engineering Book Series, Volume 2, 288pp.

Young, I. R. and Verhagen, L. A. (1996) The Growth of Fetch Limited Waves in Water of Finite Depth, Part 1: Total energy and peak frequency. *Coastal Engineering*, 29, 47–78.

Yousif, H. and El-Tabbany, A. (2007) Assessment of Several Interpolation Methods for Precise GPS Orbit. *The Journal of Navigation*, 60, 443-455.

Zedler, E. A. and Street, R. L. (2006) Sediment Transport over Ripples in Oscillatory Flow. *Journal of Hydraulic Engineering*, ASCE, 132(2), 180–193.

Zivanovic, S., Pavic, A. and Reynolds, P. (2005) Vibration Serviceability of Footbridge under Human-Induced Excitation: A Literature Review. *Journal of Sound and Vibration*, 279, 1–74.

# Index